Studies in History and Philosophy of Science

Volume 64

Studies in History and Philosophy of Science is a peer-reviewed book series, dedicated to the history of science and historically informed philosophy of science. The series publishes original scholarship in various related areas, including new directions in epistemology and the history of knowledge within global and colonial contexts. It includes monographs, edited collections, and translations of primary sources in the English language. These cover a broad temporal spectrum, from antiquity to modernity, and all regions of the world.

Gerardo Ienna

Genesis and Development of French Historical Epistemology

A Trajectory Toward Political Epistemology

Springer

Gerardo Ienna
Department of Communication
and Social Research
Sapienza Università di Roma
Rome, Italy

Translated by
Richard E. Burket
Italian public high school English
teacher at IISS Enrico Medi
Palermo, Italy

ISSN 1871-7381 ISSN 2215-1958 (electronic)
Studies in History and Philosophy of Science
ISBN 978-3-031-88698-0 ISBN 978-3-031-88699-7 (eBook)
https://doi.org/10.1007/978-3-031-88699-7

Translation from the Italian language edition: "Genesi e sviluppo dell'épistémologie historique. Fra epistemologia, storia e politica Lecce" by Gerardo Ienna, © Ienna, G. (2023) 2023. Published by Pensa Multimedia. All Rights Reserved.

This Springer imprint is published by the registered company Springer Nature Switzerland AG
The registered company address is: Gewerbestrasse 11, 6330 Cham, Switzerland

Preface to the English Edition

In recent decades, the methodology of historical epistemology has been experiencing a real renaissance within debates on the history and philosophy of science. Many conferences and publications have been devoted to discussing the historical-critical reconstruction of this tradition, its legacy in the context of contemporary epistemology, and its methodological relevance, which continues to inspire original research topics in these disciplines. The renewed interest in this tradition is not rooted in a merely antiquarian interest in an epistemological model now buried in the disciplinary past of philosophy and the history of science. On the contrary, it is a rather lively debate that has continued to develop in various directions and in various national contexts.

Although there has certainly been no lack of relevant publications on these topics, it is important to highlight how, in recent years, they have increasingly focused on detailed analyses of the contributions of certain authors who are part of what can be considered the "canon" of historical epistemology is important to highlight. A large number of new translations, monographs, edited volumes, essays, articles, and doctoral theses focused on bringing attention to authors such as Gaston Bachelard, Jean Cavaillès, Albert Lautman, Hélène Metzger, Georges Canguilhem, Louis Althusser, Michel Foucault, Ian Hacking, Arnold Davidson, Hans-Jörg Rheinberger, Lorraine Daston, Jurgen Renn, and others have favored the fruitful reception of this trend in many national contexts. In this perspective, the notion of historical epistemology is often mobilized as an intellectual label aimed at encompassing an increasingly vast and heterogeneous series of authors united by a similar methodological inclination, driven simultaneously by a desire to historicize the problems of "standard" epistemology and make the traditional history of science epistemologically informed. What seems to be missing, however, is a systematic reflection that attempts to follow the reverse trajectory of this trend—that is, one that tries to take stock of the canonization processes of this tradition.

This book is, in fact, part of a broader editorial and research project conducted by Massimiliano Badino, Pietro Daniel Omodeo, and I that resulted in the publication of a volume titled *Epistemologia storica. Correnti, temi e problemi* (2022), which provides a wide-ranging reconstruction of the roots, evolution, and new theoretical perspectives of historical epistemology in different cultural, historical, and national contexts. Within this project, I was responsible for writing several chapters, including two that were explicitly dedicated to the tradition of French historical epistemology [*épistémologie historique*]. The present book aims to extend what had to be limited for editorial reasons to a relatively small number of pages.

This book focuses on three main thematic nuclei that I consider fundamental for a thorough understanding of this tradition. First, I reconstruct the historical, social, and institutional roots of *épistémologie historique* by focusing on the complex process that led to the crystallization of this term within French debates. Second, I identify the main theoretical problems that allow us to appreciate the substantial internal coherence of the methodological approach of *épistémologie historique*—despite the different positions taken up by individual authors. In discussing the first two points, a correlative need to interrogate and reconstruct the sedimented canon of this tradition by including authors whose relevance in this context has been underestimated or not fully legitimated, such as Pierre Bourdieu, or by valorizing the contributions of the French philosophy of mathematics arose.

This volume also includes several appendices focused on the construction of a canon of Italian historical epistemology that, as will be seen, has had notable theoretical and thematic intersections and overlaps with the French variant. However, I leave the burden of a more in-depth clarification of the issues addressed in this book with respect to the broader panorama of debates in historical epistemology to *Epistemologia storica. Correnti, temi e problemi.*

Rome, Italy
August 2024

This volume has been corrected to include the name of the translator, Richard E. Burket.

Acknowledgments

This book was funded by the European Union's Horizon 2020 research and innovation program under a Marie Skłodowska-Curie grant agreement (n. 101026146).

My heartfelt gratitude goes to Pietro Daniel Omodeo, Charles Wolfe, and Massimiliano Badino, who had the patience to discuss the structure of this volume with me right from its initial stages. Special thanks go to my friend and colleague Giulia Gandolfi, who patiently read the Italian and English editions several times, providing important suggestions and edits. Without their help, this book would not exist. I am also grateful to Mario Castellana and Charles Alunni, who, for a decade now, have encouraged my research on French historical epistemology, giving me valuable advice and constantly fostering an intellectual exchange characterized by their profound epistemological acumen, to which I am greatly indebted. For this reason, I am particularly happy that the original Italian volume of this work was published in their editorial series *Pensée des sciences*, which, over the years, has stood out for its originality and editorial seriousness. I am also extremely grateful to Elena Gagliasso, who first opened the way for me toward a more militant perspective on the history and philosophy of science and who, since my undergraduate thesis, has never tired of reading my work. In addition, I wish to thank Marco Santoro for helping me develop a sensitivity toward the sociology of intellectual fields and for accompanying me in my first steps in the field of academic research.

The opportunity to work as a post-doc within the European research project ERC-*Early Modern Cosmology* (GA n. 725883) in 2020–2021, coordinated by Pietro Daniel Omodeo, was particularly important to my intellectual trajectory. Animated by a climate of fruitful mutual collaboration, this research group enabled me to exchange ideas, opinions, doubts, and perplexities, providing extremely varied intellectual stimulation. Thus, I thank all of them for their support and teamwork. I am particularly grateful to the Political Epistemology collective, made up of Senthil Babu, Giovanni Fava, Sascha Freyberg, Giulia Gandolfi, Lukas Meisner, Pietro Daniel Omodeo, Francesca Putignano, Giulia Rispoli, Matthias Schemmel, and Charles Wolfe, which is always developing exciting experiences and stimulating research projects.

I must also thank Étienne Anheim, Marco Beretta, Giuseppe Bianco, Jean-François Braunstein, Éric Brian, Vincent Bontems, Francesco Paolo de Ceglia, Enrico Castelli Gattinara, Annagiulia Canesso, Andrea Cavazzini, Giovanni Ciccotti, Claudio D'Aurizio, Ronan de Calan, Paola Donatiello, Jean-Louis Fabiani, Paolo Fabbri, Francesco Galofaro, Luca Guzzardi, Bruno Karsenti, Fabio Lusito, Francesco Masciani, Alvise Mattozzi, Fabrizio Palombi, Frédéric Patras, Luigigiovanni Quarta, Paolo Savoia, Emiliano Sfara, Maria Turchetto, and Matteo Vagelli, with whom I have had the privilege of discussing issues of historical epistemology over the years. To these, I add my friends and colleagues from *Réseau de recherche épistémologie historique*, with whom I had the opportunity to exchange profound reflections on my research topics on both professional and informal occasions.

Particularly illuminating for my path were the constant sociological exchanges with Matteo Bortolini, Anna Boschetti, Barbara Grüning, Carmelo Lombardo, and Marco Santoro, discussing the relationships between epistemology and social sciences with particular reference to Pierre Bourdieu's theory. Special thanks go to my friend and colleague Lorenzo Sabetta, who constantly nudged me to complete this volume.

This publication was made possible by the intense relationship of mutual exchange that I formed with these friends and colleagues; however, I would like to specify that all responsibility regarding what is presented in this text lies exclusively with its author.

My biggest thanks, however, go to my wife, Stefania, who encourages me every day, without ever skimping on her sweetness, and to my parents, Domenico and Martina, without whom nothing would have ever been possible.

Competing Interests The author has no competing interests to declare that are relevant to the content of this manuscript.

Contents

Chapter 1
The Social Genesis of an Intellectual Field

Abstract The historiographical literature dedicated to historical epistemology tends to adopt a normative approach. On the contrary, this chapter proposes a description of the uses that have been made of the label "French historical epistemology [*épistémologie historique*]" (FHE) in the French context in relation to the sociocultural context from which it emerged. This reconstruction brings its canonization process to light. The first section takes the reader inside the field of historical epistemology and provides some preliminary definitions. The second section details how a broad discussion developed around the legacy of positivism in the transitional period between the nineteenth and early twentieth centuries, resulting in various intellectual tensions that led to the formation of a polarized intellectual space within which FHE developed. I describe the international circulation of ideas that generated the emergence of different intellectual reactions to positivism and how these factors were correlated with the institutionalization dynamics for the philosophy of science in various national contexts. The third section reconstructs the emergence of the label "*épistémologie historique*" in France and how this term was used from the early twentieth century until the 1970s. The fourth section argues that the success of FHE can be tracked to the French political-cultural climate of the 1970s. This section also analyzes how the "major canon" of this tradition, consisting of Bachelard, Canguilhem, and Foucault, emerged. In the last section, I show how the crystallization of this "major canon" gradually obscured the key role played by French mathematical thought, which I therefore propose as a "minor canon" of FHE.

1.1 A Non-*Imaginabilis* Epistemology

The term *historical epistemology* usually indicates a methodology or research program aimed at constructing a historically informed theory of knowledge and/or a history of science and technology that highlights its epistemological aspects. As is already clear from this preliminary formulation, it is an area of investigation whose methodological principle is the inseparability of the philosophical and historical analysis of science and technology.

G. Ienna, *Genesis and Development of French Historical Epistemology*, Studies in History and Philosophy of Science 64, https://doi.org/10.1007/978-3-031-88699-7_1

The fundamental assumption of historical epistemology is the dynamic nature of knowledge. Scientific knowledge, like any other form of knowledge, evolves over time. For this reason, unlike other traditions, historical epistemology abandons the idea that the philosophy of science should limit itself to the analysis—according to Hans Reichenbach's famous formulation—of the "context of justification," understood as the study of connections between facts and the theories advanced to explain them.[1] Ludwik Fleck had already openly sided against this "insufficient" neo-positivist epistemology in *Genesis and Development of a Scientific Fact*, where in a famous passage he claimed that "epistemology without historical and comparative investigations is no more than an empty play on words or an *epistemologia imaginabilis*" (Fleck 1935/1979, 21; translation modified).[2] Historical epistemology therefore deals with correlating the analysis of the contents of scientific knowledge (and its philosophical implications) with their "context of discovery." The objective of this tradition is to emphasize how forms of rationality are neither universal and fixed nor disembodied with respect to the historical, cultural, and material processes that determine them. To avoid falling back into an *epistemologia imaginabilis*, epistemology and the history of science need to be in close dialogue with each other.

In this tradition, scientific reason is not the fruit of "an immaculate conception" but rather a historical product like other forms of cultural production (Bourdieu et al. 1968). Thus, epistemology—and a general theory of knowledge—cannot avoid analyzing the methods through which knowledge is transformed diachronically. It is therefore one of the possible ways—alongside other meta-scientific traditions, such as the sociology of scientific knowledge[3]—of interrogating the distinction between the "facts" and "values"[4] upon which the ideology of the axiological neutrality of scientific knowledge was built. This questioning represents its fundamental political tone. Claims of the non-neutrality of scientific knowledge usually involve at least two specific stances: On the one hand, scientific production, like other cultural forms, is influenced by the historical and social conditions in which it is produced. In this sense, science and technology are ideologically influenced by forms of cultural and economic hegemony. On the other hand, science and technology are conceived as conformations of knowledge that cooperate in the structuring of the forms that organize society, production, power, and so forth (the social function of science). From this perspective, science and technology are tools that can be used ideologically in the creation of cultural hegemony.

---

[1] On the distinction between the context of discovery and the context of justification, on which a large part of the Vienna and Berlin Circle's programs were built, see Reichenbach (1938, 6–7).

[2] The English edition translates the Latin expression "*epistemologia imaginabilis*" as "epistemology of the imagination," which I think is misleading. Fleck was not referring to an epistemology that has imagination as its object, but rather to a form of "imaginable/imaginary epistemology." I therefore prefer to retain the Latin expression contained in the original text.

[3] For some of the key works in this tradition, see Barnes (1974); Bloor (1976); Latour and Woolgar (1979); Mulkay (1979).

[4] On this point, see Elgin (1996, 2013).

In this chapter, I illustrate how historical epistemology tends to shift the investigation into the conditions of the possibility of knowledge from an a priori perspective to one rooted in the historical conditions in which scientific discourses emerge. At a methodological level, the historical-epistemological approach starts from an analysis of the historical variability of epistemic concepts, thus highlighting the relationships between the historical-cultural context and the activity of rationalizing the world. From these assumptions, several possible research trajectories emerge. Going in one direction, historical epistemology can analyze the relationships between the practices of epistemic subjects and the crystallization of concepts (practice and concepts). Alternatively, the ways that scientific objects are constituted and their relationship with the technical dimension (practice and materiality) can be investigated. Or, in yet another direction, historical epistemology can pursue a deeper understanding of the transformation of the cognitive frameworks within which experiences take shape (practices and cognitive structures).[5]

As can already be seen from these brief indications, it is an uneven field of study containing various theoretical orientations that are not always homogeneous. Despite this unevenness, some privileged sites of investigation that are common to historical epistemology can be traced. First of all, the investigation of the genesis of scientific knowledge is transformed from a search for the absolute foundations of scientific reason to a reconstruction of the historical stages that led to the crystallization of certain epistemological forms. Both scientific knowledge and the epistemologies are fragmented into a plurality of rational forms irreducible to a single formal logic, as the members of the Vienna Circle and the Berlin Circle sought to demonstrate (the problem of the nonunity of science). These positions also influence how the problem of justification is investigated. It is no longer analyzed exclusively through internal logical coherence and the adherence of theories to isolable facts. On the contrary, sociocultural influences on the validation systems of theories, as well as the historical and material nature of scientific practices, become central. Added to this are philosophical-historiographical analyses of the dynamics of the historical evolution of scientific knowledge: Is the history of science progressive? Is the transformation process of science continuous or discontinuous? How can we determine what a historical event is? Consequently, these investigations also mobilize meta-epistemological problems for historical epistemology itself, related to the methodology to be adopted in the study of scientific theories and practices.

However, these theoretical principles correspond to a tradition with very nuanced boundaries for which tracing a common genealogical path or identifying a single methodological nucleus is difficult. For these reasons, the canon of authors who fall within this tradition is widely debated.[6]

---

[5] For a general introduction to historical epistemology, see Badino, Ienna, and Omodeo (2022).

[6] There have been substantial controversies on the notion and definition of *historical epistemology*. For examples of the state of this debate, see Gingras (2010); Rheinberger (2010); Feest and Sturm (2011); Taylan (2018); Braunstein, Moya Diez, and Vagelli (2019). In addition, debates about the so-called integrated history and philosophy of science and the "historical turn" in the philosophy of science are progressively being assimilated into the context of historical epistemology. See Bird

The term *historical epistemology* has been used to encompass currents of thought and methodologies that have changed over a rather long period of time, ranging from the mid-nineteenth century to the present day. It has developed across various national academic fields, including French, English, German, and Italian, which have in turn impacted the circulation of certain authors' ideas and lines of research at a transnational level. These have led to the consolidation of terms such as *épistémologie historique*, *historical epistemology*, *historische epistemologie*, and *epistemologia storica*, each of which corresponds to a genealogy based on various inclusion/exclusion criteria. What is at stake in these genealogies is the creation of a canon that has a correlative impact on the processes and forms of the consecration (or, more precisely, canonization) of the authors implicated in it.[7]

As a methodological premise, I propose, at a preliminary level, to treat historical epistemology as an intellectual label—that is, as an object of negotiation and controversy between communities of researchers engaged in the struggle over the legitimation of its definition. In a Bourdieusian sense, this approach facilitates the emergence of the relative stakes of the actors involved in the debates on the intellectual consolidation of historical epistemology. The first step of my analysis, therefore, consists of looking at the use that has been made of this concept within certain cultural and intellectual contexts, to trace a history of the ideas that constitute its central nucleus.

The historiographical literature devoted to historical epistemology tends to adopt a normative approach. On the contrary, my objective is to propose a description of the uses that have been made of the label "French historical epistemology [*épistémologie historique*]" (FHE) in the French context in relation to the sociocultural context from which it emerged. Through this reconstruction, I will be able to highlight its canonization process. Thus, I must proceed from words to things, or to retrace the evolutionary history of the label "historical epistemology" in the French language. In this way, we can appreciate its diachronic variations and bring out how—at each stage of the historical path thus marked—"objects," "practices," and changing theories linked to different intellectual, political, and institutional stakes and stances correspond to these concepts.[8] I must therefore employ a sociohistorical epistemology aimed at understanding the dynamics of the "circulation of ideas" that have led to the specific formation of this area of research (Bourdieu 2002; Ienna 2019b).

(2008); Domski and Dickson (2010); Kitcher (2011); Schickore (2011); Arabatzis and Schickore (2012); Stadler (2014, 2017); Arabatzis and Howard (2015); Herring et al. (2019).

[7] By the term *canon*, I mean the identification of certain salient aspects common among a series of authors that would constitute a tradition of thought or an intellectual posture. The process of canonizing authors, instead, has to do with the social circulation of their ideas and the "social branding" that is attributed to them. On this point, see, for example, Bourdieu (2002); Ienna and Santoro (2016); Sapiro, Santoro, and Baert (2020).

[8] This reconstruction process is a necessary prerequisite for objectifying the same theoretical position contained in this text within a wider space of possible definitions relating to this branch of research.

My primary focus is on the type of relationship that is established between a form of "intellectual labeling" and the theoretical contents that have become distinctive in it compared to other traditions and, consequently, the types of relationships that are created between these labels and the theoretical practices of the authors considered to be its representatives or who are (willingly or unwillingly) included within it.

The term *historical epistemology* must be analyzed with caution. First of all, key authors in this tradition, such as Gaston Bachelard, Michel Foucault, Louis Althusser, and Pierre Bourdieu, never used this label. Other representative authors, such as Georges Canguilhem and Dominique Lecourt, used it as a historiographical category. Some theorists of historical epistemology have used this concept in light of their self-recognition as belonging to this label, retroactively constructing an intellectual tradition of which they would be the continuation (Ian Hacking, Arnold I. Davidson, members of the Berlin School linked to the Max Planck Institute). Moreover, Lecourt, Davidson, Hacking, Marx Wortofsy, and the Berlin School do not use this term in precisely the same way (Badino, Ienna, and Omodeo 2022). Finally, there is certainly no shortage of controversy regarding the existence of this category, from both a historiographical point of view and a methodological-programmatic point of view. Historian and sociologist of science Yves Gingras (2010), for example, has contested the validity and usefulness of the term *historical epistemology*. He attempted to undermine the intellectual label of historical epistemology by diagnosing what—in his opinion—would be the pathological aspects of the use of this term (especially referring to the group of scholars at the Max Planck Institute). In my opinion, Gingras's argument is weak and does not undermine the intellectual programs developed by the Max Planck Institute at all. He adopts an argumentative strategy aimed at questioning the existence of the label "historical epistemology" by initially "rediscovering" an older FHE (Bachelard, Canguilhem, Althusser, Foucault). In this sense, Gingras, on the basis of a methodology taken from Bourdieu and Jean-Claude Passeron (as well as the Hacking of *The Social Construction of What?*) was intent on reconstructing the genesis and international circulation of the label "historical epistemology." This objective was not achieved, and the historiographical research used as the basis for his reconstruction is completely lacking.

Moreover, how highlighting "previous uses of a label" should result in the weakening of the intellectual project of the Berliners (who are, among other things, treated as a single block without analyzing, for example, the differences between Lorraine Daston, Jürgen Renn, and Hans-Jörg Rheinberger) remains unclear. Second, Gingras questioned whether the program developed by the Berliners pertained more to the history of science than to historical epistemology. Here, he made the mistake of interpreting the role of the history of science in the program of historical epistemology as merely "ancillary" and not constitutive of the program itself, and thus, he reproduced the legitimized structure of disciplinary divisions in the scientific field. Finally, Gingras changed the focus of his investigation by moving from a reconstruction of the emergence of the label "historical epistemology" to the search for criteria that can represent the methodological canon of this intellectual program. In this regard, Gingras even further radicalized the process of searching for precursors by arguing that the authentic nucleus of research contained under the label "historical epistemology"—in its various linguistic declinations—was already entirely contained in the sociology of knowledge proposed by Mauss and Durkheim with their studies of the forms of classification. Gingras elaborated a genealogical line that, from Comte and passing through Mauss and Durkheim, arrived at the sociology of Bourdieu in which the program of a historical sociology of science would be accomplished in its entirety. In Gingras's argument, this should invalidate the project of historical epistemology (in its French as well as other national versions) because of its complete futility. I consider this argument completely ineffective and Gingras's reconstruction of the historical sources incomplete. However, I believe that the attempt to bring the tradition of FHE closer to the historical sociology of knowledge (and of science in particular) of the Durkheimian and Bourdieusian approaches is interesting and fruitful.

In the context of twentieth-century debates, some "major canons" were formed, generating a broad debate within the scientific community. Among these, FHE, with its key figures of Bachelard, Canguilhem, and Foucault, can be considered paradigmatic, and this book specifically focuses on this tradition. The reader, however, should expect not a philologically detailed treatment of the works and intricacies of the epistemological positions of the individual authors involved but rather an analysis of the general theoretical tensions of the debates in which they participated.

In the following pages, I describe how a broad debate over the legacy of positivism emerged in the transition between the end of the nineteenth century and the beginning of the twentieth. In particular, I focus on the emergence of the historical-critical method in Germany and its links to the French academic field. Indeed, I believe that this can be considered the first intellectual label based on a research program that claimed the inseparability of historical and epistemological analyses of knowledge. For this reason, I believe that the historical-critical method can be considered the most authoritative antecedent of FHE. Again, my goal is to highlight the circulation of intellectual programs and labels rather than focus on the contributions of individual authors.

Second, I closely follow how the first occurrences of the term *épistémologie historique* appeared and how these took on different tones depending on the intellectual climate within which they emerged. In particular, I argue that the intellectual consolidation of this label—as well as its success—can be traced back to the interaction between two main structural factors. On the one hand was the crystallization of the Bachelard–Canguilhem line within the French academic context thanks to Canguilhem's philosophical teachings; on the other was the emergence of the 1968 movements in France and the considerable attention that French intellectuals paid to Marxism and historical materialism. Finally, I highlight how the spread of Bachelard and Canguilhem's ideas in the Parisian academic milieu of the 1960s and 1970s generated a twist or bend—which I see as a rupture, despite the presence of some elements of continuity—in the debates that generated a shift of focus away from the physicomathematical sciences and toward the life sciences, along with the human and social sciences. This process generated a "major canon" composed of authors such as Bachelard, Canguilhem, Althusser, Foucault, and Bourdieu. At the same time, it overshadowed the tradition of the *pensée des mathématiques* of authors such as Jean Cavaillès, Albert Lautman, Bachelard himself, and Jean-Toussaint Desanti, which I propose to identify as a "minor canon" of FHE.

1.2 A Disputed Legacy?

This section details how a broad discussion developed around the legacy of positivism in the transitional period between the nineteenth and early twentieth centuries, resulting in various intellectual tensions that led to the formation of a polarized intellectual space within which FHE developed. I describe the circulation of ideas that generated the emergence of different intellectual reactions to positivism and how these factors were correlated with the dynamics of the institutionalization of the philosophy of science in various national contexts.

An early tendency common to debates within both historical epistemology and the integrated history and philosophy of science (iHPS) involved a return to the predisciplinary stage in which the dialogue between epistemology and the history of science was completely organic as it had not yet been conditioned by their institutionalization as distinct disciplinary branches. In this sense, the prehistory of historical epistemology coincided with the same conceptual genesis as that of the philosophy of science, which, moreover, is a relatively recent term, which emerged simultaneously in France and England in the first half of the nineteenth century.[9] In French, physicist André-Marie Ampère was responsible for inserting the concept of *philosophie des sciences* into the philosophical vocabulary with the 1834 publication of his *Essai sur la philosophie des sciences, ou exposition analytique d'une classification naturelle des toutes le connaissances humaines*. That the expression was rendered in the plural is no coincidence given that, for Ampère, the objective of this particular discipline was to classify the sciences to investigate their specificity (with respect both to each other and, individually and collectively, to the faculties of the intellect) (Lecourt 2001, 13). These were the same years in which Auguste Comte worked on his monumental *Cours de Philosophie Positive*, which he published

[9]For something like the "philosophy of science" to emerge, both philosophy and science had to be conceived as two relatively autonomous forms of knowledge. Robert K. Merton has highlighted how the concept of scientist is a relatively recent invention. Only in the seventeenth century was the term coined, shifting from the English *artist* to *scientist*, to avoid using the generic term *philosopher* to indicate that body of knowledge that had been fragmented into disciplinary specificities to which there was a corresponding social division of research work (Merton 1988). The process by which philosophy and science gained autonomy was, in fact, the nineteenth-century result of the long historical process of the "Scientific Revolution," which led to the emergence of epistemic thresholds capable of functioning as principles of social inclusion/exclusion and professional activities functional to the distinction and specialization of forms of knowledge (Bourdieu 2001). Before the Scientific Revolution, the social figures of the "scientist" and the "philosopher" largely overlapped. Only through the emergence of their social distinction did the idea that philosophical knowledge could deal with science—starting from an external point of view at that point—become legitimate.

between 1830 and 1842. Even for Comte, the primary objective of the *philosophie des sciences* was to identify the principles for classifying human knowledge, their mutual connections, and their hierarchical order. In the Francophone context, Comte was also one of the first authors to clearly state the need to integrate epistemological reflection with the historical reconstruction of scientific thought (and vice versa), and for this reason, he is widely considered one of the "fathers" of FHE (Braunstein 2002b).

In the English-speaking world, William Whewell—a constant reference point in contemporary debates within iHPS—coined the phrase *philosophy of science* with the 1840 publication of *The Philosophy of Inductive Sciences, Founded Upon Their History*. Although the title includes the plural form, it is also found in the singular within the text. This tension clearly reflects the author's dual objective: on the one hand, to classify the various types of scientific knowledge and, on the other, to seek, through historical investigation, the common elements of the multifaceted "discoveries" that characterized the Scientific Revolution and that thus represented rationality.

In the transition between the nineteenth and twentieth centuries, positivism had established itself as an essential point of reference in debates around science. Comtian thought had been a driving force for the professionalization of the philosophy and history of science (but also of sociology). From an institutional point of view, the first university chair in philosophy of science was created in Zurich in 1870, under the title of "inductive philosophy," for the German philosopher and sociologist Friedrich Albert Lange. This position was later occupied by the neo-Kantian Wilhem Windelband and, subsequently, by the father of empiriocriticism, Richard Avenarius (Schnädelbach 1984, 243; Gundlach 2012, 267; Russo Krauss 2019, 15). This chair had been established with the specific aim of putting the hard sciences in communication with philosophy, without the latter's having to lose its autonomy (therefore, in a certain continuity with positivism and neo-Kantianism). Subsequently, in 1895, physicist, philosopher, and historian Ernst Mach was appointed to a chair in Geschichte und Theorie der induktiven Wissenschaften at the University of Vienna, and thanks to his fame, the philosophy of science began to spread and become institutionalized (Lorenzano 2010, 5).

In France, the first chair in this field was established in 1909 with the name histoire de la philosophie dans ses rapports avec les sciences exactes and was occupied by Gaston Milhaud, an intellectual heir of Comtian positivism. This chair was later occupied by Abel Rey (a proponent of an absolute positivism) and, subsequently, by Bachelard and Canguilhem, with the name épistémologie et histoire des sciences.

Even the history of science obtained its institutional autonomy as a discipline only at the end of the nineteenth century in France. In 1832, Comte sent a letter to François Guizot (at the time the minister of public instruction) to officially request the creation of the first chair in histoire générale des sciences at the Collège de France (Petit 1995). After a direct refusal, Comte proposed the position at least twice more (in 1846 and 1848, after the revolution), but it would be created only more than 30 years after his death. This chair was established in 1892 and was occupied by

Comte's main disciple, Pierre Laffitte. However, Laffitte was unable to fully realize the scope of this task, leaving it instead to another Comtian, Paul Tannery, to provide the history of science with a truly scientific form (Sarton 1913, 10; Braunstein 2008).[10] In 1932, Rey founded the Institut d'histoire des sciences et des techniques at the Faculty of Letters at the Sorbonne, whose directorship was later held by Bachelard and Canguilhem.[11]

In this period, a series of intellectual currents emerged over a short time, including pragmatism, conventionalism, empiriocriticism, and instrumentalism. These, though distancing themselves from positivism, remained linked in some aspects to its contentious legacy. In particular, Comtian positivism was at the origin of the variant of epistemology that took root in German-speaking countries under the name of the "historical-critical method," which to my knowledge was the first and clearest example of intellectual labeling in this area—with a precise research program underlying it—aimed at underlining the inseparability of historical and epistemological analyses of science.[12] Despite the relative differences between these traditions, the historical-critical method can be considered not only in direct continuity

[10]Even in the Anglophone tradition, a direct connection to the positivist program can be traced: George Sarton, a Belgian author who emigrated to the US in 1915, published a pioneering work in three volumes titled *Introduction to History of Science* between 1927 and 1947, establishing himself as the father of the history of science, understood as a professional discipline in the US. In his famous opening article of the journal *Isis*, he declared continuity with Comte's project of an *histoire générale des sciences* (Sarton 1913, 10). Sarton was concerned by the opposition that had hardened between humanistic and scientific cultures, and in this sense, he saw in the history of science an opportunity to create common cultural ground (Dear 2009).

[11]For a detailed analysis of the founding of this institute, see Braunstein (2015).

[12]Ronan de Calan has highlighted how the label "historical-critical method" originated in the context of biblical exegesis during the Reformation. The historical-critical method is, in fact, an area of research that aims to create a form of positive knowledge about the Bible, in contrast to the allegorical readings provided by the Church Fathers, the Latin Vulgate taken from the translation of Saint Jerome, and the scholastic interpretations of the Doctors of Church (de Calan 2020). This research method, characterized by a critical-philosophical posture with respect to historical sources, began to circulate over the centuries in other disciplinary contexts, including comparative grammar, political economy, law, and philosophy.

To the points that de Calan raises, I also add that, my still-partial research has shown that the first occurrences of the term *historical epistemology* in English can be dated to the end of the 1800s. Also in this case, these came from the field of theological and religious studies. The first text to use the term *historical epistemology* was Henry H. Davis's 1898 text in *The American Journal of Theology* titled "Origen's Theory of Knowledge." The second appearance of the term, in 1913, also in the US, was in John A. Singmaster's "Current Theological Thought," published in *The Lutheran Quarterly*. The text in question is a long "survey" of approaches to the study of biblical texts. Such occurrences in the field of theological and religious studies therefore may not be entirely random, because they create a further point of contact with the historical-critical method. However, this point merits further investigation. Despite these suggestive hints, I think that the formulation of the historical-critical method in the context of the history and philosophy of science can be found in the historicization of Kantian criticism, which is at the basis, for example, of the philosophical positioning of many authors of the time, especially including Lange, who was the first to use the term in this context.

with but also as an essential element in understanding the development of FHE (de Calan 2020).

The emergence of the historical-critical method as a tool for investigating the structure of scientific concepts from both a historical point of view and an epistemic point of view was favored by the intertwining of multiple cultural and institutional factors in the German academic and scientific fields. The first to explicitly use this term was Lange in *Geschichte des Materialismus* (1866). This text was one of the first manifestations of the neo-Kantian current that Lange himself contributed to founding in Marburg (subsequently developed by writers such as Hermann Cohen, Paul Natorp, and Ernst Cassirer[13]).

In the second part of *Geschichte des Materialismus*, Lange proposed using the historical-critical method to study the principles of physics, which the author intended as a new way of reconstructing the history of the ideas that had generated the transformations in this discipline. In particular, Lange criticized the naive conception that modern materialists had of concepts such as force or energy, as well as the positivistic naivety with which scientists conceived the historical evolution of their work. Both disciplinary and synthetic histories of science were already largely consolidated. Likewise, historiographical reconstructions of the evolution of physics were widespread—among physicists, presenting theorems from a historical point of view was tradition—but these were limited to the reconstruction of the experimental history of physics theories. The major gap highlighted by Lange was the lack of a history of the ideas and conceptions that had made the theorization and correlative experimentation of physicists possible.

De Calan (2020) has provided a detailed reconstruction of the fundamental stages through which this methodology came to permeate the field of the history and philosophy of science at an institutional level. In 1847, the independent and simultaneous discovery of the principle of the conservation of energy, by Julius Robert von Mayer and Hermann von Helmholtz in Germany and by James P. Joule in England, reopened disputes on the foundations of physics. Indeed, in the interpretation offered by de Calan, this scientific event led to the investigation of whether this principle could be considered a confirmation of the mechanistic paradigm (synthesized by the mechanics of Joseph-Louis Lagrange) or put it in crisis, which implied a questioning of the definition of certain central concepts for physics, such as matter, energy, force, and movement.[14]

On the basis of these questions, in 1869—shortly after the publication of Lange's work on materialism—the University of Göttingen announced a competition for a

[13]In particular, I point out that through Lange's support (against that of Kuno Fischer), Cohen managed to obtain the qualification that allowed him, upon Lange's death in 1876, to succeed him in his post. Cohen and Natorp, in particular, recognized Lange as an inspiration, especially from a historical-epistemological point of view (Patton 2004). In the 1914 edition of Lange's text was an introduction by Cohen in which he indicated the need to integrate a reconstruction of the principles of mechanics from both historical and philosophical points of view.

[14]On this point, see Thomas Kuhn's proposed reconstruction in "Energy Conservation as an Example of Simultaneous Discovery" (1977, 66–104).

prize named in memory of philosopher Friedrich E. Beneke for an essay on the critical history of the principles of mechanics, for which complex and detailed submission instructions were written by philosopher Hermann Lotze and physicist Eugen Weber.[15] The objective of the competition was to promote research projects investigating the formulation of the principles of mechanics in the period between Galileo and more recent principles of energy conservation. The first part of the submissions (i.e., the historical part) was to include a reconstruction of how, when, and under what impulse the principles of mechanics had been formulated and modified (or unified). The critical (i.e., epistemological) part of the work, instead, would investigate whether each of these principles of mechanics corresponded to logical foundations—and related mathematical formulations—or whether these foundations were instead the result of universally valid experiences acquired over time. Within this historical-critical reconstruction, the role played by the philosophical conceptions of the various historical eras in the physicomathematical work of formulating the principles of mechanics was also to be highlighted.

The winner was Eugen Dühring, who is known today almost exclusively for his dispute with Engels[16] but who, at the time, was a relatively well known and politically active figure in Germany. With respect to de Calan's reconstruction, I emphasize that Dühring was far from a novice in this methodological tradition. In 1869, the year of the competition, Düring had already published *Kritische Geschichte der Philosophie von Ihren Anfängen bis zur Gegenwart*, showing his mastery of the historical-critical methodology. Alongside this work, he also published *Kritische Geschichte der Nationalökonomie und des Sozialismus* in 1871 and, already in 1872, *Kritische Geschichte der allgemeinen Principien der Mechanik*, an improved version of the essay presented for the Göttingen competition.[17]

---

[15]For a full reading of the submission guidelines, see de Calan (2020).

[16]In 1877, Engels published *Herr Eugen Dühring's Revolution in Science* (better known by the title *Anti-Dühring*), which included the first formulation of the dialectical materialism, which he further developed in *Dialectics of Nature*. Engels sought to demonstrate that dialectics was a scientific procedure—indeed, in some reviews, Dühring had criticized Marx's positions in *Capital* by attacking the scientific nature of the dialectical method. Whereas in the initial formulation of historical materialism in *The German Ideology*, Marx and Engels had limited the application of the dialectical method to social and historical facts, with the *Dialectics of Nature*, Engels went beyond this limitation, arguing that even the physical-natural world is regulated by laws of dialectical transformation.

[17]In the preface to the 1883 first edition of *The Science of Mechanics*, Mach referred to Dühring's works while partly distancing himself from them: "if Duehring's [sic] estimable work, *Kritische Geschichte der Principien der Mechanik* (Berlin 1873), did not particularly influence me, it was that at the time of its appearance, my ideas had been not only worked out but actually published. Nevertheless, the reader will, at least on the destructive side, find many points of agreement between Dühring's criticisms and those here expressed" (1883/1893, ix). These kinds of observations suggest a space for articulating positions in relation to the common stakes of the definition of the historical-critical method aimed at clarifying the problems of the foundations of the physics of the time through an investigation that was, at the same time, epistemological.

Inspired by this competition, many works devoted to reconstructing the principles of mechanics from a historical-critical point of view were published in Germany in the immediately following years. In 1872, Hermann Klein—who also participated in the Göttingen competition—published *Die Prinzipien der Mechanik, historisch und kritisch dargestellt*. In 1873, physicist and philosopher Mach also began to engage in this debate by publishing his first contribution to critical history, titled *Die Geschichte und die Wurzel des Satzes von der Erhaltung der Arbeit*,[18] followed by *The Science of Mechanics* in 1883, which is the best-known example of the canonical historical-critical method in the study of this science (Mach 1872; 1883). Indeed, upon this text, Mach's philosophical authority as a historian and philosopher of science was built, and he obtained the prestigious chair in Geschichte und Theorie der induktiven Wissenschaften in 1895 (Lecourt 2001, 25).

In Mach's formulation, the historical-critical method had a double objective: the first was a detailed reconstruction of the historical and epistemological evolution of the concepts of the history of science (specifically physics in this case) to show, through their genealogy, that they were part of a broader conceptual apparatus within which they acquired a specific sense. The historical-critical method, therefore, restored the precise meanings of scientific concepts in relation to their original contexts of formalization and then focused on subsequent diachronic transformations. The second aim, instead, had to do with the practical applicability of these notions. Thanks to the clarification made through the genealogy of these concepts, their original metaphysical residue, which needed to be eradicated from contemporary scientific activity, could be stressed.[19] According to Mach, contemporary physics required this activity of conceptual clarification, which drew attention to the inadequate interpretations and theoretical assumptions upon which previous scientific activity had been based. Failure to conduct this type of investigation would risk of reproducing the underlying metaphysical apparatus in the practice of scientific research. It was, therefore, a historically informed epistemology in all respects, with its theoretical objectives and methodological program already mature and fully formed.

Mach's thought was a cornerstone upon which a broader debate emerged and from which various intellectual trajectories developed. The aforementioned Avenarius had, in fact, formulated the program of empiriocriticism based on a

[18] In his preface to the first edition of *The Science of Mechanics*, Mach declared that "The elucidations which I here offer are, in part, substantially contained in my treatise, *Die Geschichte und die Wurzel des Satzes von der Erhaltung der Arbeit* (Prague, Calve, 1872)" (1883/1893, viii).

[19] In his preface to the first edition, Mach wrote that "The present volume is not a treatise on the application of the principles of mechanics. Its aim is to clear up ideas, expose the real significance of the matter, and get rid of metaphysical obscurities […]. If the reader's interest is […] to know how the principles of mechanics have been ascertained, from what sources they take their origin, and how far they can be regarded as permanent acquisitions, he will find, I hope, in these pages some enlightenment. All this, the positive and physical essence of mechanics, which makes its chief and highest interest for a student of nature, is in existing treatises completely buried and concealed beneath a mass of technical considerations" (1883/1893, vii).

philosophy of pure experience that was elevated to the status of a rigorous science. The central thesis of this approach—like Mach's—was the impossibility of giving a materialistic or idealistic interpretation of the process of coming into knowledge because this precedes the distinction between the physical and the mental. For both Avenarius and Mach, sensations were the fundamental elements of knowledge through which an empiricist epistemology, updated to the state of nineteenth-century science, could be constructed. Both things and thoughts were seen as the result of networks of sensations that took on meaning based on the relationships among them; the ego itself was nothing more than a complex of memories, feelings, and sensations, all attributable to the same body.

Paradoxically, however, studies of the reception of Mach's work have focused mainly on the circulation of his philosophical conceptions relating to the formulation of phenomenalist monism, the economy of thought, and the psychology of research elaborated in works such as *Knowledge and Error* and *The Analysis of Sensations and Relation of the Physical to the Psychical* (Mach 1905, 1906). However, his contribution to the historiographical debates on *The Science of Mechanics* has been partly overlooked. At that time, though, following the success of Mach's work, the number of publications referring to the historical-critical method grew exponentially, consolidating a true stylistic canon in the history of physics.[20]

During these transformational phases in the debates between the end of the nineteenth and the beginning of the twentieth century, different forms of the rearticulation and critical revision of the initial positivist program emerged. These trends in German-speaking countries also began to spread to France with the arrival of hybrid figures at the intellectual crossroads between science and philosophy who energized what was called the *critique des sciences* (Castellana 2020) by some of its proponents. These events constituted what Enrico Castelli Gattinara called the "intellectual background" for the double articulation of "epistemology and history" upon which the epistemological turning point, guided by Bachelardian thought, was then built (Castelli Gattinara 1996, 80).

Some scientist-philosophers—such as Henri Poincaré, Milhaud, Tannery, Édouard Le Roy, Pierre Duhem, and Émile Meyerson—and philosopher-scientists—such as Émile Boutroux, Henri Bergson, and Léon Brunschvicg—began to debate the foundations of the science, reconstructing their epistemological properties and, in some cases, their historical aspects as well.[21] Mario Castellana, in fact, has noted how "Their contributions, especially focused on the new mechanics and the proliferation of different branches within the field of mathematics, were

[20] A large part of the German-based production continued to follow the institutional indications proposed by the University of Göttingen for the Beneke prize (which also had subsequent editions). Examples can be found in some of the works of Gustav Ludwig Lange, Max Planck, Georg Helm, Joseph Petzoldt, and others (de Calan 2020). For broader insights into the reception and varied uses of Mach's work, see Gori (2018) and Stadler (2019).

[21] For a reconstruction of the hybridization of science and philosophy in this phase, see Leroux (2009, 13–99).

crucial for the 'deconstruction' of positivism and the consequent development of the *critique des sciences*" (Castellana 2020, 29).

As both Charles Alunni and Castellana have pointed out, even in the French context the roots of this interdisciplinary dialogue were partly due to institutional reasons arising from university teaching reforms, which, on the basis of the German model, were supposed to encourage exchanges and discussion among various branches of knowledge (Alunni 2018; Castellana 2020, 32). Moreover, as emphasized at the beginning of this chapter, these authors were absolutely crucial to the dynamics through which the philosophy and the history of science were professionalized and intellectually consolidated in the French context and through which the first chairs in these disciplines were institutionalized. Castellana also noted that the *critique des sciences* has not received adequate attention in the critical literature, because both at the pedagogical level and in its international circulation, it has largely been made to coincide "both historically and conceptually" with the "conventionalist movement […] for the radical criticism of every form of theoretical essentialism" (Castellana 2020, 27). In this regard, Castellana highlighted the specificities of the *critique des sciences*:

> the *critique des sciences*, both in the French and Mach's German contexts […] was also that movement of epistemological thought which, from an analysis of the processes of critical revision that occurred in the science, derived a "method" for studying them by giving a more appropriate place to another of the "essential pillars of scientific discourse," namely "the historicity of science." (Castellana 2020, 28)

On the contrary, Jean Piaget's *Logique et connaissance scientifique* (1967) provided authoritative testimony to the fact that the perception of Mach's reception in the Francophone debates was different. Here, Piaget used the label "historical-critical method" to identify the tradition that, starting from the 1970s, would come to be recognized as FHE:

> The historical-critical method has provided, and continues to provide, an important stimulus for French-speaking authors. The body of work by Duhem, Milhaud, Boutroux, Brunschvigc, Meyerson, Raymond, Koyré, Gaston and Suzanne Bachelard, Canguilhem, and so forth, attests to its fruitfulness in all domains, which extend from the epistemology of mathematics to that of biology. (Piaget 1967, 165)

This generation of philosopher-scientists and scientist-philosophers was oriented primarily around three journals: the *Revue philosophique de la France et de l'étranger*, the *Revue de métaphysique et de morale* and the *Revue de synthèse* (Castelli Gattinara 1996, 114–152; Soulié 2009, 213–249; Alunni 2018, 425–438; Castellana 2020). These were characterized by intense editorial activity focused on the relationship between philosophy and the natural sciences, and consequently, they functioned as the main conduits for circulating ideas in the field of the *critique des sciences*.

Despite this series of convergences, the panorama—to use Le Roy's expression—of the French "new positivism" (Le Roy 1901) was fragmented into different theoretical positions (Brenner 2015a). The epistemological theme of conventionalism took on different tones between Poincaré's first formulations, those developed by his student Duhem (known as the "Duhem-Quine" thesis), Milhaud's extensions, and Le Roy's further radicalizations. Poincaré, for example, devoted a part of *La valeur de la science* to a critical examination of Le Roy's positions, which he labeled "nominalism" (Poincaré 1905).

In spite of this, Mach's thought remained an important interest for and influence on the entire generation of French authors belonging to the *critique des sciences* who saw a point of reference in the historical-critical method. In particular, notable intersections existed between Mach and the different versions of positivism proposed by authors such as Poincaré and Duhem. Mach's *The Science of Mechanics* was translated into French for the first time in 1903. In that same year, Duhem had also published a study of mechanics titled *L'évolution de la mécanique*, which would go on to be considered the equivalent of Mach's text in France. Duhem was asked to review Mach's work, and Mach, after hearing about it, got in touch with him. From there, as Brenner has described, an intense correspondence developed between the two authors as they read, commented on, and reviewed each other's work, mutually developing parallel theoretical trajectories from various points of view (Brenner 2003, 109). Poincaré, in *Science et méthode*, appreciated how Mach's concept of the economy of thought could be applied not only to the study of physical sciences but also to mathematics[22] (Poincaré 1920, 26; Brenner 2003, 115). Rey was also moving in this direction. In *La théorie physique chez le physiciens contemporains* (1907), he addressed controversies relating to how the value of the mechanistic model in physics could be determined (Leroux 2009, 43)—i.e., the same type of problem underlying the emergence of the historical-critical method since the original Göttingen competition. For his part, Brunschvicg, while discussing Lagrange's work in *Les étapes de la philosophie mathématique* (1912), necessarily referred to Mach's study on mechanics (Brunschvicg 1912, 247–248). Meyerson also addressed Machian philosophical perspectives several times in *Identity and Reality* (1908), comparing them to Comte's (Meyerson 1908). Mach's historical-critical method was a central point of reference, especially for Maximilien Winter (one of the founders of the *Revue de métaphysique et de morale*), who explicitly stated the following:

[22] Despite this convergence, elements of theoretical opposition between these authors must be pointed out. Duhem was a Catholic and conservative thinker who progressively focused on the historiographical study of those elements that would demonstrate a continuity between medieval and early modern thought (the continuism thesis, which will be explored in depth in the next chapter). Beyond the initial points of contact, Duhem continued his work as a historian of science while Mach, instead, turned toward epistemological analyses and the development of a theory of knowledge. Among other things, Mach had a progressive orientation and was a point of reference for Austro-Marxism as well as for the nonorthodox Marxism of authors such as Bogdanov (in relative opposition to the beliefs of Duhem). Furthermore, whereas in *The Science of Mechanics* Mach had investigated the processes of forming the principles of mechanics, following a trajectory that from primary experiences and practical knowledge reached high-level epistemological formulations, Duhem and Poincaré for their part instead highlighted the breaking points between common sense and scientific theory.

> In this study, we will not limit ourselves to simply following a historical method. Inspired by Mach's example, we will try to apply to mathematics the critical and historical method developed by this illustrious thinker in the study of the formation of the principles of mechanics; for this reason, we will try to show the evolution of ideas and, in addition to any historical considerations, we will highlight how the elementary notions were formed in their logical concatenation. This method is legitimate because a completely objective history of science is impossible; moreover, the complete publication of everything that appears would be the only truly objective historical method, a method whose absurdity is immediately obvious. Critiques will always intervene to discern the important works that constitute fundamental scientific theories.
>
> But we will nevertheless deviate a little from Mach's approach in the sense that we will not support a specific philosophical thesis […]. We will be more cautious and limit ourselves to showing how, starting from a certain moment in its evolution, science has developed in order to see how scientific theories grow and transform. (Winter 1911, 144–145)

These are just some of the many possible examples of the convergence between the German-Austrian historical-critical research program and the revision of positivism taken up by influential French thinkers. However, the French *critique des sciences* was not the only one that referred to Mach's intellectual project. The different ways that his declination of the positivist program was received became, indeed, the theoretical crossroads at which major theoretical conflicts emerged and which went on to form the bases for the main epistemological positions of the early twentieth century.[23] Among these positions, the reasons behind the formation of what has been called the analytic–continental divide,[24] which corresponded largely

[23] For example, in the Soviet context, Mach had been the main object of contention in the dispute between Lenin and the Russian Machians led by Alexander Bogdanov (Lenin 1909; Bogdanov 1910; Tagliagambe and Rispoli 2016, 19–32). Bogdanov had formulated an empiriomonist philosophy aimed at bringing together the results of the science of the time, the theory of knowledge advanced by empiriocriticism, and the theses of Marxist materialism. On the other hand, for the Lenin of *Materialism and Empirio-criticism*, Mach's thought led to a bourgeois philosophy and epistemology because it was based on the devaluation of the function of materiality in knowledge and, therefore, in complete contrast with Marxist materialism. A declination of Mach's positivist model also circulated in American philosophical debates, exerting an influence on the current of American pragmatism (Holton 1992). For detailed analyses of the various ways that Mach's philosophy was received—though without dedicating much space to the historical-critical method—see Gori (2018) and Stadler (2019).

[24] Among the explanations that philosophers have provided for the polarization of the space of contemporary philosophy, the distinction between analytic philosophy and continental philosophy seems to be crucial for studying the emergence of FHE. The claim of a so-called analytic–continental divide stems, in fact, from a categorization based on a wide variety of factors, including the geographical-cultural niches in which certain styles of thought originated, techniques used, problems addressed, areas of cultural influence in which these variations are most widely accepted, etc. For further discussion on this distinction, see D'Agostini (1997, 2014). These criteria are relatively opaque from an analytical-conceptual point of view but well inscribed in the disciplinary common sense through which the productive and reproductive effects of such taxonomies are generated, which is inscribed in and, at the same time, determines the contemporary philosophical habitus. In this sense, using both historical-philosophical and quantitative tools of a sociological nature, I am examining whether this distinction between these two varieties could correspond to two research paradigms in the Kuhnian sense of the term (Levy 2003; Petrovich 2018).

to the double legacy of the Machian program, are particularly important to explore: on the one hand, the continuing trajectory of historical-critical analysis; on the other, the consolidation of the neo-positivist program. According to the interpretation proposed by Simon Critchley (1997), the genesis of this bifurcation was linked primarily to two ways of thematizing and reacting to the so-called crisis of reason, which characterized the transition between the end of the nineteenth and the beginning of the twentieth century. Piaget summarized the crisis the following way: "at present, the progressive differentiation of knowledge means that the main epistemological innovations arise from scientists' reflections on the conditions of knowledge in their disciplines: this happens especially during crises that force a revision of principles and methods" (Piaget 1967, 79).

An entire generation of intellectuals was, in fact, dealing with the ethical-moral crisis of a world that had been torn apart by two world wars and, at the same time, with the crisis of the so-called hard sciences (as well as their relationship to the human and social sciences). According to this interpretation, analytic philosophy reacted to this "crisis" by seeking a unifying principle in the logical language common to all the sciences from which the program of logical positivism had been generated and which represented the precursors upon which this paradigm was founded. Continental epistemology—of which historical epistemology was an essential element—instead took the problematic nature of the crisis as its own founding criterion.[25] The reflection of this epistemological variant, rather than seeking a priori unifying principles as a possible solution to this critical state, followed the trajectory of a reflection on the various sciences, starting precisely from their intrinsic plurality, historicity, and social contingency.

Mach's proposal to purge metaphysical residue from scientific knowledge was a central point for the neo-positivist program for the scientific foundation of philosophy developed by the Vienna Circle.[26] That the group initially called itself the "Ernst

[25] Among the various interpretations of how reactions to this crisis spurred the development of historical epistemology, I particularly suggest Castelli Gattinara's, which focused on the "double articulation between epistemology and history" (1996); Luca Guzzardi's, which emphasized how, in the transition between the nineteenth and twentieth centuries, physics and psychology called into question the dominant notions of subject and object (2010); and Castellana's, along with Gaspare Polizzi's, which focused on the origins of the divide between analytic and continental philosophy, dating back to the clash at the Paris congresses in 1935 and 1937 (Polizzi 1982; Castellana 1990).

[26] Between 1922 and 1936, a group of scholars from various disciplines met regularly in Vienna and proposed a version of positivism that was up to date with recent scientific discoveries and contemporary philosophical debates. This group, known as the Vienna Circle, was inspired by the empiriocriticism of Mach and Avenarius (from which the professionalization of the philosophy of science had begun). In fact, in 1924, Moritz Schlick obtained the chair previously held by Mach in the philosophy of inductive sciences around which a heterogeneous group of scientists, philosophers, and intellectuals from various disciplines had formed. In addition to Schlick himself—who organized the Thursday afternoon meetings—the regular participants were Hans Hahn, Philipp Frank, Otto Neurath, Rudolf Carnap, Herbert Feigl, Richard von Mises, Karl Menger, Kurt Gödel, Friedrich Waismann, Felix Kaufmann, Viktor Kraft, and Edgar Zilsel. Alongside these were also occasional participants such as Alfred Tarski, Hans Reichenbach, Carl Gustav Hempel, Willard Van Orman Quine, Ernest Nagel, Alfred Jules Ayer, Oskar Morgenstern, and Frank P. Ramsey. Ludwig

Mach Association" and that one of the protagonists was Moritz Schlick, who occupied the Viennese chair previously held by Mach and Ludwig Boltzmann, are not coincidental (Stadler 2007).

However, the dialogue between certain French authors and Mach did not go in only one direction. In fact, Mach promoted the diffusion of some important French works in German. In particular, Mach advocated for Duhem's thought, promoting the German translations of *La théorie physique son objet et sa structure*, for which he had written the preface, and *L'évolution de la mécanique*, translated by Philipp Frank, a member of the Vienna Circle (Brenner 2003, 113). In addition, Poincaré's *La valeur de la science* was first reviewed in 1907 and then translated into German in 1915 by Hans Hahn, another member of the Vienna Circle (Brenner 2003, 119).

The common purpose of French conventionalism and Mach's thought was widely accepted by the members of the Vienna Circle. Duhem and Poincaré were mentioned several times as intellectual points of reference by Hahn, Frank, and Rudolf Carnap in the Circle's famous manifesto, titled *The Scientific Conception of the World* (Hahn, Neurath, and Carnap 1929/1973). Rey's role as cultural mediator must also be considered as the main point of reference for the Vienna Circle in France. His *La théorie physique chez le physiciens contemporains* was translated into German in 1908 (only a year after its French publication), and Rey—despite his reticence regarding an epistemological program centered on logic—promoted a series of French translations of texts by Frank, Neurath, and Schlick for the *Revue de synthèse* and invited Reichenbach to contribute to the *Revue philosophique de la France et de l'étranger*. Alongside these authors, others, such as Marcel Boll and Louis Rougier, proposed various research projects aimed at making the Viennese program known in France. For his part, Jules Vuillemin translated various works by Schlick, Carnap, Reichenbach, and Frank. Both Cavaillès and Bachelard reviewed works by Reichenbach, Hahn, and Popper. Reichenbach's French translations, in particular, would come to be at the center of the controversies in the history and philosophy of physics among Bachelard, Meyerson, Brunschvicg, and Rey (Castellana 2019, 60–61). The dynamic of this international circulation of ideas led to at least two myths. On the one hand, in the Viennese reception of Duhem, Poincaré, and Rey, the specific differences between these authors were flattened due to their common connection to conventionalism.[27] On the other hand, rhetoric regarding French

Wittgenstein and Karl Popper also had close contact with the Vienna Circle, with occasional informal meetings. This Viennese philosophical tradition, also called the Unity of Science movement, proposed an epistemological position (politically committed to the left) whose main objectives were to eliminate metaphysics and to unify science under a single logical language.

[27] In fact, Castellana supported this point: "through awareness of Rey's work *La théorie de la physique chez les physiciens contemporains*, Poincaré and Duhem were reused within the scientific philosophy proposed by the Vienna Circle; many 'conventionalist' theses, in fact, were crucial for the development of some basic theses of logical empiricism, and the relationships between these two research traditions can be better understood within a history of the philosophy of science. But Rey proposed a unitary reading of Duhem and Poincaré's positions, accepted uncritically by the neo-positivist literature, which did not exactly correspond to the reality of French philosophy of

imperviousness to the logicist program was largely consolidated in epistemological debates (Castellana 2019, 60).

The rupture between the neo-positivist program and the historicist one proposed by some of the representatives of French philosophical culture—albeit with significant internal differences in orientation—was made explicit in the immediately following years.[28] Starting in 1900, the first international philosophy congresses were held, favoring the faster international circulation of ideas. Thanks to these events, the main philosophical trends solidified, leading to the emergence of a more complex space of theoretical polarization in the field.[29] Looking at the authors and titles of the presentations of the sessions dedicated to the history and philosophy of science exposes a significant Francophone philosophical cultural hegemony that began to be challenged only between the end of the 1920s and the early 1930s.[30]

---

science, where there was a clear difference between the philosophical-scientific research programs of Duhem, Poincaré, and Rey" (Castellana 2019, 59–60).

[28] Recently, commentators such as Brenner have called into question this radical break between Francophone epistemological culture and Anglophone analytic culture (Brenner 2015b). This position was supported by a recognition of common intellectual influences spanning the two traditions. This was the case of authors like Poincaré, who simultaneously inspired Bachelard, Bertrand Russell, and Carnap (Brenner 2015b, 211). Brenner also showed how the hostility toward logical empiricism on the part of authors such as Duhem, Rey, and Bachelard was less marked than the successors of this tradition wanted to convey. A second aspect stressed by Brenner was related to the participation of some authors in both research paradigms. As has been mentioned, certain authors who participated in the historical epistemology debates, such as Hacking and Davidson, had solid analytic training. These authors developed epistemological conceptions capable of arousing interest on both sides of the analytic–continental divide. Alongside these, Brenner also emphasized how a series of French authors, including Vuillemin and Gilles-Gaston Granger, accepted analytic philosophy with interest, developing original perspectives while not remaining completely outside of the debates of FHE (Brenner 2006, 119–120). On the reception of analytic philosophy in France, see Pudal (2004). On the peculiar forms of the French analytic tradition, see Grange (2014). These attempts to build historiographical and theoretical bridges between these traditions, however, have remained in the minority in historiographical debates.

[29] The creation of international philosophy congresses was a clear indicator of the social transformation of philosophical practices toward full professionalization. As noted by Jean-Louis Fabiani, this process began between the end of the nineteenth century and the beginning of the twentieth century. Much philosophical activity was performed by nonacademics. The philosophical field had an open and heteronomous structure compared to that of other fields of cultural production. Progressively, "philosophy became more and more a purely professorial affair" (Fabiani 1988, 28). As Christophe Charle noted, this process was intrinsically linked to the autonomization of the academic field (Charle 1983, 85). As Soulié has rightly pointed out, the creation of these international conferences played a role in the process of deindividualizing philosophical activity and creating an international community. Such congresses transformed the governance and social regulation of philosophical production, pushing the intellectual field to acquire a community rule excluding amateurism and leading to the progressive specialization and fragmentation of knowledge through a process of social division of labor that—though not organized in collective terms—would represent a clear threat to the reflective unity of philosophical knowledge (Soulié 2009, 113–151).

[30] The historical reconstruction of the theoretical orientations that emerged from international philosophy congresses would require an extensive treatment that cannot be accomplished within

The two main congresses of note were the First Congress of Scientific Philosophy[31] in Paris in 1935 and the International Congress of Philosophy in Paris in 1937, known as the Congrès Descartes.

Despite the short time interval, the landscape of epistemological debates had already started to change. In the wake of the movement begun with the Vienna Circle regarding the neo-positivist program, various intellectual groups were already being formed, including the Berlin Circle, the Polish logicians of the Lviv-Warsaw School, the Göttingen School, and the Münster School. At the same time, a new generation of French epistemologists was also emerging—authors such as Winter, Ferdinand Gonseth, Bachelard, Cavaillès, and Lautman—who were acquiring ever-greater importance. Furthermore, the high esteem in which certain Italians, the most representative of which was Federigo Enriques, were held in international debates should not be underestimated.

The idea of organizing the First Congress of Scientific Philosophy dated back to a conversation between Reichenbach and Rougier in Berlin in 1932. This idea was further developed during the International Congress of Philosophy in Prague in 1934 (Soulez 2015; Bourdeau 2018). Rougier—a key figure in the reception of the neo-positivist program in France—had maintained a close relationship with members of both the Vienna Circle and the Berlin Circle since at least 1931 (Lambert 2015; Bourdeau, Heinzmann, and Wagner 2018, 3–4). Rey, who along with Duhem and Poincaré was considered a key point of reference for the emerging movement for the unity of science, was also among the French scholars who promoted the organization of this conference, offering the support of the Centre international de

the scope of this book. Thus, I limit myself to some essential aspects. The first five congresses, in 1900 in Paris (chaired by Émile Boutroux), 1904 in Geneva (chaired by Jean-Jacques Gourd), 1909 in Heidelberg (chaired by Wilhelm Windelband), 1911 in Bologna (chaired by Federigo Enriques), and the 1924 meeting in Naples (chaired by Guido Della Valle), were dominated by French-speaking philosophers and historians of science in dialogue and interaction with the Germans and Italians. At subsequent conferences—in 1926 in Cambridge, Massachusetts (chaired by Nicholas Murray Butler); 1930 in Oxford; 1934 in Prague (chaired by Zdeněk Smetáček); and 1937 in Paris, titled Congrès Descartes (chaired by Brunschvicg and Raymond Bayer)—and afterwards, the presence of neo-positivism and logicism progressively consolidated. As Castellana pointed out, these congresses were linked to others: "As we know, the community of scholars at the Congress of physicists and mathematicians held in Prague in 1929 was exposed to the theses of logical empiricism: in this meeting, it was decided to organize annual *Tagungen*, dedicated specifically to the epistemology of the exact sciences. At the important meeting in Königsberg in 1930, where Carnap, Heyting, and von Neumann, respectively, presented their theses on logicism, intuitionism, and formalism, Weisman introduced logical neo-positivism. Thus, at the eighth International Congress of Philosophy held in Prague in 1934, the neo-empiricist movement participated together in the 'Unity of Science' session and, on this occasion, it was decided to organize the first Congrès International de Philosophie Scientifique the following year in Paris. After the first *Tagungen* of 1929 and 1930, other International Congresses for the Unity of Science were held in Copenhagen in 1936, Cambridge in 1938, and the last one, before the Second World War, in Harvard in 1939" (Castellana 2019, 41). On these congresses, see also Bianco (2014).

[31] Others followed in the next few years: in Copenhagen in 1936, Paris in 1937, Cambridge in 1938, Harvard in 1939, and Chicago in 1941 (Bourdeau, Heinzmann, and Wagner 2018).

synthèse and, above all, the Institut d'histoire des sciences et des techniques (Brenner 2018).

Despite the already substantial presence of the Viennese group at another series of conferences, the Parisian meeting of 1935 was the first occasion in which the theses of this new epistemology were presented at an international level (see Polizzi 1983). The conference was organized in various sessions: (I) *Philosophie scientifique et empirisme logique*[32]; (II) *Unité de la science*[33]; (III) *Langage et pseudo-problèmes*[34]; (IV) *Induction et probabilités*[35]; (V) *Logique et expérience*[36]; (VI) *Philosophie des mathématiques*[37]; (VII) *Logique*[38]; and (VIII) *Histoire de la logique et de l'empirisme logique.*[39] This meeting was particularly significant because it gave space to the a wide range of theoretical perspectives. As can be seen from the names of the participants reported in the notes, the conference was attended by members of the main groups that drove the epistemological turn of neo-positivism and logical empiricism, including the Vienna Circle and the Berlin Circle as well as the Lviv-Warsaw School, the Göttingen School, and the Münster School, whose official publication outlet was the journal *Erkenntnis*, founded in 1930 by Reichenbach and Carnap.

One of the first aspects that the conference clearly revealed was the fragmentation of the original neo-positivist program. The unity of intent expressed in *The Scientific*

---

[32] This volume of the conference proceedings included texts by Rougier, Enriques, Frank, Russell, Reichenbach, M. Ajdukiewicz, Charles W. Morris, Carnap, Neurath, Kotabinski, Adam Wiegner, and Léon Chwistek (Rougier, Neurath, and Lautman 1936a).

[33] This volume of the conference proceedings included texts by Frank, P. Lecomte du Noüv, Egon Brunswik, M. Gibrat, Neurath, C. G. Hempel and Paul Oppenheim, Emil J. Walther, Carnap, and Morris (Rougier, Neurath, and Lautman 1936b).

[34] This volume of the conference proceedings included texts by Alfred Tarski, Marja Kokoszyn, Louis Massignon, Paul Masson Oursel, Jules Richard, M. Chevalley, Allessandro Padoa, Thomas Greenwood, Rougier, Georges Matisse, Ferbert Feigl, and Jules Vuillemin (Rougier, Neurath, and Lautman 1936c).

[35] This volume of the conference proceedings included texts by Reichenbach, Schlick, Carnap, Bruno de Finetti, Zygmunt Zawirski, and Janina Hosiasson (Rougier, Neurath, and Lautman 1936d).

[36] This volume of the conference proceedings included texts by Ajdukiewicz, Cornelius Benjamin, Paul Renaud, Gérard Petiau, Jean-Louis Destouches, Jacques Métadier, Edouard Habermann, Chwistek, M. R. B. Braithwaite, M. E. Tranekjaer Rasmussen, and M. Grelling (Rougier, Neurath, and Lautman 1936e).

[37] This volume of the conference proceedings included texts by Gonseth, Lautman, Goustave Juvet, Georges Buligand, Destouches, Basilio Mania, M. St. Jaskowski, Arnold Reymond, M. Albr. Becker, and M. Schrecker (Rougier, Neurath, and Lautman 1936f).

[38] This volume of the conference proceedings included texts by Tarski, Olaf Helmer, Eugeniu Sperantia, Adolphe Lindenbaum, F. Bachman, Alessandro Padoa, G. Molfitano, A. Honnelaitre and A. Bollengier, and Hugo Bergmann (Rougier, Neurath, and Lautman 1936g).

[39] This volume of the conference proceedings included texts by Heinrich Scholz, M. Jasinowki, Antoinette Reymond and Friedrich Bachmann, Alessandro Padoa, Einar Tegen, Walter Hollitscher, Alfred J. Ayer, P. Zervos, Jörgen Jörgensen, Frank, Rougier, and Fritz Heinemann (Rougier, Neurath, and Lautman 1936h).

Conception of the World—which had been further consolidated with the first conferences on the topic—was giving way to the emergence of a polarized space of internal positions within this movement (relating to both the orientation of the individual authors and the various schools). For example, differences in orientation and misunderstandings had already arisen between Rougier and Neurath while planning the conference (Bourdeau 2018). The members of the Berlin Circle entered into theoretical tension with the Viennese, but their affinity with the Göttingen School was strengthened (Benis-Sinaceur 2018), and at the same time, the Polish Lviv-Warsaw School gained momentum at the international level (Miskiewicz 2018).

The rise of Nazism caused the initial nucleus of the Vienna Circle to disintegrate because many of its members had already begun the expatriation that took them to the United States. Only Schlick chose to remain in Austria, and the year after this conference, he was killed on the steps of the University of Vienna by a Nazi student. This issue is far from irrelevant because it generated a transformation of the theoretical structures of the movement. Schlick's murder brought the initial program of the Vienna Circle to a definitive end, giving rise to the phase that Carnap called the "liberalization of logical empiricism" (Feigl 1968; Castellana 2019, 53). Most of the members of the Vienna Circle were united by their affiliation with Austro-Marxism, which, following their US migration, was significantly weakened. This dismantling of the political dimension of the initial neo-positivist program was also accompanied by contact with the US empiricist tradition, especially for Carnap, Neurath, and Reichenbach. Thus, paradoxically, at the precise moment in which the unity of the positivist movement began to fragment, it acquired full international notoriety. The 1935 conference in Paris was the first opportunity for wide-ranging discussions in which the authors at the center of the epistemological turning point of neo-positivism and logical empiricism could confront each other in their different trajectories and internal polarizations. However, exponents of Franco-Italian culture also participated in the conference, in particular Enriques, Cavaillès, Lautman, Gonseth, and Jean-Louis Destouches, who opposed these intellectual trends in their contributions. Effectively, a double polarization emerged from the 1935 Paris conference: both internal polarization within the neo-positivist movement and external opposition.

However, at the Congrès Descartes of 1937, the clash between the French tradition and logical empiricism took on a clearer outline. This meeting was organized by Brunschvicg and Raymond Bayer, who during the 1934 Prague conference had arranged to bring the conference to Paris. Thanks to the international notoriety that they had acquired from the 1935 conference, representatives of various currents of logical empiricism, such as the former Viennese group, the Berlin Circle, the Lviv-Warsaw School, etc., were invited to this meeting. Before Schlick's death, he had already prepared a speech titled "The Vienna School and Traditional Philosophy" (Schlick 1979).

In the context of the epistemological sessions, Rougier and Vuillemin succeeded in gaining approval for one devoted explicitly to the unity of science. Interestingly, this session was coordinated by Bachelard—"the silent guest of the congress"—who preferred not to intervene, due to his total disagreement—as will be seen in detail in the next section—with the very idea of the unity of science[40] (Alunni and Brian 2002, 130; Alunni 2018, 182). Castellana has noted how the internal structure of this session "was set up according to the French historical tradition" (Castellana 2019, 46). The six sessions were divided as follows: (I) *Le problème de la raison* (Enriques); (II) *L'Unité de la Science* (Carnap, Neurath, Reichenbach, Schlick, Rougier, and Gonseth); (III) *L'unité de la méthode* (Kotabinski); (IV) *Formation de la Science*; (V) *La méthode de l'histoire*; and (VI) *L'unité de la science dans l'histoire de la pensée* (Bayer 1937b, 1937c). Lautman, instead, coordinated the session dedicated to logic and mathematics—where Hempel, Tarski, Oppenheim, Cavaillès, and Lautman himself, among others, presented their talks (Bayer 1937d). Destouches chaired the session focused on the philosophy of physics, *Causalité et déterminisme*—whose participants included Louis De Broglie, Frank, Vuillemin, and Destouches himself (Bayer 1937e). A large part of the conference was, obviously, dedicated to Cartesian studies. Among the participants in the session dedicated to *La méthode et les mathématiques* were Milhaud and Rey; Alexandre Koyré presented in the session dedicated to *La physique*; and Canguilhem presented in the session on *La morale et la pratique* (Bayer 1937a).

During the conference, clear opposing positions were articulated between the two traditions, which were in competition for the legitimacy of the legacy of positivism and its Machian and conventionalist evolutions despite their coming from the same cultural background. The attempt to redefine the relationship between science and philosophy was a common aspect of both the tradition of logical empiricism and the emerging neo-rationalism, marked by a strong historicist and antilogicist tendency, which was largely French but also included the Italian Enriques. Even the theme of the unity of science was certainly not unknown to authors such as Enriques, Cavaillès, Lautman, and Gonseth, but rather, to them, it had a different version from the results of logical empiricism. Thus, during this conference, one of the first signs of what would take the form of the analytic–continental divide in the following years emerged:

> Two years earlier, the First Congress of Scientific Philosophy had established the success and international scope of Viennese logical empiricism, initiating a theoretical turning point within this movement. However, the Congrès Descartes radicalized the irreconcilability of the neo-positivist positions of the initial project of a *Wissenschaftliche Weltauffassung* (Carnap), created an openness to a variety of much less homogeneous philosophical positions, and became the place for an official encounter between two very different conceptual traditions and styles: one provided a preview of what would become analytic philosophy, destined to become a standard current of the twentieth century; the other (Italian-French) shared fundamental principles in the attempt to re-inscribe the various sciences and their history within a neo-rationalist project of critical epistemology. (Alunni and Brian 2002, 131)

[40]Bachelard had participated in the 1934 conference in Prague with a paper titled "Critique préliminaire du concept de frontière épistémologique," in which he had presented the theoretical foundations of his epistemological break with common sense.

Both of these traditions measured themselves against the same research topics: Einstein's theory of relativity; the emergence of quantum mechanics; the logical-mathematical thinking of David Hilbert, Kurt Gödel, and Gerhard Gentzen; set theory; and so forth. But the ensuing philosophical reactions went in radically opposite directions. In particular, the essential point of divergence between these traditions was related to the interpretation attributed to the nature, development, and structure of mathematical thought. Mathematics had always been considered the purest scientific activity, and consequently, any position taken regarding it would certainly generate an impact on the broader understanding of epistemology.

The logicist epistemological line of authors, such as Gottlob Frege, Russell, and Carnap, solidified around their definition of the concepts of pure mathematics in purely logical terms, especially those assumed to be primitive. From this followed the possibility of deriving the truths of pure mathematics (and in particular those considered most evident), starting from purely logical principles and using completely explicit reasoning methods. In this reductionist sense, mathematics was nothing more than a part of logic.

In the dynamics of the Parisian conference of 1937, Enriques is a central figure for understanding a series of ruptures that shaped epistemological debate for the first half of the twentieth century. At the time of the conference, Enriques was one of the most prominent figures on the international philosophical scene. He had already published *Problems of Science* in 1906, *Razionalismo e storicismo* in 1909, *Scienza e razionalismo* in 1912, *The Historic Development of Logic* in 1922, *Il significato della storia del pensiero scientifico* in 1934, and other important works in which he had sought a "synthetic or critical" vision of the science (Enriques 2016) as well as laid the foundations for a historical-epistemological study of scientific activity. For this reason, Enriques was recognized by both factions as a precursor of neo-positivism as well as neo-rationalism. In the manifesto of the Vienna Circle, Enriques was mentioned as one of the authors of reference in the study of the foundations, aims, and methods of empirical science (Hahn, Neurat, and Carnap 1929/1973, 304). Similarly, at the first conference on scientific philosophy in 1935, Enriques was considered to be "at a similar level to Bertrand Russell. Both appear as the true fathers of the movement of logical empiricism and modern scientific philosophy" (Stöltzner 1998, 463). On the other hand, in 1934, Bachelard recognized Enriques as the author of "a reversal of the epistemological perspective" (Bachelard 1934/1984, 50) and later continued to refer to him in other important works, such as *Essai sur la connaissance approchée* and *The Formation of the Scientific Mind*:

> the neo-positivists' reading of Enriques's program was an external, almost self-justifying reading of their own theoretical choices, traceable in some of their precursors […]. Moreover, precisely in these years, Enriques distanced himself from logical neo-positivism in his reviews of the main works of the protagonists of the neo-empiricist movement and in his interventions at international conferences. (Castellana 2019, 66)

At previous international conferences, Enriques had already clearly emphasized how necessary the historical dimension was for every possible epistemological investigation.[41] In his 1937 presentation "Il problema della ragione," he outlined the fundamental characteristics of a "new epistemology" aimed at understanding the unfolding of reason in the history of scientific thought through historical-critical reconstruction. The axis of epistemological interest shifted from the fundamentals of science to the analysis of the contents of scientific theories. In complete contrast to the logicist program described above, it was not a matter of reducing mathematics to its logical foundations but rather a matter of looking at the historical evolution of mathematics as a source of epistemological reflection. Thus, the "new epistemology" was founded on the historical criticism of scientific concepts. This essential point was well understood by French-speaking epistemologists such as Bachelard, Cavaillès, Lautman, and Gonseth, with whom a programmatic harmony could be established. At this intersection, the main coordinates of the method of FHE were established.

In fact, at the same conference, the French-speaking authors also focused their presentations on showing how mathematical thinking is distinguished by its own creativity, characterized by a dialectical activity that is irreducible to logic. Gonseth, in the wake of Enriques's *Problems of Science*, focused his talk on the idea that the meaning of mathematical notions is never closed and fixed once and for all. Meaning, in fact, depends on the state of knowledge itself, which is by its nature in constant "evolution, extension, or withdrawal." The knowledge contained in mathematical concepts is, therefore, always partial, imperfect, approximate, and summary (Gonseth 1937, 64). Cavaillès devoted his paper "Réflexions sur le fondement des mathématiques" to proposing a modified formalist theory that did not lead to logicalistic outcomes. He focused on how mathematical activities can be founded by reducing them to something else. In this sense, mathematics is not just "intuitions" but also an "original historical becoming":

> The initial objects of mathematics are neither whole numbers, nor elements of a special intuition, but the result of concrete attempts to organize the action of consciousness in the world. At every moment of history, mathematical data is the system of objects on which current mathematics is based. (Cavaillès 1937, 137)

In "Sur la réalité inhérente aux théories mathématiques," Lautman did not hesitate to define philosophical reflections confusing mathematical philosophy with the study of logical formalisms as "flawed conceptions." This theoretical posture would lead, he claimed, to the support of a tautological character of mathematics (Lautman 1937/2011, 27).

The presentations at the conference were therefore oriented toward deconstructing the trajectory of logical empiricism. Bachelard, Gonseth, Cavaillès, and Lautman had already been working on an epistemology that recognized in mathematics a condition of irreducibility to any form of logical reductionism (I will return to this point in sections 1.5 and 2.5). In the wake of the debates at the Congrès Descartes, Gonseth, starting from 1938, organized a series of conferences and meetings called Entretiens de Zurich, in which authors such as Bachelard,

[41] On this point, see Castellana (2016b, 2019, 63–92).

Cavaillès, Lautman, Henri Lebesgue, Destouche, Enriques, Jan Łukasiewicz, and Piaget participated.[42] In addition to the publication of a series of texts, this group also founded a journal, *Dialectica. Revue Internationale de philosophie de la Connaissance*, in 1947 (Castellana 2019, 47). The Congrès Descartes was the last large public opportunity for the international philosophical community to discuss these issues before the start of the Second World War. The devastating effects of the war had significant repercussions on the field of epistemology: Authors such as Cavaillès and Lautman died in the French resistance, and Enriques died in 1946. The philosophical panorama therefore had, in part, new protagonists.

The program set forth for FHE was thus shaped against this cultural and institutional background. The epistemological production of the following years led to the development of two main trajectories. The first, which is examined in the next two sections, followed a line of continuity between authors such as Bachelard, Canguilhem, Foucault, Althusser, Bourdieu and another long series of authors that, together, make up what I call the "major canon" of FHE. The second trajectory, to which sect. 1.5 of this chapter is devoted, instead sought to extend philosophical reflection to the field of the philosophy of mathematics, forming what I call the "minor canon" of FHE.

1.3 The Emergence of the Label "*Épistémologie Historique*"

In light of what has been said so far, I must examine how and when the label "*épistémologie historique*" "emerged" in the French context. Various sources have claimed that the first to use the term was Dominique Lecourt, a student of Althusser at the École Normale Supérieure in Paris (in 1965) and of Canguilhem at the Sorbonne. In 1969, Lecourt published his *mémoire de maîtrise* (master's thesis), titled *L'épistémologie historique de Gaston Bachelard*, in which the term appears in the title itself. Lecourt overtly claimed authorship of the expression, claiming that

> Having been told that this point of lexicological history is the subject of today's debate, I would like to emphasize that I owe the expression *épistémologie historique* to Georges Canguilhem, who used it to designate Gaston Bachelard's philosophy of science; this expression appeared for the first time in the title of my *mémoire de maîtrise*, published with a preface by Canguilhem in 1969.
>
> A few years later, I discovered that the same term had been used in English, quite independently, by Marx Wartofsky[43] […] to characterize his own Marxist-tinged methodology in the history of science. (Lecourt 2007, 30)

[42] On the dynamics of the formation of the so-called Zurich School, see the 2005 special issue of the *Revue de synthèse* titled *Sciences et philosophie au XXe siècle. L'École de Zurich et le programme surrationaliste*, 126 (2).

[43] Wartofsky was the first author to use the term *historical epistemology* in English, giving it a distinctive twist. His privileged points of reference were Piaget's genetic epistemology and Marx's historical materialism, which Wartofsky used as tools to investigate how Kantian a priori emerged

Further proof of Lecourt's authorship of this expression can be found in an anecdote reported by Jean Gayon, who recounted that when Canguilhem was revising Lecourt's text in his capacity as supervisor, he suggested the following correction: "Not *épistémologie historique*, but *histoire épistémologique*" (Gayon 2003, 53; Bontems 2006, 139–140). In this sense, there had clearly been a disagreement between Canguilhem and Lecourt regarding the intellectual label to describe the specific theoretical program developed by Bachelard compared to other contemporary trends in epistemology. This symbolic negotiation was definitively won by Lecourt, who kept *L'épistémologie historique de Gaston Bachelard* as the title of his book, for which Canguilhem wrote a laudatory preface. A few years later, Lecourt published his second book, titled *Pour une critique de l'épistémologie: Bachelard, Canguilhem, Foucault*, in which, for the first time, these three authors were identified as representatives of a common intellectual movement. The Bachelard–Canguilhem–Foucault triptych would eventually come together to form the most canonical representation of FHE (Lecourt 1972; Braunstein 2002a).

In reality, however, the term *épistémologie historique* was already in use and becoming widespread both in France and in other national and linguistic contexts, even though, in some cases, it was used only occasionally (here, I refer to only those of Francophone origin). A few years earlier, for example, Georges Gusdorf had used this term in *Introduction aux sciences humaines* to refer to the German historicist tradition from Wilhelm Dilthey onward, personified, among others, by Heinrich Rickert and Max Weber (Gusdorf 1974, 412, 521). But even more significant was that—beyond the anecdote reported by Gayon—Canguilhem himself had already used the term *épistémologie historique* on various occasions and with different nuances.[44] In 1965, in fact, he had published the second edition of the volume *Knowledge of Life*, which included a series of previously unpublished texts and lectures from 1946 to 1952. In particular, in the essay "The Living and Its Milieu" (dated between 1946/1947), Canguilhem, analyzing the relationship between Georges-Louis Leclerc de Buffon and Jean-Baptiste Lamarck, wrote:

> as teacher and precursor to Lamarck in his theory of milieu, Buffon appears at the convergence of the two components of the theory: the mechanical and the anthropo-geographical. Here is posed a problem of *epistemology and of the historical psychology of knowledge*, a problem whose scope greatly exceeds the present example. Shouldn't we interpret the fact that two or more guiding ideas combine at a certain moment to form a single theory as a sign that—in the final analysis and despite their apparent differences—they have a common origin, whose meaning and very existence we forget when we consider separately their disjointed parts? (Canguilhem 1952/2008, 101; emphasis added)

from historical and social conditions related to practice. With the label "historical epistemology," two strands of research converged: the study of the role of scientific thought patterns, representations, and practices in the construction of scientific theories and an analysis of the relationship between history and practice. On this point, see Wartofsky (1968, 1979, 1983).

[44] On four occasions, Canguilhem has been found to have used the term *épistémologie historique* (see Limoges 2012, 57; Méthot 2013, 116). In one of these, he used this term to indicate Kuhn's project of integrating the history and philosophy of science.

In this passage, Canguilhem used the expression "historical epistemology and psychology of knowledge" to refer to a methodology relating to the study of the formation of scientific knowledge. The reference to psychology is certainly not surprising if we remember that the subtitle of Bachelard's *The Formation of the Scientific Mind*, published in 1938, was *A Contribution to a Psychoanalysis of Objective Knowledge* (Bachelard 1938/2002). The second occasion in which Canguilhem explicitly used the term *épistémologie historique* was in a 1963 article on the reconstruction of Bachelard's relationship with the history of science:

> We must understand Bachelard's precise position in relation to the history of science. In one sense, he never really does it. In another sense, he never really stops doing it. If the history of science consists in taking inventory of the variants in subsequent editions of a treatise, Bachelard is not a historian of science. If the history of science consists in making sensible—and at the same time intelligible—the difficult, contested, recovered, and rectified construction of knowledge, then Bachelard's epistemology is a history of science that is always in progress. Hence his interest in errors, horrors, disorders, in everything that represents the fringes of history that are not covered by historical epistemology. [...]
>
> In short, until philosophy provided the history of science with the key concept of epistemological obstacle, epistemology risked being the victim of an overly candid history of science, "which almost never restores the obscurities of thought," causing us to "mistake all the glimmers of the past for lights." The epistemologist is thus led towards a historical psychology of the scientific spirit. (Canguilhem 1994, 178)

Whereas in the first of the two quotes, the label "historical epistemology [and psychology] of knowledge" positively represented a way of analyzing knowledge, adopted by Canguilhem himself against the background of a Bachelardian reference, in the second, the situation is completely reversed. In fact, in this passage, he used this expression to describe a research program opposed to Bachelard's. *Épistémologie historique*, according to this use by Canguilhem, was a history of science that is "overly candid," which reasons exclusively in terms of a positive development of knowledge:

> The historian proceeds from the origins towards the present in such a way that contemporary science is always, to a certain extent, foretold in the past. The epistemologist proceeds towards the beginning, so that only a part of what was taken for science yesterday is, to some extent, still found to be grounded in the present. (Canguilhem 1994, 178)

On the contrary, the Bachelardian project, defined by Canguilhem as the "historical psychology of scientific knowledge," consisted in dwelling on "errors," on "epistemological obstacles," and on moments of "rupture"[45]:

> But, as a counterpart, attention to epistemological obstacles will allow the history of science to be authentically a history of thought. It will protect the historian from the false objectivity which consists in creating an inventory of all the texts in which the same word appears in a certain era, or in different eras [...]. The same word does not express the same concept. It is necessary to simultaneously reconstruct the synthesis in which the concept is inserted, that is

[45] Following current specialist usage in English, I use *rupture* to translate the French term *rupture* used by Bachelard and *break* to translate the French term *coupure* used by Althusser.

> to say, both the conceptual context and the guiding intention of the experiences or observations. (Canguilhem 1994, 177)

Though Canguilhem's theoretical approach remained the same, he clearly used the same expression to connote different intellectual positions. So, when Canguilhem referred to *épistémologie historique* in negative terms, to whom was he actually referring?

As has been rightly pointed out by Braunstein and Brenner (Braunstein 2012, 35; Brenner 2016, 160), Rey[46]—Bachelard's direct predecessor as professor at the Sorbonne—was actually responsible for the first formulation of this concept within the Francophone scene. Thus, Canguilhem could have been referring only to Rey when he used the term *épistémologie historique*[47] with a negative meaning. In 1907, in the introduction to his doctoral thesis (directed by Émile Boutroux) published under the title *La théorie physique chez les physiciens contemporains*, Rey wrote:

> Philosophy, especially American philosophy, has given the name epistemology to transcendental research into the principles and general conditions of science. This word, reduced to a positive meaning, could happily designate documentary research, the set of historical observations necessary for a precise, comprehensive vision of the various sciences, a positive science of sciences. And the historical research relating to the general spirit of each science would, then, be one of the main elements of this *épistémologie historique*. (Rey 1907, 13)

The label "*épistémologie historique*," in this context, indicated a research program that clearly distinguished itself from the Anglo-American logicist program relating to the search for the principles and general conditions of scientific knowledge. Rey did not make any explicit reference to the authors whom he had in mind when he expressed his opinion on this debate, but he referred to a previous article of his (Rey 1907, 11), in which John Stuart Mill, Alexander Bain, and Charles Renouvier were mentioned. As Brenner rightly emphasized, Rey opposed the logicist program developed by these authors, proposing instead a historical approach aligned with the Comtian line of thought[48] developed by Tannery, Milhaud, and Duhem (Brenner 2016, 161). As we have seen, however, shortly afterward, Rey established himself as one of the main Francophone intellectual points of reference for the Vienna Circle. On the one hand, logical positivism had found a fruitful source of inspiration in his and Duhem's work. On the other hand, although Rey had expressed reluctance regarding an epistemological program centered on logic, he

[46] As noted previously, Rey was of significant institutional importance in the field of French epistemology. A professor at the Sorbonne in the history and philosophy of science, Rey was also the founder and director of the Institut d'histoire des sciences et des techniques in 1932 (Braunstein 2015). After Rey, Bachelard and Canguilhem would each take on the dual role of the Sorbonne professorship and director of the institute. Bachelard renamed this chair épistémologie et histoire des sciences.

[47] This thesis is further supported by the fact that at the beginning of Canguilhem's contribution, Rey is cited as a direct precursor of Bachelard (Canguilhem 1952; Brenner 2016, 168).

[48] Rey, extending Comte's program with some radicalizations, proposed an intellectual model that he called absolute positivism (Rey 1909).

had personally facilitated the translation of texts by Frank, Neurath, and Schlick into French for the *Revue de synthèse* and had worked hard on the organization of the 1935 Paris Congress of Scientific Philosophy.

The continuities between Rey's ideas and those of Bachelard are notable, such that many of Bachelard's theses were partly contained in nuce in Rey's texts. Despite this, Canguilhem tended to emphasize Bachelard's moments of rupture, in relation to a discontinuous conception of the developmental process of scientific thought, compared to the continuous conception supported by Rey.[49] In this sense, Canguilhem's hesitation and ambiguity in the quotations above become clear. Although he wanted to keep history and epistemology united, in line with the principles of Bachelardian epistemology, he hesitated to use the label "*épistémologie historique*" because Rey had already used it in elaborating a continuist historiographical research program against which Bachelardian philosophy had been established.

1.4 Intellectual and Institutional Context

From the point of view of a historical socio-epistemology of the intellectual label "*épistémologie historique*," what changed during the 60 years between Rey's first formulation and the now famous one by Lecourt must be investigated. Moreover, why exactly Lecourt's formulation ended up being established as an intellectual label must be asked. What factors led Canguilhem, despite his initial reluctance, to accept[50] the term *épistémologie historique* proposed by Lecourt in 1969? What intellectual and institutional changes had occurred between 1907 and 1968 in the field of French science studies? In fact, the label "*épistémologie historique*" seemingly did not have widespread success before the advent of 1968.

An initial terminological clarification is warranted here. To understand the truly innovative scope of the circulation of the label "*épistémologie historique*," I must stress that it is, for all intents and purposes, redundant in French. The word *épistémologie* was introduced into contemporary French in the early 1900s, quickly acquiring the meaning of the "philosophy of science" (Brenner 2016, 162). Meyerson was one of the first authors to use it in this sense, from the first lines of the foreword of his 1908 work *Identity and Reality*: "The present book belongs, by reason of its method, to the domain of the philosophy of science or *épistémologie*, to

[49] Indeed, we should consider that the 1963 text in which Canguilhem took a position "against" Rey's historical epistemology was published the year after Bachelard's death. The text is, therefore, of a celebratory nature and aimed at highlighting the innovative aspects of Bachelard's thought compared to the intellectual context preceding him.

[50] This was an implicit acceptance, demonstrated by the fact that Canguilhem not only approved the definitive version of Lecourt's thesis but also promoted its publication as a book with a glowing preface.

use a word sufficiently appropriate and now becoming current" (Meyerson 1908/1930, 5). In the preface to the second edition, he would add:

> Following a program outlined but not implemented by A. Comte, we wanted to identify, *a posteriori*, the *a priori* principles that guide our thinking in its pursuit of reality. This is why we analyze science, not to extract what have been considered its *results* (as materialists and "natural philosophers" have often done), nor to draw inspiration from its methods (as positivists claim), but to consider it as the raw material of work, as a tangible example of human thought and its development. (Meyerson 1951, viii)

André Lalande, in his famous *Vocabulaire critique et technique de philosophie* of 1926, emphasized that the word *épistémologie*

> refers to the philosophy of science, but in a very precise sense. It is not strictly the study of scientific methods, which is the object of Methodology and is part of Logic. Nor is it a synthesis or a hypothetical anticipation of scientific laws [...]. It is essentially the critical study of the principles, hypotheses, and results of the various sciences, intended to determine their logical (not psychological) origin, their value, and objective scope. (Lalande 1926, 293)

Moreover, Lalande highlighted the substantial difference between the French word *épistémologie* and the English word *epistemology*:

> The English word epistemology is very frequently used (contrary to its etymology) to designate what we call the "theory of knowledge" or "gnoseology." [...] In French, one should correctly say only philosophy of science, in the way it was defined above, and philosophical history of the sciences. (Lalande 1926, 293)

This word in French, therefore, designates a particular way of practicing the philosophy of science—understood neither as methodology nor as logic but rather as a critical study of the principles, hypotheses, and results of the different sciences—which takes the form of a philosophical history of science. Similarly, Pierre Jacob emphasized the substantial difference between the English and French expressions while also focusing on another aspect implicit in Lalande's definition. In French, the regionalist and pluralistic approach led to the now common expression *philosophie* or *histoire des sciences*, expressed in the plural, as opposed to the expression *philosophy or history of science* in the singular.

> 1) In the English expression philosophy of science, the genitive is in the singular and announces the idea of the unity of the scientific process, while in French the expression "philosophy of science" is rarely used, and the usual expression "philosophy of sciences" suggests the idea of the plurality of scientific disciplines. 2) In contemporary English, the term epistemology refers to the theory of knowledge in general, while in contemporary French the term *épistémologie* indicates the study of solely scientific knowledge. (Jacob 1991, 285–286)

In light of this, clearly, French epistemology can only be a historical epistemology, or a philosophical history of the sciences, as this intellectual orientation is already largely crystallized in the semantics of the term as it came to be, and continues to be, used in France. The list of authors who can be included in the category of FHE, understood in a broad sense, includes almost the entire French epistemological tradition as it has been historically informed, in terms of both theoretical orientation and the internal dynamics of the French intellectual field.

From a historiographical point of view, many have attempted the arduous task of tracing common features, as well as lines of intellectual affiliation within this tradition, to identify its salient characteristics. Below, I present just a few exemplary cases out of a wide range of possibilities.[51]

One issue with universal agreement is the centrality of Bachelard to this philosophical context. On the one hand, some have emphasized how historical epistemology tends to coincide with the Bachelardian legacy in epistemology (which has also crossed French borders). Thus, a line of intellectual influence could be traced that starts from Bachelard, would pass down to Canguilhem, and then would spread through Althusser, Foucault, Gilbert Simondon, Desanti, Michel Serres, Hacking, Davidson, and so forth (Lecourt 1972; Davidson 2001; Braunstein 2002a; Hacking 2002; Bontems 2006, 2010; Chimisso 2008). On the other hand, an attempt has also been made to trace lines of continuity between Bachelard (and some of his contemporaries) and the earlier epistemological tradition: Comte, Duhem, Meyerson, Rey, Brunschvicg, Cavaillès, Lautman, Koyré (Vinti 1977; Redondi 1978; Braunstein 2002b, 2008; Brenner 2003; Cavazzini and Gualandi 2006; Chimisso 2008). In this way, a long-term view of this tradition has been established, which has been described as a "French style of epistemology" (Braunstein 2002a) and which holds together both pre- and post-Bachelardian perspectives (Jacob 1991; Braunstein 2002a, 2008; Chimisso 2008; Taylan 2018).

Others have highlighted certain thematic connections, investigating the specificities of the forms of rationality specific to each science from a regionalist approach (see Jacob 1991). Alunni and Castellana, for example, have foregrounded on various occasions the line of continuity in French mathematical thought, whose greatest representatives were Winter, Brunschvicg, Bachelard, Lautman, Cavaillès, Gonseth, Desanti, Gilles Châtelet and, among contemporary thinkers, Alain Badiou. This internal line of the so-called *pensée des mathematiques* or, with a clearly Bachelardian reference, *école surrationaliste*, was not limited exclusively to France but rather also included international figures such as Hermann Weyl, Wolfgang Pauli, Enriques, Ettore Majorana, and Fernando Zalamea because of their theoretical orientations (Alunni 2018). This interpretative model has also had notable effects on the (re)reading of Bachelard's work itself in light of mathematical thought (Castellana 1974, 1990, 2019, 2020, 2021; Palombi 2017; Alunni 2018; Ienna 2019a, 2021b). In the field of the life sciences, Canguilhem's students, such as Foucault, François Dagognet, Camille Limoges, Claude Debru, and Gayon, were those who continued his approach (Jacob 1991; Bontems 2010). For his part, Castelli Gattinara can be credited with having brought attention to the "double articulation" between epistemology and history in the French interwar intellectual context, tracing interconnections between FHE and historical schools such as that of Henri Berr's Centre International de Synthèse or Marc Bloch and Lucien Febvre's *Annales* (Castelli Gattinara 1996). In contrast, authors such as Gingras, Éric Brian, and

[51] The following examples are clearly simplifications, useful for making a brief synthesis, with respect to the perspectives developed by each of the scholars included in this reconstruction.

Terry Shinn have tried to maintain a Bachelardian sociology within Bourdieu's framework (Bontems 2010). More recently, Castellana has focused on the fundamental characteristics of a female tradition of Francophone epistemology, following a path that connects Suzanne Bachelard, Hélène Metzger, and Simone Weil (Castellana 2018). In the dynamics characterizing this "major canon," however, the intellectual and political climate within which Lecourt's formulation became such an effective term needs to be foregrounded. In my opinion, the consolidation of this label came about because of the deliberate redundancy of the term *épistémologie historique*[52] and its relations with the French intellectual field of the time.

The period between the second half of the 1950s and the end of the 1960s was marked by great political ferment. In 1956, the Twentieth Congress of the Communist Party of the Soviet Union was held, during which Nikita Khrushchev denounced the "cult of personality" that his predecessor, Joseph Stalin, had established, inaugurating a process of de-Stalinization. This event created the need for a broad internal questioning within the major communist parties and organizations of the time. This discussion simultaneously stimulated the emergence of new theoretical debates. From an intellectual point of view, the interdisciplinary current of structuralism was emerging in France in this period, with a strong critique of humanistic philosophies "in their different variants of personalism, existentialism, phenomenology, and Marxist humanism" (Pianciola 1979, 115). In this context, Althusser and his group at the École Normale Supérieure (including Jacques Rancière, Étienne Balibar, Pierre Macherey, Roger Establet, Badiou, Michel Fichant, Michel Pécheux, and François Regnault) worked on a critique of Marxist humanism rooted in a deep epistemological reflection. Desanti, a philosopher of mathematics and student of Cavaillès, who was strongly committed to the construction of a materialist epistemology (of a Marxist nature) largely inspired by Bachelard (Desanti 1975), certainly played a decisive role in Althusser's intellectual training. Althusser's works published between 1960 and 1965 (*For Marx* and *Reading Capital*, with Balibar), as well as his courses at the École Normale (from 1965 to 1968), were ultimately aimed at proposing a "left-wing" criticism of "Stalinism" and a new reading of Marx's thought that was diametrically opposed to the widely shared version based on Roger Garaudy's humanism:

> But the "criticism of Stalin's errors" was formulated at the Twentieth Congress in terms such that there inevitably followed what we must call an unleashing of *bourgeois* ideological and philosophical themes within the Communist Parties themselves. This was the case above all among Communist intellectuals, but it also touched certain leaders and even certain leaderships. Why?
>
> Because the "criticism of Stalin's errors" (some of which—and rather a lot—turned out to be crimes) was made in a non-Marxist way. [...]

[52] As previously noted, the term *épistémologie* had been in the French language only since the beginning of the 1900s. In this sense, when Rey used the term *épistémologie historique* in contrast to its Anglophone meaning, it had not yet achieved this sense of redundancy, which emerged only in later years.

> If you take Communist philosophers and other Communist "intellectuals" and set them officially on a bourgeois ideological and philosophical line, in order to "criticize" a regime under which they (among others) have suffered deeply, you must not be surprised when the same Communist philosophers and intellectuals quite naturally take the road of bourgeois philosophy. It has been opened up right in front of them! You must not be surprised when they make up their own little bourgeois Marxist philosophy of the Rights of Man, exalting Man and his Rights, the first of which is *liberty*, whose reverse is *alienation*. It is quite natural for them to lean on Marx's early works—that is what they are there for—and then on humanism in all its forms! Shall it be Garaudy's socialist humanism, the pure humanism of John Lewis, the "true" or "real" socialism of others, or even (why not?) "scientific" humanism itself? (Althusser 1973/1976, 75–76)

In short, Althusser saw Marxist humanism as a bourgeois philosophy that would give a "right-wing" reading to the problems denounced at the Twentieth Congress and, through its extra-Marxist theoretical approach, would itself act as a bourgeois ideology.[53] To carry out this critical operation, however, Althusser needed a strong epistemological framework, so starting from 1960, he began to apply the principles of Bachelardian epistemology to his reading of Marx's work. Between 1960 and 1965, Althusser published the texts that make up the collection *For Marx* (1965), and during the first months of 1965, the famous seminar *Reading Capital* (Althusser and Balibar 1965a, 1965b) was held in collaboration with Rancière, Macherey, and Balibar. Starting from 1967, Althusser delivered courses on the philosophy of science, published with the Bachelardian title of *Philosophy and the Spontaneous Philosophy of the Scientists*, which was followed by a series of other modules taught by some of his collaborators. The program consisted of the following[54]: (1) Althusser, *La philosophie et les sciences*; (2) Macherey, *L'objet de la science*; (3) Pécheux, *Pratique sociale et histoire des sciences*; (4) Fichant, *Épistémologie et histoire des sciences*; (5) Regnault, *Y a-t-il des précurseurs dans les sciences?*; (6) Balibar, *La méthode expérimentale*; and (7) Badiou, *Qu'est-ce qu'un modèle?* These were the years that theoretically prepared the French for the events of 1968[55] (Althusser et al. 1967). Introducing his philosophy of science course, Althusser wrote:

[53] For more on this issue, see Althusser and Balibar (1965a) and Althusser (1973).

[54] Of these, only *Sur l'histoire des sciences*, by Fichant and Pécheux, and *Le concept de modèle*, by Badiou, were actually published.

[55] Summing up this period of political-intellectual ferment, Althusser wrote: "A good deal of water has flowed under the bridge of history since 1960. The Workers' Movement has lived through many important events: the heroic and victorious resistance of the Vietnamese people against the most powerful imperialism in the world; the Proletarian Cultural Revolution in China (1966–1969); the greatest workers' strike in world history (ten million workers on strike for a month) in May 1968 in France—a strike which was 'preceded' and 'accompanied' by a deep ideological revolt among French students and petty-bourgeois intellectuals; the occupation of Czechoslovakia by the armies of the other countries of the Warsaw Pact; the war in Ireland, etc. The Cultural Revolution, May 1968 and the occupation of Czechoslovakia have had political and ideological repercussions in the whole of the capitalist world" (Althusser 1973/1976, 35–36).

> This experiment, inaugurated by the present exposition and continued by the interventions of Pierre Macherey, Étienne Balibar, François Regnault, Michel Pêcheux, Michel Fichant and Alain Badiou, was to last up to the eve of the great events of 1968.
>
> The texts of the lectures were immediately mimeographed and soon began to circulate. Later, on the initiative of students, certain of them were even reproduced in the provinces (Nice, Nantes). (Althusser 1967/1990, 71)

As mentioned, starting in 1965, Lecourt had been Althusser's student at the École Normale, where he fully experienced the intellectual ferment at the intersection of Marxism and epistemology. When he formulated the intellectual label "*épistémologie historique*" to describe Bachelard's thought, he did so with a clear political intent, (re)creating this expression on the basis of historical materialism:

> I wrote my first books on Bachelard and epistemology to nourish a political reflection which, in fact, I have never abandoned. In what sense can we say (like Althusser) that Marx founded a "science of history"? What was the value of the notion of "*coupure épistémologique*" [epistemological break] that he had proposed to answer this question? Was it justified to speak, like many others, of "political science" alongside the nascent "human and social sciences," intoxicated with formalism and enamored of absolute relativism? (Lecourt and Godin 2014, 220)

In fact, in the Althusserian program, the formulation of historical materialism played—as we will see in the following pages—the role of a real stake on which to base the rereading of Marx's work. Even in his doctoral work under the direction of Dagognet, Lecourt maintained the same political-epistemological emphasis, which was no coincidence. Introducing the resulting text, titled *For a Critique of Epistemology*, Lecourt declared that

> The texts you are about to read have a history, and to know it will doubtless help you to appreciate their true import. This history hangs on an encounter—that of Marxism and epistemology—which may itself appear as the fruit of a double historical "accident." (Lecourt 1972/1975, 119)
>
> My aim will be different: I should like to cast light on the apparent paradox of the encounter that has taken place here between strictly specialized works in the history of the sciences and the theoretical preoccupations of the Marxist-Leninist philosophers grouped around Louis Althusser. (Lecourt 1972/1975, 162)

Therefore, separating an analysis of the emergence of the FHE label from the French political-intellectual climate of the time is clearly impossible.[56]

Certainly, the success enjoyed by Lecourt's text on Bachelard's *épistémologie historique* was due largely to its being the first systematic introduction in book form to the thought of this extremely important figure in French culture of the time. But in turn, the centrality of a particular reading of Bachelard within the French academic field was brought about by certain institutional contingencies. Of course, thanks to his Sorbonne teachings, Bachelard had had the opportunity to spread his ideas,

---

[56] As Bontems has rightly pointed out, Lecourt subsequently distanced himself from this model of a Bachelardo-Althusserianism and, instead, moved toward a Bachelardo-Canguilhemism (Bontems 2010, 190–191).

creating a strong discontinuity compared to previous conceptions of both epistemology and the history of science in the French-speaking world as a large group of students and admirers of his work spread throughout the French academic field. But a clarification is necessary in this regard: When discussing FHE, I am not referring to Bachelard's work as such but rather to a certain reception of his thought.[57]

Canguilhem once again played an ambiguous role in this context, having taken over the chair in 'épistémologie et histoire des sciences' at the Sorbonne (in 1955) and the direction of the Institut d'histoire des sciences et des techniques from Bachelard. We should also remember that Canguilhem held other important institutional roles in the field of French philosophy, such as inspecteur général de philosophie (1948–1955) and président du Jury d'Agrégation (1964–1968), and participated in commissions of the Centre national de la recherche scientifique (CNRS). This institutional centrality facilitated his influence over an entire generation of the French intellectual field. Through this unique position, Canguilhem implemented a real process of the institutionalization and consecration of the thought of his predecessor[58] but, to a large extent, from the point of view of his own historical-epistemological conceptions. This operation was oriented toward Bachelardian elements that Canguilhem had integrated into his system of thought. In the introduction to the first English edition of Canguilhem's *The Normal and the Pathological*, Foucault clearly stated:

> But, take away Canguilhem, and you will no longer understand very much about a whole series of discussions that took place among French Marxists; nor will you grasp what is specific about sociologists such as Pierre Bourdieu, Robert Castel, Jean-Claude Passeron, what makes them so distinctive in the field of sociology. (Foucault 1985/2003, 465)[59]

For his part, Bourdieu perfectly described the institutional role played by the *homo academicus* Canguilhem in the generational transition within the French philosophical field between the hegemonic period of Sartre's existentialism to the

[57] The most recent interpretations of both Bachelard's and Canguilhem's works highlight the elements of discontinuity between them. Some scholars have emphasized how a certain Canguilhemian interpretation has led to the obscuring of some aspects of Bachelard's philosophy—if not their outright misrepresentation. On the other hand, Canguilhem's work has, for a long time, been subjected to a distorted interpretation because it was analyzed under a lens that saw him as "Bachelard's student" and "Foucault's teacher," reducing his contribution to that of a generational mediator. Especially with the publication of his complete works, a critical (re)reading of his intellectual positioning is being carried out. Likewise, the materials contained in the Georges Canguilhem collection preserved at the CAPHÉS (École Normale de Paris) also support this thesis. See, for example, Sfara (2016, 2018).

[58] Bontems has called this process of Canguilhem's transmission of Bachelard's epistemology a "Bachelardization of French philosophy" (Bontems 2010, 181).

[59] As mentioned, the first version of the text can be dated to 1978 on the occasion of the first English translation of *The Normal and the Pathological*. The French version was published in 1985 in the *Revue de métaphysique et de morale* in an issue on Canguilhem. Foucault, exhausted by his illness, only had time to revise the original text. This was the last text authorized by Foucault for publication.

emergence of FHE and structuralism. In this context, the new way of conceiving the relationship between science and philosophy played a decisive role:

> The complexity of the relationship between the two generations, and the subterranean complicities between members of the same generation, are seen clearly in the almost universal recognition accorded to Georges Canguilhem. A fellow student at the École Normale of Sartre and Aron, from whom he was separated by his lower-class provincial origin, Canguilhem could be evoked by the occupants of opposite positions in the university field. As an exemplary *homo academicus* [...] but as the advocate of a tradition of history of science and epistemology who, at the height of the triumph of existentialism, represented the last bastion of seriousness and rigour, he was consecrated, with Gaston Bachelard, as the *maître à penser* of philosophers more remote from the heart of the academic tradition, such as Althusser, Foucault and some others; it was as if his central yet minor position in the university field and the quite rare, even exotic dispositions that had predisposed him to occupy it had designated him to play the role of a totemic emblem for all those who sought to break with the dominant model and who constituted themselves as an "invisible college" by rallying around him. (Bourdieu 1997/2000, 38)

Canguilhem was thus a "totemic emblem" of the "invisible college"[60] that was being established around him. He had met Althusser in 1948 through Gusdorf. Canguilhem, despite having participated in the resistance (alongside figures such as Cavaillès and Lautman), was known for his liberal and anticommunist political views (Bontems 2010, 188–189). On the other hand, Althusser was not particularly knowledgeable in the field of the history of science. Despite their divergence, Canguilhem declared: "And it so happens that I did not seem too reactionary to him, nor did he seem too stubborn or too narrow-minded. So we managed to get along pretty well" (Canguilhem 1998, 127). Their mutual understanding was built around the fact that Canguilhem had recognized a certain debt toward Althusser, from whom he had taken the notion of "scientific ideology" (developed in his courses at the Sorbonne[61]) (see Canguilhem 1977). Moreover, as has been said, Althusser was interested in developing Bachelardian reflections relating to "epistemological rupture" as an interpretative key for demonstrating the scientific nature of the historical materialism formulated by Marx in *The German Ideology*. Considerable circulation occurred between the Sorbonne and the École Normale Supérieure as scholars from the latter, including Foucault, Macherey, Balibar, Lecourt, and Bourdieu, came into close contact with Canguilhem by following his courses at the Sorbonne.

[60] In the sociology of science and knowledge, the concept of invisible college was introduced by Derek De Solla Price (on the basis of an expression used by Robert Boyle). With this term, Price meant the informal circuit of relationships between researchers who link different research institutions. This informal network permits those who are recognized in a certain field to periodically come into contact with other protagonists within that scientific community, creating a circle of symbolic legitimation. The concept of invisible college was later taken up and developed by Crane (1972).

[61] Between 1967 and 1968, Canguilhem recognized his debt to Althusser by dedicating an entire course to the notion of ideology (especially if one considers his liberal political orientation), which is certainly no coincidence.

What is more important to emphasize, however, is that with the events of 1968, Canguilhem, Althusser, Bourdieu, and Foucault simultaneously recognized the absolute centrality of Bachelard's thought (especially in the version proposed by Canguilhem's reading) in the respective contexts of the five most authoritative and legitimized cultural institutions of the era.[62] Canguilhem and Althusser had assumed central roles at the Sorbonne and the École Normale Supérieure, respectively. At the same time, Foucault's notoriety was already becoming widespread due to the publication of works such as *Mental Illness and Psychology* (1954), *Madness and Civilization* (1961), *The Birth of the Clinic* (1963), *The Order of Things* (1966), and *Archaeology of Knowledge* (1969), thanks to which he was elected to the Collège de France (with the support of Vuillemin) in 1969. At the same time, Foucault was responsible for forming the philosophy department of the Université de Vincennes, an experimental university founded following the French May 1968 in response to student demands. Intellectuals of the highest importance were part of this institution, including some who were very close to the FHE approach, such as Serres, Badiou, Rancière, Regnault, Balibar, François Châtelet, and, among the sociologists, Jean-Claude Passeron (Bourdieu's famous collaborator and coauthor) (Soulié 1998).

Starting from 1964, Bourdieu was based at the sixth section of the École Pratique des Hautes Études,[63] where he taught, among others, a course on the epistemology of the social sciences in collaboration with Passeron. A large group of scholars was also forming around Bourdieu. Among these was Jean-Claude Chamboredon (professor at the École Normale Supérieure), with whom Bourdieu and Passeron wrote an epistemology textbook for the social sciences with a strong Bachelardian-Canguilhemian structure, published in 1968 (Bourdieu et al. 1968; Bourdieu and Krais 2005). Both Bourdieu and Passeron had had philosophical training from which they later distanced themselves to devote themselves to sociology. Starting from 1961, they began collaborating, publishing important works in the field of the sociology of education and culture that influenced the political-intellectual climate of 1968 France. Especially with *The Inheritors: French Students and Their Relation to Culture* (1964)—and its related in-depth analysis in *Reproduction in Education, Society and Culture* (1970)—these two authors provided the key elements for a criticism of the reproduction of social inequalities within educational systems.

Through this process, FHE became the style of epistemology of reference for the nascent current of French structuralism in which authors such as Althusser, Foucault, Bourdieu, Piaget,[64] and others were inscribed, creating a large overlap between these intellectual waves. Bachelardian thought has been recognized as an "authentically

[62] I do not wish to flatten the specific differences that distinguish these thinkers, who were indeed in opposition to one another. In those years, after an initial convergence, both Foucault and Bourdieu were already beginning to demonstrate a certain detachment from the Althusserian perspective (both had trained at the École Normale Supérieure in Paris), developing their own individual positions.

[63] In 1975, this section of the École Pratique des Hautes Études (EPHE) broke away and became an autonomous institution with the name École des Hautes Études en Sciences Sociales (EHESS).

[64] For more on Piaget's relationship to this tradition, see Castellana (2016a).

structural method, without structuralism," or as a structuralism ahead of its time, which is no coincidence (Castellana 1974, 150).

1.5 The Minor Canon: *Pensée Des Mathématiques* [Mathematical Thought]

As seen in the previous section, Canguilhem's political-institutional success at the Sorbonne, with his connection to the Marxist debates developed at the École Normale Supérieure thanks to Althusser, was at the root of the intellectual consolidation of the FHE label, understood as an epistemological reflection created on the basis of the Marxist concept of historical materialism. The appearance of the term *épistémologie historique* was linked, on the one hand, to the attempt to describe a method of theoretical and historical analysis of the development of forms of knowledge and, on the other, to the consequent process of the canonization of the authors who contributed to this epistemological reflection. Although Bachelardo-Canguilhemism allowed the original and fruitful opening of FHE toward the life sciences, Marxism, the human sciences, and the social sciences, it also generated a process of the obfuscation of the epistemological tradition dedicated to the study of physicomathematical sciences.

As seen at the end of sect. 1.3 of this chapter, the intellectual origins of the program of historicizing epistemology—in contrast to the intellectual stance of logical empiricism—were already present in French epistemology in the first half of the twentieth century. This type of research, which led to the proposal of a *pensée des mathématiques* [mathematical thought], was the cornerstone upon which what I call the "minor canon" of FHE developed in France as a sort of underground current in the history of French epistemology.

As highlighted in previous pages, a wide variety of "minor canons" of historical epistemology in general and of FHE in particular can be identified. However, what makes the variant of mathematical thought particularly interesting compared to others is that it was at the center of the debate that gave rise to the opposition between a logicist and a priori epistemology and one, instead, that was involved in studying concrete scientific production. The reconstruction of this line of thought is therefore fundamental to understanding the position of the program of historical epistemology, both in the historical context in which it emerged and in its contemporary development. Its legacy is that it opened the way for the development of an alternative epistemology to what had become the dominant point of view in English-speaking countries. Furthermore, the development of mathematical thought took on further strategic value in highlighting how FHE was not only characterized by a shift toward the biological, social, and human sciences but, on the contrary, could also continue to be profitably applied to the physicomathematical context (as confirmed by contemporary developments with Hacking, the Berliners of the Max Planck Institute, and the debates within iHPS) (Badino, Ienna, and Omodeo 2022).

The intellectual origins of this tradition, like those of the major canon, can be traced to the thought of Comte, who, alongside better-known expressions such as *philosophie scientifique* and *philosophie des sciences*, also used the term *philosophie mathématique* several times.[65] This expression was then taken up by Brunschvicg in *Les étapes de la philosophie mathématique* in 1912, giving it the specific meaning of a historical-critical reflection aimed at bringing out the philosophy implicit in mathematical thought. Brunschvicg was also among the founders of the *Revue de métaphysique et de morale* around which the first nucleus of reflections relating to contemporary mathematics developed. In particular, Winter's thought took shape in this context, one of the most lucid precursors to subsequent developments of the French philosophy of mathematics. Brunschvicg's teaching was central to the training of the following generation of scholars, including Bachelard (whose complementary doctoral thesis was directed by Brunschvig), Cavaillès, and Lautman (who, instead, wrote their main theses with him). In 1935, the collective of mathematicians known as Bourbaki was also launched, in which Cavaillès and Lautman in particular were deeply involved. At the same time, as noted previously, international collaborations with Gonseth and the Zurich School, and with the Italian tradition through Enriques, were of key importance. Desanti, Gilles Châtelet, Alexander Grothendieck, Badiou, and others continued writing about mathematical thought during the 1970s. Another aspect—beyond that of historicizing epistemology—that united the minor and major canons was certainly the political-ideological radicality of its authors, which, in various cases, was also reflected in their ways of conceiving mathematical practices.

The minority status of this tradition is obviously not due to its lack of philosophical relevance—which, on the contrary, is currently being reassessed and valorized. Rather, it was the victim of a double obfuscation created by both specific academic events and historical-political contingencies that marked the period between the two world wars. In the context of debates among French academics, the development of Bachelardo-Canguilhemism shifted the axis of epistemological research from the physicomathematical sciences toward objects of study that were still completely unexplored at the time. The migration of the main representatives of neo-positivism to the United States certainly favored the international circulation of their ideas. On an international level, in fact, the perspective of logical empiricism made substantial inroads, progressively establishing itself as the standard perspective in epistemological debates. It was therefore in a double state of minority with respect to the developments of both French and transnational debates. At the same time, the premature deaths of authors such as Cavaillès and Lautman, who were shot by the Nazis during the French resistance; the death of Enriques in 1946; and the end of

[65] Actually, this expression had already been coined in the context of the debates that developed in the scientific field between the seventeenth and eighteenth centuries, together with expressions such as mechanical philosophy, chemical philosophy, biological philosophy, etc. (Castellana 2020, 13). On Comtian uses, see Castellana (2020, 12), who reconstructed the passages in which this expression appeared in the *Cours de philosophie positive* and other minor writings.

Bachelard's teaching career in 1955 (as well as his turning down a chair at the Collège de France) generated serious repercussions for these debates.

The process of canonizing the tradition of mathematical thought within historical epistemology developed through a Franco-Italian intellectual exchange that has developed in recent decades in the context of the *Revue de synthèse*, focused largely on new interpretations of Bachelardian thought.[66] Even in this case, the approach has been both historiographical and theoretical-epistemological. Both mathematicians and philosophers have been engaged in this debate, seeking sources, stimuli, and precursors for the contemporary development of a historical epistemology of mathematics in the key stages of development of twentieth-century mathematical thought.

Among the mathematicians who have participated most actively in this debate are Frédéric Patras (2001, 2013, 2014, 2017), René Guitart (2015), Yves André (2017a, 2017b) Alunni and André (2015), and the Colombian Fernando Zalamea (2006, 2016, 2018, 2019) Lautman (2006, 2016). In this context, Alunni and Castellana in particular have focused on the coherence and theoretical consistency of the tradition of mathematical thought. Both of these authors have highlighted the fundamental stages of the development of this "new epistemology" (Castellana 1990, 2019) or "surrationalist school" (Alunni 2018). Alunni and Castellana edited the critical editions of the texts of key authors in this tradition, such as Enriques, Bachelard, Lautman, and Gilles Châtelet (Bachelard 1933a, 1933b; Châtelet 1993; Enriques 2016; Lautman 2016), through which they brought out—to use Alunni's expression—"*les harmonies cachées*" [the hidden harmonies] within European mathematical culture. Alongside this work of dissemination, they have published numerous essays, volumes, and special journal issues aimed at building a canon of the historical epistemology of Franco-Italian mathematics. Both authors refer to the concept of *surrationalism*—a term coined by Bachelard that represents the constantly dialectical, critical, and historical character of scientific activity—to indicate the quintessence of this epistemological perspective, thus recognizing Bachelard as the center of gravity of this network of intellectual affiliations (Castellana 1974; Alunni 2018; Castellana 2021). Among the issues at stake in this reading is also the tendency to reinterpret Bachelard's legacy by going beyond and deconstructing the historical-theoretical conjuncture of Bachelardo-Canguilhemism. For my part, I prefer to call this tradition the "minor canon" precisely to highlight the doubly

[66] In recent years, the *Revue de synthèse* in particular has proposed a debate on Bachelard's relationship with mathematics, one that is based on the fruitful collaboration between Charles Alunni and Éric Brian (see, for example, the following issues: 1999, vol. 120, n. 1, *Pensée des sciences*; 2001, vol. 122, n. 1, *Object d'échelle*; 2005, vol. 126, n. 2, *Science et philosophie au XXème siècle. L'École de Zurich et le program surrationaliste*; 2013, vol. 134, n. 1, *Ettore Majorana. De la légende à la science*; 2015, vol. 136 n. 2, *Philosophie et Mathématique*; 2017, vol. 138 n. 1–4, *Philosophie contemporaine de mathématiciens: Évariste Galois, Gian-Carlo Rota, Gilles Châtelet*). In the chapter "Le bachelardisme" of his book on this author, Vincent Bontemps underlined the role played in this debate by Alunni and Brian. Among the Italian authors who work (or have worked) in this direction are Castellana, Castelli Gattinara, Carlo Vinti, Fabrizio Palombi, and Andrea Cavazzini (for further details, see *Revue de synthèse*, 2011, vol. 132, n. 2, *Épistémologie historique italienne*).

subordinate character of this new epistemology, or surrationalist school, compared to the major canon of FHE and the analytic philosophical tradition.[67]

References

———. 1967. *Philosophie et philosophie spontanée des savants*. Paris: Maspero. *Philosophy and the spontaneous philosophy of the scientists, and other essays*, trans. Warren Montag, ed. Gregory Elliott. London: Verso, 1990.

———. 1973. *Réponse à John Lewis*. Paris: Maspero. *Reply to John Lewis*, trans. Grahame lock. In *Essays in self-criticism*, 35–77. London: New Left Books. 1976.

Althusser, Louis, and Étienne Balibar. 1965a. *Lire le capital, v. 1*. Paris: Maspero. *Reading capital*, trans. Ben Brewster. London: Verso, 1970.

———. 1965b. *Lire le capital, v. 2*. Paris: Maspero. *Reading capital*, trans. Ben Brewster. London: Verso, 1970.

Althusser, Louis, Pierre Macherey, Michel Pécheux, Michel Fichant, François Regnault, Étienne Balibar, and Alain Badiou. 1967. *Cours de philosophie pour scientifiques*, 169. Paris: École Normale Supérieure. MS.

Alunni, Charles. 2018. *Spectres de Bachelard. Gaston Bachelard et l'écoles surrationaliste*. Paris: Hermann.

Alunni, Charles, and Yves André, eds. 2015. *Federigo Enriques o le armonie nascoste della cultura europea. Tra scienza e filosofia*. Pisa: Edizioni della Normale.

Alunni, Charles, and Éric Brian. 2002. La mémoire des gestes de science et ses enjeux. *Actes de la recherche en sciences sociales* 141-142 (1): 127–134.

André, Yves. 2017a. Ambiguïté, infini, transcendance. Réflexions sur l'Évolution de la notion de nombre, après une lecture de Gilles Châtelet. *Revue de Synthèse* 138 (1–4): 261–278. https://doi.org/10.1007/s11873-000-0000-12.

———. 2017b. "Quels sont vos grands auteurs?" Réflexions sur l'écriture et le style en mathématique. *Revue de Synthèse* 138 (1–4): 471–486. https://doi.org/10.1007/s11873-000-0000-24.

Arabatzis, Theodore, and Don Howard. 2015. Introduction: Integrated history and philosophy of science in practice. *Studies in History and Philosophy of Science Part A* 50:1–3. https://doi.org/10.1016/j.shpsa.2014.10.002.

Arabatzis, Theodore, and Jutta Schickore. 2012. Introduction: Ways of integrating history and philosophy of science. *Perspectives on Science* 20 (4): 395–408.

Bachelard, Gaston. 1933a. Physique et métaphysique. In *Septimana Spinozana. Acta conventus oecumenici in memoriam Benedicti de Spinoza*. The Hague: Curis Societatis Spinozanae Edita.

———. 1933b. *Métaphysique des mathématiques. Eds. Charles Alunni and Gerardo Ienna. Paris: Hermann. 2021. Originally published as physique et métaphysique, in Septimana Spinozana. Acta conventus oecumenici in memoriam Benedicti de Spinoza*. The Hague: Curis Societatis Spinozanae Edita, 1933.

———. 1934. *Le Nouvel esprit scientifique. Paris: Alcan. The new scientific Spirit.*, trans. Arthur Goldhammer. Boston: Beacon Press, 1984.

———. 1938. *La formation de l'esprit scientifique. Contribution à Une psychanalyse de la connaissance objective. Paris: Vrin. The formation of the scientific mind: A contribution to a*

[67] On this point, see the publications in which I have stressed the strategic importance of the study of this perspective for the debates dedicated to the major canon of FHE (Ienna 2016, 2019a, 2019b, 2021a).

psychoanalysis of objective knowledge., trans. Mary McAllester Jones. Manchester: Clinamen Press, 2002.

Badino, Massimiliano, Gerardo Ienna, and Pietro D. Omodeo. 2022. *Epistemologia Storica: Correnti, temi e problemi*. Rome: Carocci.

Barnes, Barry. 1974. *Scientific knowledge and sociological theory*. London/New York: Routledge.

Bayer, Raymond. 1937a. Travaux du IXe Congrès International de Philosophie. In *Études Cartésiennes II*, vol. 2. Paris: Hermann.

———. 1937b. Travaux du IXe Congrès International de Philosophie. In *L'Unité de la Science: la Méthode et les méthodes I*, vol. 4. Paris: Hermann.

———. 1937c. Travaux du IXe Congrès International de Philosophie. In *L'Unité de la Science: la Méthode et les méthodes II*, vol. 5. Paris: Hermann.

———. 1937d. Travaux du IXe Congrès International de Philosophie. In *Logique et Mathématiques*, vol. 6. Paris: Heinemann.

———. 1937e. Travaux du IXe Congrès International de Philosophie. In *Causalité et Déterminisme*, vol. 7. Paris: Hermann.

Benis-Sinaceur, Hourya. 2018. Philosophie scientifique: origines et interprétations. Hans Reichenbach et le groupe de Berlin. *Philosophia Scientiæ. Travaux d'histoire et de philosophie des sciences* 22-23 (October): 33–76. https://doi.org/10.4000/philosophiascientiae.1553.

Bianco, Giuseppe. 2014. Le long et monotone chapelet de l'Esprit universel. Disciplinarisation et internationalisation dans les congrès de philosophie. *Revue de Métaphysique et de Morale* 84 (4): 483–497.

Bird, Alexander. 2008. The historical turn in the philosophy of science. In *The Routledge companion to philosophy of science*, ed. Stathis Psillos and Martin Curd, 67–77. London/New York: Routledge.

Bloor, David. 1976. *Knowledge and social imagery*. Chicago: Chicago University Press.

Bogdanov, Alexander. 1910. *Padenie velikogo fetishizma (Sovremennyj krizis ideologii). Vera i nauka (o knige V. Il'ina "Materializm i empiriokriticizm")*. Moscow: S. Dorovatovskogo and A. Charushnikova Publishers.

Bontems, Vincent. 2006. L'actualité de l'épistémologie historique. *Revue d'histoire des sciences* 59 (1): 137–147.

———. 2010. *Bachelard*. Paris: Les Belles Lettres.

Bourdeau, Michel. 2018. La préparation du congrès de 1935 dans la correspondance d'Otto Neurath et de Louis Rougier. *Philosophia Scientiæ. Travaux d'histoire et de philosophie des sciences* 22-23 (October): 17–32. https://doi.org/10.4000/philosophiascientiae.1546.

Bourdeau, Michel, Gerhard Heinzmann, and Pierre Wagner. 2018. Introduction. *Philosophia Scientiæ. Travaux d'histoire et de philosophie des sciences* 22-23 (October): 3–15.

Bourdieu, Pierre, Jean-Claude Chamboredon, and Jean-Claude Passeron. 1968. Le métier de sociologue. Paris: Mouton Bordas. The craft of sociology: Epistemological preliminaries, trans. Richard Nice, ed. Beate Krais. Berlin: de Gruyter, 1991.

Bourdieu, Pierre. 1997. *Méditations pascaliennes. Paris: Éditions du Seuil. Pascalian Meditations.*, trans. Richard Nice. Stanford: Stanford University Press, 2000.

———. 2001. *Science de la science et Réflexivité. Paris: Raisons d'agir. Science of Science and Reflexivity.*, trans. Richard Nice. Cambridge: Polity Press, 2004.

———. 2002. Les conditions sociales de la circulation internationale des idées. *Actes de la recherche en sciences sociales* 145 (5): 3–8.

Bourdieu, Pierre, and Beate Krais. 2005. Entretien avec Pierre Bourdieu—Recueilli par Beate Krais en décembre 1988. In *Le métier de sociologue*, by Pierre Bourdieu, Jean-Claude Chamboredon, and Jean-Claude Passeron, 5th ed., v-xix. Berlin/New York: Mouton de Gruyter.

Braunstein, Jean-François. 2002a. Bachelard, Canguilhem, Foucault. Le "style français" en épistémologie. In *Les philosophes et la science*, ed. Pierre Wagner, 920–963. Paris: Gallimard.

———. 2002b. La philosophie des sciences d'Auguste Comte. In *Les philosophes et la science*, ed. Pierre Wagner, 787–822. Paris: Gallimard.

———. 2008. *L'histoire des sciences. Méthodes, styles et controverses*. Paris: Vrin.

———. 2012. Historical epistemology old and new. In *Epistemology and history from Bachelard and Canguilhem to Today's history of science*, 33–40. Berlin: Max Planck Institute for the History of Science.
———. 2015. Abel Rey et les débuts de l'Institut d'histoire des sciences et des techniques (1932-1940). In *L'épistémologie française, 1830–1970*, ed. Michel Bitbol and Jean Gayon, 163–180. Paris: Éditions Matériologiques.
Braunstein, Jean-François, Iván Moya Diez, and Matteo Vagelli, eds. 2019. *L'épistémologie historique: Histoire et méthodes*. Paris: Éditions de la Sorbonne. https://doi.org/10.4000/books.psorbonne.39172.
Brenner, Anastasios. 2003. *Les origines françaises de la philosophie des sciences*. Paris: PUF.
———. 2006. Quelle épistémologie historique?: Kuhn, Feyerabend, Hacking et l'école bachelardienne. *Revue de Métaphysique et de Morale* 49 (1): 113–125. https://doi.org/10.3917/rmm.061.0113.
———. 2015a. Un "positivisme nouveau" en France au début du XXe siècle (Milhaud, Le Roy, Duhem, Poincaré). In *L'épistémologie française, 1830–1970*, ed. Michel Bitbol and Jean Gayon, 17–30. Paris: Éditions Matériologiques.
———. 2015b. Is there a cultural barrier between historical epistemology and analytic philosophy of science? *International Studies in the Philosophy of Science* 29 (2): 201–214. https://doi.org/10.1080/02698595.2015.1119421.
———. 2016. L'épistémologie historique d'Abel Rey. *Revue de metaphysique et de morale* 2:159–176.
———. 2018. From scientific philosophy to absolute positivism: Abel Rey and the Vienna circle. *Philosophia Scientiæ. Travaux d'histoire et de Philosophie Des Sciences* 22-23 (October): 77–95. https://doi.org/10.4000/philosophiascientiae.1562.
Brunschvicg, Léon. 1912. *Les étapes de la philosophie mathématique*. Paris: Alcan.
Canguilhem, Georges. 1952. *La connaissance de la vie. Paris: Vrin. Knowledge of life*., trans. Stefanos Geroulanos and Daniela Ginsburg, eds. Paola Marrati and Todd Meyers. New York: Fordham University Press, 2008.
———. 1977. *Idéologie et rationalité dans l'histoire des sciences de la vie*. Paris: Vrin. trans. Arthur Goldhammer. Cambridge, MA: MIT Press
———. 1994. *Études d'histoire et de philosophie des sciences concernant les vivants et la vie*. Paris: Vrin.
———. 1998. Entretien avec Georges Canguilhem. In *Actes du Xe Colloque de la Société internationale d'histoire de la psychiatrie et de la psychanalyse*, ed. François Bing, Jean-François Braunstein, and Élisabeth Roudinesco, 121–135. Le Plessis-Robinson: Synthélabo.
Castellana, Mario. 1974. *Il surrazionalismo di Gaston Bachelard*. Naples: Glaux.
———. 1990. *Alle origini della nuova epistemologia: il Congrès Descartes del 1937*. Lecce: Il protagora.
———. 2016a. Dal metodo storico-critico all'epistemologia genetica: Federigo Enriques e Jean Piaget. In *L'epistemologia genetica, by Jean Piaget*, ed. Mauro Ceruti, Luisa Damiano, and Mario Castellana, 173–201. Rome: Studium.
———, ed. 2016b. *Introduzione. In L'anima religiosa della scienza, 5–17*. Rome: Castelvecchi.
———. 2018. *Cuori pensanti in epistemologia*. Rome: Castelvecchi.
———. 2019. *Federigo Enriques e la Nuova Epistemologia*. Lecce: Pensa MultiMedia.
———, ed. 2020. *Il contributo di Maximilien Winter alla critique des sciences. In Il metodo storico-critico per una nuova filosofia delle matematiche, 9–85*. Milan: Meltemi.
———. 2021. In *Il surrazionalismo di Gaston Bachelard con due saggi inediti*, ed. Pietro Console. Lecce: Milella.
Castelli Gattinara, Enrico. 1996. *Epistemologia e storia. Un pensiero all'apertura nella Francia fra le due guerre mondiali*. Milan: FrancoAngeli.
Cavaillès, Jean. 1937. Réflexions sur le fondement des mathématiques. In *Travaux du IXe Congrès International de Philosophie, vol. 6, Logique et Mathématiques*, 136–139. Paris: Hermann.

Cavazzini, Andrea, and Alberto Gualandi, eds. 2006. *L'epistemologia francese e il problema del "trascendentale storico"*. Macerata: Quodlibet.

Charle, Christophe. 1983. Le champ universitaire parisien à la fin du 19ème siècle. *Actes de la Recherche en Sciences Sociales* 47 (1): 77–89. https://doi.org/10.3406/arss.1983.2189.

Châtelet, Gilles. 1993. *Les Enjeux du mobile. Mathématique, physique, philosophie. Paris: Seuil. Figuring space: Philosophy, mathematics and physics.*, trans. Robert shore and Muriel Zagha. Dordrecht: Kluwer, 2000.

Chimisso, Cristina. 2008. *Writing the history of the mind: Philosophy and science in France, 1900 to 1960s*. Aldershot, Hants, England/Burlington, VT: Ashgate.

Crane, Diana. 1972. *Invisible colleges: Diffusion of knowledge in scientific communities*. Chicago: University of Chicago Press.

Critchley, Simon. 1997. What is continental philosophy? *International Journal of Philosophical Studies* 5 (3): 347–363. https://doi.org/10.1080/09672559708570862.

D'Agostini, Franca. 1997. *Analitici e continentali: guida alla filosofia degli ultimi trent'anni*. Milan: Raffaello Cortina.

———. 2014. Analitici e continentali: un progetto fallito? *Bollettino Filosofico* 29:73–90. https://doi.org/10.6093/1593-7178/2837.

Davidson, Arnold. 2001. *The emergence of sexuality: Historical epistemology and the formation of concepts*. Cambridge: Harvard University Press.

Dear, Peter. 2009. The history of science and the history of the sciences: George Sarton, *Isis*, and the two cultures. *Isis* 100 (1): 89–93. https://doi.org/10.1086/597570.

de Calan, Ronan. 2020. L'histoire critique des sciences: un chapitre dans l'histoire des idées. In *Histoire et historiens des idées: Figures, méthodes, problèmes*, ed. David Simonetta and Alexandre de Vitry, 203–220. Paris: Collège de France. http://books.openedition.org/cdf/9922.

Desanti, Jean-Toussaint. 1975. *La philosophie silencieuse; ou critique des philosophies de la science*. Paris: Seuil.

Domski, Mary, and Michael Dickson. 2010. *Discourse on a new method: Reinvigorating the marriage of history and philosophy of science*. Chicago-La Salle: Open Court.

Elgin, Catherine. 1996. The relativity of fact and the objectivity of value. *The Harvard Review of Philosophy* 6 (1): 4–15. https://doi.org/10.5840/harvardreview1996611.

———. 2013. Fact–Value Distinction. In *International encyclopedia of ethics*, ed. Hugh Lafollette. Malden, MA: Wiley.

Enriques, Federigo. 2016. In *L'anima religiosa della scienza*, ed. Mario Castellana. Rome: Castelvecchi.

Fabiani, Jean-Louis. 1988. *Les philosophes de la république*. Paris: Édition de Minuit.

Feest, Uljana, and Thomas Sturm. 2011. What (good) is historical epistemology? Editors' Introduction. *Erkenntnis* 75 (3): 285–302. https://doi.org/10.1007/s10670-011-9345-4.

Feigl, Herbert. 1968. The Wiener Kreis in America. In *The intellectual migration, Europe and America, 1930–1960*, ed. Donald Flemin and Bernard Bailyn, 630–674. Cambridge, MA: Harvard University Press.

Fleck, Ludwik. 1935. *Entstehung und Entwicklung einer wissenschaftlichen Tatsache. Frankfurt am Main: Suhrkamp. Genesis and development of a scientific fact.*, trans. Fred Bradley and Thaddeus J. Trenn, eds. Thaddeus J. Trenn and Robert K. Merton. Chicago: University of Chicago Press, 1979.

Foucault, Michel. 1985. La vie: L'experience et la science. *Revue de méthaphysique et de morale* 90 (January–march): 3–14. Life: Experience and science. In *Aesthetics, method, and epistemology: Essential works of Foucault 1954–1984*, vol. 2., trans. Robert Hurley, ed. James D. Faubion, 465–478. New York: New Press. 2003.

Gayon, Jean. 2003. Bachelard et l'histoire des sciences. In *Bachelard et l'épistémologie française*, ed. Jean-Jacques Wunenburger, 51–113. Paris: PUF.

Gingras, Yves. 2010. Naming without necessity: On the genealogy and uses of the label "historical epistemology". *Revue de Synthèse* 131 (3): 439–454. https://doi.org/10.1007/s11873-010-0124-1.

Gonseth, Ferdinand. 1937. L'unité de la connaissance scientifique. In *Travaux du IXe Congrès International de Philosophie, vol. 4: L'Unité de la Science: la Méthode et les méthodes I*, 58–69. Paris: Hermann.

Gori, Pietro, ed. 2018. *Ernst Mach: tra scienza e filosofia*. Pisa: Edizioni ETS.

Grange, Juliette. 2014. De la philosophie française des sciences à la "philosophie analytique" à la française. *Cites* 58 (2): 11–37.

Guitart, René. 2015. Deux problèmes en vue d'une épistémologie transitive des mathématiques. *Revue de Synthèse* 136 (1–2): 237–279.

Gundlach, Horst U. K. 2012. Germany. In *The Oxford handbook of the history of psychology: Global perspectives*, ed. David B. Baker, 255–289. Oxford: Oxford University Press.

Gusdorf, Georges. 1974. *Introduction aux sciences humaines. Essai critique sur leurs origines et leur développement*. 2nd ed. Paris: Les Éditions Ophrys.

Hacking, Ian. 2002. *Historical ontology*. Cambridge MA: Harvard University Press.

Hahn, Hans, Otto Neurat, and Rudolf Carnap. 1929. *Wissenschaftliche Weltauffassung. Der wiener Kreis, Veröffentlichnungen des Vereines Ernst Mach. Vienna: Arthur wolf Verlag. The scientific conception of the world: The Vienna circle, in empiricism and sociology, 299–318.*, trans. Paul Foulkes and Marie Neurath, eds. Marie Neurath and Robert S. Cohen. Dordrecht: D. Reidel, 1973.

Herring, Emily, Kevin Matthew Jones, Konstantin S. Kiprijanov, and Laura M. Sellers, eds. 2019. *The past, present, and future of integrated history and philosophy of science*. Abingdon/ New York: Routledge.

Holton, Gerald. 1992. Ernst Mach and the fortunes of positivism in America. *Isis* 83 (1): 27–60.

Ienna, Gerardo. 2016. Presentazione. In *Metafisica della matematica*, ed. Charles Alunni and Gerardo Ienna, 5–18. Rome: Castelvecchi.

———. 2019a. *Natura constructa* et phénoménotechnique. Spinozisme et pensée des mathématiques chez Gaston Bachelard. In *L'épistémologie historique: Histoire et méthodes*, ed. Jean-François Braunstein, Iván Moya Diez, and Matteo Vagelli, 43–58. Paris: Éditions de la Sorbonne.

———. 2019b. *Science and Technology Studies. Socio-epistemologia storica delle negoziazioni disciplinari*. Doctoral thesis. Bologna: Alma Mater Studiorum Università di Bologna.

———. 2021a. Corrazionalismo. La dimensione sociale dell'epistemologia bachelardiana. In *Il surrazionalismo di Gaston Bachelard con due saggi inediti*, ed. Mario Castellana and Pietro Console, 353–385. Lecce: Milella.

———. 2021b. Métaphysique des mathématiques. Bachelard et le spinozisme mathématisant, trans. Charles Alunni. In *Métaphysique des mathématiques, by Gaston Bachelard*, ed. Charles Alunni and Gerardo Ienna, 23–48. Paris: Hermann.

Ienna, Gerardo, and Marco Santoro. 2016. Per una scienza delle relazioni internazionali in campo culturale. *Studi culturali* 13 (1): 61–69.

Jacob, Pierre. 1991. Il regionalismo epistemologico: una tendenza della filosofia contemporanea delle scienze in Francia. *Rivista di filosofia* 82 (2): 279–300.

Kitcher, Philip. 2011. Epistemology without history is blind. *Erkenntnis* 75 (3): 505–524. https://doi.org/10.1007/s10670-011-9334-7.

Kuhn, Thomas S. 1977. *The essential tension. Selected studies in scientific tradition and change*. University of Chicago Press.

Lalande, André. 1926. *Vocabulaire critique et technique de philosophie*. Paris: Alcan.

Lambert, Jacques. 2015. L'épistémologie française et le Cercle de Vienne: Louis Rougier. In *L'épistémologie française, 1830–1970*, ed. Michel Bitbol and Jean Gayon, 91–107. Paris: Éditions Matériologiques.

Latour, Bruno, and Steve Woolgar. 1979. *Laboratory life*. 2nd ed. Princeton: Princeton University Press.

Lautman, Albert. 1937. De la réalité inhérente aux théories mathématiques. In *Les mathématiques, les idées et le réel physique*. Paris: Vrin.

———. 2006. *Les mathématiques, les idées et le réel physique. Paris: Vrin. Mathematics, ideas and the physical real.*, trans. Simon B. Duffy, ed. Fernando Zalamea. London/New York: Continuum, 2011.

———. 2016. *La matematica come resistenza.*, trans. and ed. Mario Castellana. Rome, Castelvecchi.

Le Roy, Édouard. 1901. Un positivisme nouveau. *Revue de Métaphysique et de Morale* 9 (2): 138–153.

Lecourt, Dominique. 1972. *Pour Une critique de l'épistémologie: Bachelard, Canguilhem, Foucault. Paris: Maspero. For a critique of epistemology. In Marxism and epistemology: Bachelard, Canguilhem, Foucault.*, trans. Ben Brewster, 117–216. London: New Left Books, 1975.

———. 2001. *La Philosophie des sciences.* Paris: Presses Universitaires de France - PUF.

———. 2007. Georges Canguilhem le philosophe. In *Canguilhem: histoire des sciences et politiques du vivant*, ed. Jean-François Braunstein, 27–44. Paris: PUF.

Lecourt, Dominique, and Cristian Godin. 2014. Dominique Lecourt entretien avec Cristian Godin. *Cités* 48:217–222.

Lenin, Vladimir Ilyich. 1909. *Materializm i empiriokrititsizm: kriticheskie zametki ob odnoi reaksionnoi filosofii.* Moscow: Zveno. Materialism and Empiro-Criticism: Critical Comments on a Reactionary Philosophy, trans. Abraham Fineberg. In V.I. Lenin Collected Works, vol. 14, ed. Clemens Dutt, 17–362. Moscow: Progress Publishers, 1962.

Leroux, Jean. 2009. Une histoire comparée de la philosophie des sciences. *Volume I.* In *Aux sources du Cercle de Vienne.* Quebec: Presses de l'Université Laval.

Levy, Neil. 2003. Analytic and continental philosophy: Explaining the differences. *Metaphilosophy* 34 (3): 284–304. https://doi.org/10.1111/1467-9973.00274.

Limoges, Camille. 2012. L'épistémologie historique dans l'itinéraire intellectuel de Georges Canguilhem. In *Epistemology and history from Bachelard and Canguilhem to Today's history of science*, ed. Henning Schmidgen et al., 53–66. Berlin: Max Planck Institute for the History of Science.

Lorenzano, Pablo. 2010. Logic and methodology of science: An introduction to philosophy of science. In *History and philosophy of science and technology - volume I*, ed. Pablo Lorenzano, Hans-Jörg Rheinberger, Eduardo Ortiz, and Carlos Delfino Galles, 1–50. London: EOLSS Publications.

Mach, Ernst. 1883. *Die Mechanik in ihrer Entwicklung. Historisch-kritisch dargestellt. Leipzig: Brockhaus, 1912. The science of mechanics: A critical and historical exposition of its principles.*, trans. Thomas J. McCormack. Chicago: Open Court Publishing, 1893.

———. 1905. *Erkenntnis und Irrtum: Skizzen zur Psychologie der Forschung. Vienna: JA Barth. Knowledge and error: Sketches on the psychology of enquiry.*, trans. Thomas J. McCormack. Dordrecht: D. Reidel, 1976.

———. 1906. *Die analyse der Empfindungen und das Verhältnis des Physichen zum Psychischen. Fischer: Jena. The analysis of sensations and relation of the physical to the psychical.*, trans. C. M. Williams. Chicago: Open Court Publishing, 1914.

Merton, Robert K. 1988. Le molteplici origini e il carattere epiceno del termine inglese "Scientist". *Scientia: L 'Immagine e il Mondo* 123:279–293.

Méthot, Pierre-Olivier. 2013. On the genealogy of concepts and experimental practices: Rethinking Georges Canguilhem's historical epistemology. *Studies in History and Philosophy of Science Part A* 44 (1): 112–123. https://doi.org/10.1016/j.shpsa.2012.06.002.

Meyerson, Émile. 1908. *Identité et realité. Paris: Alcan. Identity and reality, trans. Kate Lowenberg.* London: George Allen & Unwin, Ltd. 1930.

———. 1951. *Identité et realité.* 2nd ed. Paris: Alcan.

Miskiewicz, Wioletta A. 2018. Between semantics and logistics. The polish delegation at the congress for scientific philosophy in Paris 1935. *Philosophia Scientiæ. Travaux d'histoire et de Philosophie Des Sciences* 22-23 (October): 97–118. https://doi.org/10.4000/philosophiascientiae.1574.

Mulkay, Michael. 1979. *Science and the sociology of knowledge*. London/Boston: G. Allen & Unwin.
Palombi, Fabrizio. 2017. *Elogio dell'astrazione. Gaston Bachelard e la filosofia della matematica*. Milan: Mimesis.
Patras, Frédéric. 2001. *La pensée mathématique contemporaine*. Paris: PUF.
———. 2013. Mathématiques et herméneutique. *Archives de Philosophie* 76 (2): 217–238.
———. 2014. *La possibilité des nombres*. Paris: PUF.
———. 2017. De la combinatoire algébrique à la phénoménologie. *Revue de Synthèse* 138 (1–4): 151–175. https://doi.org/10.1007/s11873-000-0000-08.
Patton, Lydia. 2004. *Hermann Cohen's history and philosophy of science*. Doctoral thesis. Ottawa: Library and Archives Canada. Bibliothèque et Archives Canada.
Petit, Annie. 1995. L'héritage du positivisme dans la création de la chaire d'histoire générale des sciences au Collège de France. *Revue d'histoire des sciences* 48 (4): 521–556.
Petrovich, Eugenio. 2018. Accumulation of knowledge in Para-scientific areas: The case of analytic philosophy. *Scientometrics* 116 (2): 1123–1151. https://doi.org/10.1007/s11192-018-2796-5.
Piaget, Jean. 1967. *Logique et connaissance scientifique*. Paris: Gallimard.
Pianciola, Cesare. 1979. *Filosofia e politica nel pensiero francese del dopoguerra*. Turin: Loescher.
Poincaré, Henri. 1905. *La valeur de la science*. Paris: Flammarion.
———. 1920. *Science et méthode*. Paris: Flammarion.
Polizzi, Gaspare, ed. 1982. *La filosofia scientifica a congresso: Parigi 1935*. Livorno: Belforte Editore Libraio.
Pudal, Romein. 2004. La difficile réception de la philosophie analytique en France. *Revue d'Histoire des Sciences Humaines* 11 (2): 69–100.
Redondi, Pietro. 1978. *Epistemologia e storia della scienza: le svolte teoriche da Duhem a Bachelard*. Milan: Feltrinelli.
Reichenbach, Hans. 1938. *Experience and prediction. An analysis of the foundations and the structure of knowledge*. Chicago: University of Chicago Press.
Rey, Abel. 1907. *La théorie physique chez le physiciens contemporains*. Paris: Alcan.
———. 1909. Vers le positivisme absolu. *Revue philosophique de la France et de l'étranger* 67: 461–479.
Rheinberger, Hans-Jörg. 2010. *On historicizing epistemology*. Stanford, CA: Stanford University Press.
Rougier, Louis, Otto Neurath, and Albert Lautman, eds. 1936a. *Actes du Congrès international de philosophie scientifique, Sorbonne, Paris 1935, vol. 1. Philosophie scientifique et empirisme logique*. Paris: Hermann.
———. 1936b. Actes du Congrès international de philosophie scientifique, Sorbonne, Paris 1935, vol. 2. In *Unité de la science*. Paris: Hermann.
———. 1936c. Actes du Congrès international de philosophie scientifique, Sorbonne, Paris 1935, vol. 3. In *Langage et pseudo-problèmes*. Paris: Hermann.
———. 1936d. Actes du Congrès international de philosophie scientifique, Sorbonne, Paris 1935. In *Induction et probabilité*, vol. 4. Paris: Heinemann.
———. 1936e. Actes du Congrès international de philosophie scientifique, Sorbonne, Paris 1935. In *Logique et expérience*, vol. 5. Paris: Hermann.
———. 1936f. Actes du Congrès international de philosophie scientifique, Sorbonne, Paris 1935. In *Philosophie des mathématiques*, vol. 6. Paris: Hermann.
———. 1936g. Actes du Congrès international de philosophie scientifique, Sorbonne, Paris 1935. In *Logique*, vol. 7. Paris: Hermann.
———. 1936h. Actes du Congrès international de philosophie scientifique, Sorbonne, Paris 1935, vol. 8. In *Histoire de la logique et de la philosophie scientifique*. Paris: Heinemann.
Russo Krauss, Chiara. 2019. *Wundt, Avenarius, and scientific psychology: A debate at the turn of the twentieth century*. Cham: Palgrave Macmillan. https://doi.org/10.1007/978-3-030-12637-7.

Sapiro, Gisèle, Marco Santoro, and Patrick Baert, eds. 2020. *Ideas on the move in the social sciences and humanities: The international circulation of paradigms and theorists*. Cham: Palgrave Macmillan.

Sarton, George. 1913. L'Histoire de la Science. *Isis* 1 (1): 3–46. https://doi.org/10.1086/357728.

Schickore, Jutta. 2011. More thoughts on HPS: Another 20 years later. *Perspectives on Science* 19 (4): 453–481. https://doi.org/10.1162/POSC_a_00049.

Schlick, Moritz. 1979. "The Vienna school and traditional philosophy." In Philosophical papers Vol. 2 (1925–1936), trans. Peter Heath, et al., eds. Henk L. Mulder and Barbara van de Velde-Schlick, 491–498. Dordrecht: Reidel.

Schnädelbach, Herbert. 1984. *Philosophy in Germany 1831–1933*. Cambridge, UK: Cambridge University Press.

Sfara, Emiliano. 2016. *Una filosofia della prassi organismi, arte e visione in Georges Canguilhem*. Turin: Nuova Trauben.

———. 2018. *Georges Canguilhem inédit Essai sur une philosophie de l'action*. Paris: L'Harmattan.

Soulez, Antonia. 2015. La réception du Cercle de Vienne aux congrès de 1935 et 1937 à Paris ou le "style Neurath". In *L'épistémologie française, 1830–1970*, ed. Michel Bitbol and Jean Gayon, 31–66. Paris: Éditions Matériologiques.

Soulié, Charles. 1998. «Histoire du département de philosophie de Paris VIII. Le destin d'une institution d'avantgarde». *Histoire de l'éducation 77*(1):47–69.

Soulié, Charles. 2009. *Les philosophes en République L'aventure intellectuelle de la Revue de métaphysique et de morale et de la Société française de philosophie (1891–1914)*. Rennes: Presses Universitaires de Rennes.

Stadler, Friedrich. 2007. The Vienna circle: Context, profile, and development. In *The Cambridge companion to logical empiricism*, ed. Alan Richardson and Thomas Uebel, 13–40. Cambridge, UK: Cambridge University Press. https://doi.org/10.1017/CCOL0521791782.002.

———. 2014. History and philosophy of science: Between description and construction. In *New directions in the philosophy of science*, ed. Maria Carla Galavotti, Dennis Dieks, Wenceslao J. Gonzalez, Stephan Hartmann, Thomas Uebel, and Marcel Weber, 747–767. Cham: Springer International Publishing. https://doi.org/10.1007/978-3-319-04382-1_52.

———, ed. 2017. *Integrated history and philosophy of science: Problems, perspectives, and case studies*. Cham: Springer International Publishing. https://doi.org/10.1007/978-3-319-53258-5.

———, ed. 2019. *Ernst Mach—Life, work, influence*. Cham: Springer International Publishing.

Stöltzner, Michael. 1998. Federigo Enriques e l'enciclopedia neurathiana. *Rivista di Storia della Filosofia* 53 (3): 463–494.

Tagliagambe, Silvano, and Giulia Rispoli. 2016. *La divergenza nella rivoluzione: filosofia, scienza e teologia in Russia (1920–1940)*. Brescia: Editrice La scuola.

Taylan, Ferhat. 2018. *Concepts et rationalités. Héritages de l'épistémologie historique, de Meyerson à Foucault*. Paris: Éditions Matériologiques.

Vinti, Carlo. 1977. *L'epistemologia francese contemporanea. Per un razionalismo aperto*. Rome: Città Nuova Editrice.

Wartofsky, Marx W. 1968. *Conceptual foundations of scientific thought*. London: Macmillan.

———. 1979. *Models*. Cham: Springer.

———. 1983. From genetic epistemology to historical epistemology: Kant, Marx, and Piaget. In *Piaget and the foundations of knowledge*, ed. Lynn S. Liben, 1–17. New York: Psychology Press.

Winter, Maximilien. 1911. *La méthode dans la philosophie des mathématiques*. Paris: Alcan.

Zalamea, Fernando. 2006. Albert Lautman e la dialectique créatrice des mathématiques modernes. In *Les mathématiques, les idées et le réel physique*, 17-33. Vrin. Albert Lautman and the Creative dialectic of modern mathematics. In *Mathematics, ideas and the physical real*, trans. Simon B. Duffy, ed. Fernando Zalamea, xxiii-xxxvii. New York, Continuum, 2011.

———. 2016. La resistenza sintetica di Lautman., trans. Massimo De Pascale. In *La matematica come resistenza*, ed. Mario Castellana, 87–90. Rome: Castelvecchi.
———. 2018. *Philosophie synthétique de la mathématique contemporaine*. Paris: Hermann.
———. 2019. *Grothendieck. Una guía a la obra matemática y filosófica*. Bogotà: Universidad Nacional–Editorial Nomos.

Chapter 2
Twists and Turns

Abstract The best way to bring out the specificity of the intellectual program that has come together under the label "French historical epistemology" (FHE) is to focus on the theoretical positions taken up in relation to the major debates in the epistemological field as well as the historiography of scientific thought, both in France and internationally. This chapter foregrounds some distinct traits of this tradition, paying particular attention to Canguilhem's institutional and pedagogical role. The first section—focusing specifically on Bachelard, Canguilhem, and Cavaillès—shows how one of the central theses of FHE is that rationality is an emergent property of scientific-technical activity. Such a stance generates a form of epistemological materialism—that is, a perspective focused on the material features producing rationality, whether practices, discourses, or objects to be understood rigorously in their historicity. Corresponding to this is also a specific way of understanding the relationship between common sense and science that can be described in terms of constructivist realism. However, FHE was also characterized by theoretical positions regarding historiographical matters, which are addressed in the second and third sections. Indeed, compared to many of Bachelard's predecessors, he held a discontinuist conception of the history of scientific thought, triggering a real controversy that emerged within historiography between "continuists" and "discontinuists," involving Duhem, Meyerson, Alistair Crombie, Koyré, and Canguilhem. I also bring out the main characteristics of the models of "recurrent history" and "presentism" developed within the methodology of French historical research (with particular reference to the *Annales* tradition). In addition, I accentuate the specificities of FHE (focusing on Bachelard, Canguilhem, Koyré, and Metzger) compared to the Anglophone debates on Whiggism and anti-Whiggism, as well as Herbert Butterfield and his subsequent reception in Anglophone debates with, for example, Crombie, Kuhn, and the sociology of scientific knowledge. The fourth section reconstructs the thesis of epistemological regionalism advocated by Bachelard and other FHE authors as a firmly antithetical position to the "unity of science" thesis proposed by neo-positivism. Instead, the fifth section explores a specific region of rationality, namely mathematical thought, reconstructing the main theses of the "minor canon" of FHE put forward by figures such as Brunschvicg, Bachelard, Cavaillès, Lautman, Gonseth, the Bourbaki group, Desanti, and Gilles Châtelet.

G. Ienna, *Genesis and Development of French Historical Epistemology*, Studies in History and Philosophy of Science 64, https://doi.org/10.1007/978-3-031-88699-7_2

What are we to do with French historical epistemology (FHE) in light of the historical-critical reconstruction and deconstruction that was proposed in the previous chapter? Can we trace common criteria that would link the complex twists and turns that occurred in the development of twentieth-century French epistemology? The best way to bring out the specificity of the intellectual program that has come together under the FHE label is to pay attention to the theoretical positions taken up in relation to the major debates in the epistemological field as well as the historiography of scientific thought, both in France and internationally. The aim of this chapter—and the next—is therefore to highlight some distinct traits of this tradition without denying the specific differences among the authors belonging to this period of French thought.

Bachelard certainly represented a strong moment of discontinuity compared to previous debates in epistemology and the history of science, both in France and abroad. As has been said, the term *épistémologie historique* itself achieved its success and intellectual prominence precisely because of the way that it was used to describe the specificity of Bachelard's thought; therefore, most of the fundamental elements of this tradition were already present in his work, though obviously not limited to it. In the following pages, I focus on how Bachelard's thinking was transmitted, received, and disseminated in French culture thanks, in particular, to Canguilhem.[1]

In this chapter, I demonstrate how the foundations of FHE rested upon the proposal for a profound rethinking of the relationship between epistemological reflection and scientific practices. For a long time, philosophical reflection was dedicated to defining the logical, epistemological, and gnoseological spaces within which scientific knowledge could be justified through a prescriptive approach. On the contrary, FHE did not seek to delineate the requirements that scientific practices had to respect to be considered as such, determining a priori what they "ought to be." Instead, FHE developed a descriptive posture aimed at understanding how rationality is an emerging property of technical-scientific activity itself. Obviously, such a conceptualization implied a redefinition of both the philosophical status of science and the function and meaning of philosophical-epistemological reflection.

In the first section—focusing in particular on Bachelard, Canguilhem, and Cavaillès—I show how this radical change of perspective generated a form of epistemological materialism—that is, a perspective focused on the material features producing rationality, whether these are practices, discourses, or objects to be understood rigorously in their historicity. This scientific activity was described largely according to dialectical principles intrinsically inscribed both within scientific practice itself and in the historical dynamics describing its development. As seen in the previous chapter, FHE emerged within the cultural and intellectual context of

[1] As seen in the previous chapter, I term *Bachelardo-Canguilhemism* the process by which Bachelard's theses were transmitted by Canguilhem through his teaching at the Sorbonne. I have chosen this term because the circulation of Bachelard's ideas has, on various occasions, been received in the form of a hybridization with the historical-epistemological positions proposed by Canguilhem.

the political ferment of the 1960s and 1970s. In addition to the socioinstitutional aspects mentioned in the previous chapter are several theoretical reasons why Bachelard was recognized as a leading figure in this tradition. This was due, in part, to the central role played by a renewed concept of materialism in his reflection and by its application to epistemology. In addition, he interpreted science as a phenomenon regulated by a double dialectic: one internal to scientific knowledge itself and one that is historical in nature.

Alongside these elements, Bachelardo-Canguilhemism took a position against an abstract and "immaculate conception" (Bourdieu et al. 1968) of scientific reason through an analysis that explored the concreteness of scientific procedures: so-called applied rationalism. Similarly, the authors belonging to FHE were characterized by a common antiempiricist approach that presupposed the "constructed" character of the scientific object to be "conquered," against the naive image of the world provided by common sense. All these elements implied an underlying questioning of the relationship of philosophy to the sciences and, therefore, of the role played by epistemology and the history of science as particular branches of knowledge.

However, FHE was also characterized by theoretical positions regarding historiographical matters, which I address in the second and third sections of this chapter. Indeed, compared to many of his predecessors, Bachelard held a discontinuist conception of the history of scientific thought, triggering a real controversy that emerged within historiography between "continuists" and "discontinuists," in which authors such as Duhem, Meyerson, Alistair Crombie, Koyré, and Canguilhem were involved. In the following pages, I also bring out the main characteristics of the models of "recurrent history" and "presentism" developed within the methodology of French historical research (with particular reference to the *Annales* tradition). In addition, I accentuate the specificities of FHE (focusing on Bachelard, Canguilhem, Koyré, and Metzger) compared to the Anglophone debates on Whiggism and anti-Whiggism, as well as Herbert Butterfield and his subsequent reception in Anglophone debates with, for example, Crombie, Kuhn, and the sociology of scientific knowledge.

In the international context, FHE emerged in its own right particularly in contrast to neo-positivism and logical empiricism. The crisis of reason had, in fact, led both the Vienna Circle and the Berlin Circle to seek a way to unify science through the "immutable" structures of logic. FHE reacted to this crisis in a diametrically opposite way, assuming the fragmentary nature of reason as a fundamental characteristic of scientific knowledge. Thus, in the fourth section, I focus on Bachelard's support for a firmly antithetical position to the "unity of science" proposed by neo-positivism, sustaining on the contrary the intrinsically "regional" character of scientific reason. This epistemological attitude became a distinctive mark of the entire French tradition in historical epistemology, favoring investigations into specific and local forms of rationality.

Once again, Canguilhem's role in this context should not be underestimated. Indeed, whereas traditionally Duhem, Meyerson, Rey, Brunschvicg, Cavaillès, Lautman, Metzger, Koyré, and Bachelard himself had privileged the exact sciences—physics, mathematics, mathematical physics, chemistry—as their objects of inquiry, Canguilhem must be recognized for his key role in opening up French epistemological investigations to the life sciences, including medicine. From the perspective of hegemonic Bachelardo-Canguilhemism, this aspect should not be ignored, because it facilitated the emergence of an epistemology of the human and social sciences, with the contributions of authors such as Althusser, Foucault, Bourdieu, and Passeron (to whom the second section of the next chapter will be dedicated). Moreover, it resulted in the obfuscation of a specific tradition in the epistemology of mathematics that, thanks to its historicist approach, formed an alternative to the logicist program developed in the context of neo-positivism. The fifth section is dedicated to reconstructions of this "minor canon," centering on the mathematics of FHE—put forth by figures such as Brunschvicg, Bachelard, Cavaillès, Lautman, Gonseth, the Bourbaki group, Desanti, and Gilles Châtelet.

2.1 Constructivism: A Materialist and Dialectical Epistemology

The first common element within this tradition was the redefinition of the relationship between philosophy and the sciences, from which a precise epistemological posture clearly derived. As was shown in the previous chapter, Bachelardo-Canguilhemism was dealing with the huge legacy of Comtian positivism[2] at the same time that neo-positivism and logical empiricism were spreading in the English-speaking world. In particular, Canguilhem had often engaged with Comte's thought, keeping it in mind when developing his historiographical methodology. In fact, Canguilhem had devoted both an entire course in 1958–1959 and a series of essays on biology to Comte (see Canguilhem 1994a; Braunstein 2015b).

Contrary to the common image of Comtian positivism that had crystallized over time, understood as a form of absolute scientism, in reality, the author of *Cours de philosophie positive* had proposed an idea of philosophy as a "systematizer" and "organizer" of the sciences in their plurality and multiformity (see, in particular, the second lecture of Comte's volume dedicated to the classification of the sciences). Instead, Rey can be given credit for the radicalized and scientistic version of positivism (see Braunstein 2002b, 790, 2015a, 172), which he defined as "absolute": "the philosopher, if he wants to be completely positive, has nothing to add to science" (Rey 1909, 469). For his part, Comte had stated that "a science is not completely known until its history is known" (Comte 1830/2009, 24; translation modified), already anticipating an essential trait of FHE. In addition to these references to positivism, Canguilhem also based his conception of the relationship between science and philosophy largely on Bachelard's thought.

[2]Juliette Grange underlines the distance between Comte's program and that of logical neo-positivism by designating the former as a "historical positivism" (Grange 1996, 8). For a reading of Comte's work in continuity with the tradition of FHE, see also Braunstein (2002b).

Bachelard, in *Le matérialisme rationnel*, had argued that "Science does not have the philosophy it deserves" (Bachelard 1953, 20). His diagnosis was very clear: Philosophers were unable to account for the meaning of scientific reason because their goal was to seek its a priori conditions. In this case, the targets of Bachelard's polemic were Descartes—who founded reason on cogito ergo sum—as well as Kantian apriorism.

The scientist "does not claim, as he could, the extreme philosophical dignity of his incessant work; he does not highlight the philosophical meaning of the psychic revolutions necessary to experience the evolution of a particular science" (Bachelard 1953, 20). Philosophy must, instead, strive to adapt itself to the constant evolution of science. In *The Philosophy of No*, Bachelard emphasized that "science informs reason. Reason must obey science, the most highly evolved science, science in the process of evolution" (1940/1968, 122). The task of philosophy is to translate the profound spiritual revolutions brought to humanity by contemporary science into adequate conceptual categories (Alunni and Brian 2002; Castellana 2004a, 240–241). In fact, for Bachelard in *The New Scientific Spirit*, science creates philosophy and not the other way around (Bachelard 1934/1984, 3). The task of philosophy is not to determine the "ought to be" of science. They exist, work, and produce results regardless of the existence of philosophy. Rather, epistemological reflection should measure itself against the rationalist activity intrinsic to particular sciences.

Thus, for Bachelard, it is no longer a question of analyzing the conditions of scientific knowledge a priori, nor even of analyzing only the results of the sciences, but rather of observing science in action. This is a fundamental epistemological transition from the study of science as an achieved result to that of science in the process of being carried out (Bachelard 1949). Bachelard's objective was to overcome the classic dichotomies within the traditional philosophy of knowledge, such as those between realism and idealism; empiricism and conventionalism; and positivism and formalism (see Fig. 2.1). He therefore termed his approach *applied rationalism* or *technical materialism.* With these concepts, Bachelard attempted to represent the intrinsically dialectical character of reason, an open and continuously moving reason through which one can "rationally access reality by revealing its complex structure" (Bonicalzi 2007, 31).

As can be seen in Fig. 2.1, Bachelard elaborates a space of possible epistemological positions in relative opposition to one another, starting from those identified as the most extreme and moving on to the more nuanced ones: idealism vs. realism; conventionalism vs. empiricism; and formalism vs. positivism (Bachelard 1949, 5). These positions, Bachelard claimed, need to be replaced by forms of mediation and the integration of the various definitions of rationality because philosophies are always active in scientific activity, even if they are in tension with one another (Bachelard 1949, 5). The terms *applied rationalism* and *technical materialism* are completely equivalent in this formulation. They are two sides of the same coin that take on two names only by virtue of the vector line that they follow to meet in the center of the space of epistemic positionings.

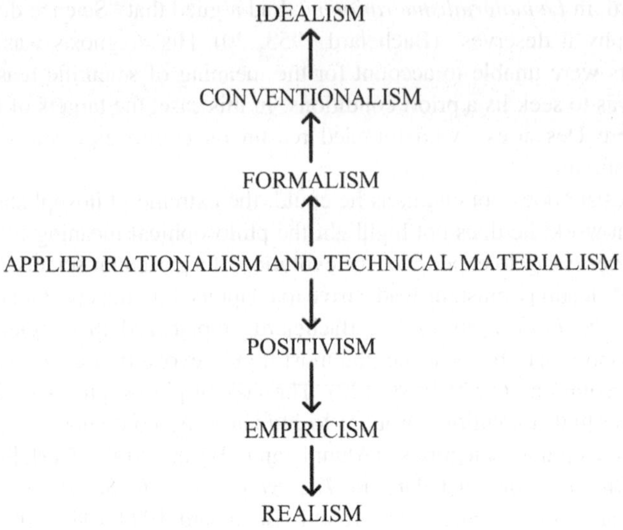

Fig. 2.1 Bachelard's diagram (1949, 5)

The term *applied rationalism* foregrounds how the essence of reason can be understood except through the concrete work that it does and its practical dynamism or understood as an emerging property of scientific activity itself that is continuously redefining its boundaries. Contrary to naive idealism, applied rationalism does not claim that nature is a mere effect of the mind but rather sees reason as the result of dialectical activity established between the knower and the known. The term *technical materialism*, instead, tends to describe the opposite epistemological vector. Unlike naive realism—based on an ideal of an objectively knowable reality and radically opposed to a knowing subject—technical materialism conceives reality as the result of an incessant dialectical process between the knower and the known, and the creative activity of scientists, using technical-scientific tools (mainly mathematical in nature), allows us to grasp the real constitution of physical reality against the illusions provided by sensory experience. For Bachelard, rationalism is always "in dialogue" and "open to dialogue" because both reason and reality are the result of a dialectical process between humans and nature. In 1953, Bachelard brought the terms *materialism* and *rationalism* (reformulated and given new meaning with the words *applied* and *technical*) together in a single formulation with the term *rational materialism*. (Bachelard 1953).

Being materialistic in epistemology means placing oneself in an intermediate space between naive idealism, which sees nature as a mere effect of the mind, and realism, which does not grant reason any productive capacity. Bachelard refused to "be confined to an immobile materialism, to a materialism without dialectics" (Bachelard 1949, 9). This materialism is dialectically shaped by the same process through which scientific activity develops in its back and forth between theory and practice as well as by its historical aspect. Being materialistic in epistemology means seeing the absolute principle of every cognitive activity in the incessant dialectic among nature, rationality, and history. Through the prism of these problematics, two specific trajectories for historical epistemology open up: The first aims to research the historical processes that brought out rationality as such, whereas the second pursues correlative research into the transformation of epistemic objects, practices, and concepts over time.

Like Bachelard, Cavaillès viewed science (particularly mathematics) as characterized by an internal, dialectical dynamism. The final sentence of Cavaillès's last work, *On Logic and the Theory of Science* (written while he was imprisoned by the Nazis), is particularly well known: "It is not a philosophy of consciousness but a philosophy of the concept which can yield a doctrine of science. The generative necessity is not that of an activity, but of a dialectic" (Cavaillès 1976/2021, 136). This cryptic phrase was then interpreted by Canguilhem as follows: "We can understand today that the enigma was worthy of being announced; Cavaillès, twenty years in advance, assigned the task recognized by philosophy today: replacing the primacy of lived or reflected consciousness with the primacy of the concept, system, or structure" (Canguilhem 1996, 29).[3] According to Cavaillès, in fact, mathematics, as a whole, forms an organic system of concepts, each of which takes on meaning in dialectical contrast with the others and with respect to their totality. The history of mathematics for Cavaillès "reveals authentic meanings in so far as it allows us to rediscover lost links—first of all to identify automatisms and sedimentations as such, and then to revivify them by plunging them back into conscious actuality" (Cavaillès 1976/2021, 133).[4]

Canguilhem had largely taken up this conception, creating an intersection between the Comtian and Bachelardian perspectives. From Comte, he derived the idea that the history of science had to be philosophically informed and at the same time had to bring organization and order to scientific experience. On the other hand, following Bachelard, for Canguilhem, the philosopher was not tasked with establishing, a priori, the extension of the concept of science. Philosophy must not intervene "before" science has done its work—with the aim of orienting and directing it—nor after a scientific result has completely crystallized (Braunstein 2002a, 927). In the texts that Canguilhem devoted to Bachelard, this aspect is

[3]In 1941, Cavaillès was called to teach at the Sorbonne, thus abandoning his course at the University of Strasbourg. Canguilhem was teaching at a high school in Clermont-Ferrand in those years, and Cavaillès convinced him to take his place in Strasbourg. The two were linked by friendship and political militancy in the French resistance. During his subsequent Sorbonne teaching, Canguilhem focused on Cavaillès's thinking, proposing an interpretation that highlighted certain aspects that would favor his reception within the intellectual convergence of Bachelardo-Canguilhemism. In particular, *On Logic and the Theory of Science* was published posthumously in 1947 thanks to Canguilhem. Canguilhem also spoke at various commemorative occasions for Cavaillès (Canguilhem 1996).

[4]Following Bachelard and Cavaillès, Desanti also worked at the intersection between mathematical thought, dialectics, and history. He traced this approach to mathematical thought as a form of materialism, particularly in *La philosophie silencieuse* (Desanti 1975, 133–153). On this point, see the last section of this chapter.

often emphasized (e.g., Canguilhem 1994a, 190–191, 200). He also commented on how these aspects have a notable impact on the methodology of the history of science.[5]

Canguilhem dedicated his classes in his first year at the Sorbonne (1955–1956) to his predecessor. In this context, the new professor underlined how, in his opinion, the three fundamental axes of Bachelardian epistemology were the "theoretical primacy of error," the "speculative devaluation of intuition," and the "position of the object as the very perspective of the idea" (Lecourt 2008, 53; Sfara 2017, 99). Bachelard, in fact, in line with the first axiom identified by his successor, had argued that

> You have to make mistakes to succeed.
>
> Thus all objectification proceeds through the elimination of subjective errors and, psychologically, the awareness of this elimination has value. This is not so much a question of fact as it is a question of law. Truth only acquires its full meaning at the end of a dispute. There can be no *prior* truth. There are only *initial* errors. (Bachelard 1970,?89)

Since his doctoral thesis—titled *Essai sur la connaissance approchée*—Bachelard had, in fact, argued that scientific knowledge is always *rectified* and the result of progressive *approximations*: "We therefore take the fundamental incompleteness of knowledge as a postulate of epistemology" (Bachelard 1927, 13), which leads to describing scientific research as an "incessant rectification of thought in the face of reality" (Bachelard 1927, 16). This theme recurs throughout Bachelard's work and is closely connected with what Canguilhem identified as the second and third axioms. In *The Philosophy of No*, Bachelard had argued that "intuitions are very useful: they serve to be destroyed" (1940/1968, 119). Intuitions, especially sensory ones and those relating to common sense, lead to error.

This expression clearly contains the deepest meaning of the notion that made Bachelard famous, the idea of "epistemological rupture." In *The Formation of the Scientific Mind*, Bachelard had declared that "the problem of scientific knowledge must be posed in terms of obstacles" (1938/2002, 25). In the economy of Bachelardian thought, "epistemological obstacles" have a stratified structure and correspond to those images and beliefs that psychologically block thought through parasitic action (Bontemps 2010, 39). Within this context, the image of the world produced by common sense, so-called primitive experience, is the first of the epistemological obstacles that must be broken with for a scientific image of the world to emerge (Bachelard 1938, 23–62).

For Bachelard, the epistemological vector goes from the rational to the real, not vice versa. It is therefore a vector that tends "essentially toward reality" (Bachelard 1934/1984, 4). The object of science is constructed through the tools of rationality, in particular through phenomenotechnics [*phénoménotechnique*][6]:

---

[5] As Camille Limoges has pointed out, this epistemological theme is more prominently discussed in the texts that Canguilhem devoted to Bachelard (Limoges 2012, 55–60, 2015,?40).

[6] On the organic relationship in Bachelard between epistemological rupture, phenomenotechnics, and epistemological regionalism, see Rheinberger (2005) and Ienna (2017).

> We understand reality to the extent that *necessity* organizes it. [...] The direction of our effort towards reality is inflexibly clear. It is a conquest, not a discovery [*trouvaille*]. Our thinking goes towards reality, it does not start from it. At no time have we found a reality that would be known through progressive abstraction; we have always had to deal with a reality that was formed through the accumulation of relationships. (Bachelard 1929, 240–241)

Phenomenotechnics is therefore the beating heart of Bachelard's technical materialism: "In all circumstances, the *immediate* must yield to the *constructed*" (Bachelard 1940/1968, 123). It is not a question of making a "phenomenology" or "phenomenography" (Bachelard 1970, 19) of common experience but rather of highlighting the epistemological process through which science "*realises* its objects" (Bachelard 1938/2002, 70).[7] It is a dialectic between concept and experimental practice that must be understood in its historicity. Phenomena "must be selected, filtered, purified, shaped by instruments; indeed, it may well be the instruments that produce the phenomenon in the first place. And instruments are nothing but theories materialized. The phenomena they produce bear the stamp of theory throughout" (Bachelard 1934/1984, 13; see also Simons 2018). Consistent with the principle of epistemological rupture, which represents the conquest of scientific knowledge over the spontaneous conquest of common sense, phenomenotechnics represents the way in which "noumenal" reality is conquered (see Bachelard 1932, 1970, 11–24): "A concept becomes scientific in so far as it becomes a technique, in so far as it is accompanied by a technique that realises" (Bachelard 1938/2002, 70). Bachelardian materialism is, therefore, based on this form of constructivist realism:

> It is rather a realism at one remove, conceived in reaction to the usual notion of reality, as a polemic against the immediate; it consists of realized reason, reason subject to experimentation. The "reality" to which this realism corresponds is not transferred into the realm of the unknowable thing-in-itself. It has a noumenal richness of quite another order. (Bachelard 1934/1984, 5–6)

Starting from the 1950s, Canguilhem also widely made use of the concepts of phenomenotechnics, epistemological rupture, and epistemological obstacle, applying them particularly to the history of the life sciences—a field in which he had already been working since the 1940s with his doctoral thesis in medicine, titled *Essai sur quelques problèmes concernant le normal et le pathologique*.[8] In *La formation du concept de réflexe aux XVIIe et XVIIIe siècles*—his 1955 doctoral thesis in philosophy directed by Bachelard—Canguilhem foregrounded how, contrary to a popular belief that had spread in the nineteenth century, Cartesianism was not a precursor of the modern physiological theory of the nervous reflex. The former, however, in its historical context, not only had not created favorable terrain for the development of a reflex theory but also represented a real epistemological obstacle to it. Canguilhem proposed to shift the axis of historiographical attention to the work of Thomas Willis, emphasizing how reflex theory emerged from practical knowledge.

---

[7] On the relation between epistemology and ontology in Bachelard, see Gaukroger (1976, 206–212).

[8] This was the title of the 1943 first edition of Canguilhem's celebrated work *The Normal and the Pathological*.

In fact, Canguilhem stressed that when "Descartes formulates a theory of medicine outside of any medical practice, he proceeds from the normal to the pathological. Willis proceeds from the pathological to the normal" (Canguilhem 1955, 59). In the epistemological reflection supporting historical reconstruction, Canguilhem also made frequent use of the Bachelardian concept of phenomenotechnics, though with some original nuances. As in Bachelard, he used this concept primarily to show that the scientific object is always constructed—that is, not independent of its conceptualization. On this basis, Canguilhem further reflected on the materialization of theories into technical tools through which scientific objects are constructed according to the sphere of conceptual production that informs them:

> This is where a truly significant difference appears between the concept of reflex in 1850 and that of 1800. In 1800, reflex was undoubtedly a "good" concept, but it was still good for nothing. It was discussed, but nothing was known because nothing was done about it. It was only written about in books. In 1850, the concept of reflex was found in books as well as in laboratories, in the form of exploration and demonstration devices built on this concept and which would not have existed without it. (Canguilhem 1955,?161)

With the consolidation of Bachelardo-Canguilhemism, the perspective of rational materialism was translated into terms of what has recently been defined as a "discursive materialism" (Benoit 2019a, b). I take up this expression here—with a somewhat different meaning—to identify that "twist" or "turn" that leads to understanding the sphere of scientific conceptual production as the result of a combined action of factors that are both internal and external to the field of scientific production.[9] This epistemological position proposed a historical analysis of the conceptual transformations of science in light of their belonging to discursive orders that condition their conceptual shifts. Materiality is to be found in the capacity of the order of discourse to produce—in terms of a constructivist epistemology—scientific objects and to have an effect on the regulation of scientific practices. In this sense, the development of discursive orders is material to the extent that it is not simply a reflection of a historical situation but rather a concrete and active part of the process of constructing reality by virtue of its material and nonideal effectiveness.[10]

For Canguilhem, the raw materials for a historian of science are scientific concepts and not the natural objects that science deals with. Taking up the methodological reflections of historian of chemistry Hélène Metzger,[11] Canguilhem argued that historical epistemology (or epistemological history) had the objective of

[9] See also Gaukroger's seminal work interpreting Bachelardian epistemology from a "discursive" perspective (Gaukroger 1976).

[10] Obviously, Foucault's perspective would be built on this idea.

[11] In particular, Canguilhem referred here to *La genèse de la science des cristaux* (Metzger 1918). Metzger was at the center of the process of historicizing epistemology both in France and in English-speaking countries. Kuhn explicitly mentioned Metzger in the introduction to *The Structure of Scientific Revolutions*, thus drawing the attention of English-speaking scholars to her contributions. More recently, she has been canonized and is rightly considered one of the pillars upon which FHE developed (Metzger 1938, 1987; Freudenthal 1989; Chimisso 2019). On this point, see also Sect. 3.1 of Chap. 3.

producing an analysis of discourses relating to the nature of scientific objects. The object of the history of science and that of science itself are irreducible to each other, and confusing them with each other is a serious mistake: "the history of sciences is the history of an object which is a history, which has a history, whereas science is the science of an object which is not history, which does not have a history" (Canguilhem 1994b/2005, 202). From this perspective, we can identify three classes of objects. The natural object in itself, beyond all the discourses produced about it; the object of science, which in the Bachelardian sense is the product of a phenomenotechnical construction that represents an achievement against the primary sensory experience of this object; and, lastly, the object of the history of science, which is, instead, the very historicity of the scientific discourse on the natural object (Rheinberger 2010, 67):

> The object of historical discourse is, in effect, the historicity of scientific discourse, inasmuch as this historicity represents the carrying out of an internally law-governed project, but one which is traversed by accidents, retarded or deflected by obstacles, interrupted by crises, i.e. moments of judgment and of truth. (Canguilhem 1994b/2005,?203)

In this sense, "the history of sciences is not a science and its object is not a scientific object" (Canguilhem 1994b/2005, 206). This aspect is fundamental to understanding the type of relationship that historical-epistemological reflection maintains with science itself:

> without reference to an epistemology, a theory of knowledge would be a meditation on the void, and without relation to a history of sciences an epistemology would be a less important labor which was completely superfluous to the science of which it pretends to speak. (Canguilhem 1994b/2005,?200)

But if Bachelard's was a historical epistemology (*épistémologie historique*), Canguilhem's was an epistemological history (*histoire épistémologique*) (see Lecourt 1972). In this sense, Canguilhem reinforced some Bachelardian intuitions regarding the methodology of the history of science. The idea of epistemological rupture required Canguilhem to rethink the history of science as a discontinuous history of "conceptual filiations." It was no longer a story aimed at merely reconstructing the biographies of scientists or the list of theories they had expounded (Canguilhem 1994a, 184). For Canguilhem, the objective was rather to develop a historiographical model that would be useful for understanding evolutions and epistemological transformations through the close observation of the complex and intricate trajectory traced by scientific concepts between cultural and disciplinary fields, in space and over time (Méthot 2013,?114).

Canguilhem's method consisted of treating scientific concepts not as already-established facts but as tools through which facts are constructed or as the concrete material substance of scientific discourse (Benoit 2019a, 104–105). These represent the essential elements that make up the history of scientific thought. This perspective gives materiality to the dialectical-discursive sphere that makes up the intricate mesh within which scientific concepts take shape.

Canguilhem's approach aimed to define genealogies capable of accounting for how concepts migrate between different cultural contexts or scientific paradigms, thus producing transformations and shifts in meaning (Chimisso 2003; Méthot 2013). For example, in "The Living and Its Milieu"—an essay in the collection *Knowledge of Life*—he sought to reconstruct the transformations that the concept of "milieu" had undergone, starting from 1800 up to the mid-twentieth century, crossing various disciplinary contexts, including physics, biology, geography, psychology, technology, economics, and social history (Canguilhem 1952/2008, 98–120). Another exemplary case in which Canguilhem detailed the radical changes in meaning and use according to context was, once again, that of "reflex" in its use in "vitalist" and "mechanistic" theories (Canguilhem 1955).

In much of twentieth-century epistemological reflection, scientific concepts were relegated to being exclusively "functions of theories." Their meaning had always been derived from the scientific theory of reference and its way of accounting for phenomena. As Gary Gutting (2001) effectively illustrated, Canguilhem made an important methodological inversion by proposing a clear distinction between the theoretical interpretation of phenomena through concepts and their theoretical explanation. Accordingly, "a given set of concepts provides the preliminary descriptions of a phenomenon that allow the formulation of questions about how to explain it" (Gutting 2001, 229). Different theories can, in fact, use the same set of basic concepts while proposing divergent explanations of phenomenal reality. The basic concept of motion that supports the Aristotelian theory of natural places and the Galilean theory of falling bodies is the same: What changes is the theory that explains the action. From this, Canguilhem deduced the "theoretical polyvalence" of concepts, or their ability to function—albeit with relative differences—even in very different theoretical contexts (Gutting 2001,?229).

I concur with Gutting in noting that Canguilhem's change in perspective generated a shift in meaning in how he used the concepts of epistemological rupture and epistemological obstacle. Through his historical-genealogical reconstructions, Canguilhem—while maintaining the idea of a clear separation between science and common sense—observed how some concepts can migrate from one scientific or cultural context to another with relative independence compared to theories that call them into question. In this sense, "the same piece of scientific work may be an obstacle in terms of the theoretical context in which it is formulated and a creative breakthrough in terms of some of its conceptual content" (Gutting 2001,?230).

2.2 Historical and Epistemological Discontinuism

As noted above, one of the fundamental features of FHE is how it accounts for the fragmentation of scientific reason. In this section, I analyze how this fragmentation has been observed in the description of scientific knowledge on a diachronic level. In this sense, the key point to be understood is how the historical evolution of scientific knowledge occurs. This problem, indeed, gave rise to the controversy between continuists and discontinuists, one of the most vivid and representative disputes in the field of epistemology and the history of science. This debate emerged around the

1930s in France and then spread to international contexts thanks to the way that Bachelard and Koyré distanced themselves from their predecessors. What was at stake was the determination of whether the history of science was seen as a continuous succession of cumulative successes or interspersed with moments of discontinuity and ruptures. This was an internal question within the methodology of the history of science that, however, had notable epistemological implications.

On the one hand, authors such as Duhem, Meyerson, and Rey (anticipated, in some respects, by Comte and his students Lafitte and Tannery) argued that the process of scientific development proceeds without interruptions and is characterized by a progressive tension. Duhem in particular was responsible for formulating this approach. Duhem was a Catholic and conservative thinker who increasingly focused his attention on the historiographical study of those elements demonstrating continuity between medieval and early modern thought:

> the mechanical and physical science of which the present day is so proud comes to us through an uninterrupted sequence of almost imperceptible refinements from the doctrines professed within the Schools of the Middle Ages. The so-called intellectual revolutions consisted, in most cases, of nothing but an evolution developing over long periods of time. The so-called Renaissances were frequently nothing but unjust and sterile reactions. Finally, respect for tradition is an essential condition for all scientific progress. (Duhem 1905/1991,? 9)

As Fichant emphasized, this position is not exempt from ideological bias: "Duhem's work dedicated to the study of ancient and medieval science aimed to eliminate the myth of the Renaissance, of the *ex nihilo* constitution of modern [*classique*[12]] scientific knowledge through the rejection of religious philosophies and theology" (Fichant and Pêcheux 1969, 72). For Duhem, the task of the historian of science was to identify lines of continuity marking the gradual evolution of scientific thought throughout history.[13] This objective was pursued through careful research into the intellectual antecedents, precursors, and correspondences of certain modern ideas with medieval and ancient ones. For example, one of Duhem's goals was to show how the concept of inertia that emerged with Galilean and Newtonian physics had its roots in the medieval physics of *impetus* and the discussions that took place within the Mertonian school of Oxford—which, in turn, were attributable to even-older reflections. In this sense, the existence of the Scientific Revolution, understood as a relatively unitary historical event that had specific differences from other events of the past, was called into question.

Duhem's historiographical attitude was accompanied by an epistemological attitude. This form of continuism underlay the conventionalist epistemological model according to which physical theories were descriptions of the world aimed at saving the appearances of phenomena. For Duhem, a physical theory was nothing more than a hypothetical explanation of material reality dependent on metaphysics

---

[12] In the Francophone cultural context, the term *âge classique* usually refers to the period between the sixteenth and eighteenth centuries.

[13] For an in-depth analysis of Duhem's continuity theses, see Ariew and Barker (1992).

(Duhem 1906/1982, 10). Physics is a science founded on the development of a system of mathematical propositions that aim to provide a representation of a set of experimental laws that is as simple and complete as possible. This conventionalist position—also defined as "antirealist"—is based on the idea that the task of this discipline is to propose formal classifications of experimental laws that progressively acquire an ever greater correspondence with the natural order (see Redondi 1978, 28–29). According to this way of conceiving physics, the theoretical transformations, changing hypotheses, and failures that occur in the articulation of the history of scientific thought do not affect the relative stability of physical theory as a constantly perfectible description of the natural world: "new discoveries merely clarify the scope for applying the laws, refine the degree of precision, and specify their entire classificatory articulation" (Redondi 1978,?29).

This gives rise to Duhem's well-known thesis (known today as the Duhem–Quine thesis[14]), according to which "An experiment in physics can never condemn an isolated hypothesis but only a whole theoretical group" (Duhem 1906/1982, 183). This position assumed that a scientific theory is never completely disproved at a specific moment but rather is gradually modified over time. Therefore, it was an epistemological thesis in support of historical continuity that did not recognize the existence of radical breaks in the history of science, but rather a long process of small, gradual modifications. In this sense, for Duhem,

> The physicist can never subject an isolated hypothesis to experimental test, but only a whole group of hypotheses; when the experiment is in disagreement with his predictions, what he learns is that at least one of the hypotheses constituting this group is unacceptable and ought to be modified; but the experiment does not designate which one should be changed. (Duhem 1906/1982,?187)

For his part, Meyerson had also adopted and extended Duhem's perspective of historical continuism (Brenner 2007, 379), but in his case, the epistemological implications appear to be different. For Meyerson, the essential point was to show the continuity between common sense and science, or between the cultural and historical background and science. In this regard, Meyerson claimed that "we have seen that, from this point of view, there is an absolute continuity between science and common sense" (Meyerson 1908/1930, 419) and that "the elaboration of scientific hypotheses is the continuation of the process which creates the realities of common sense; but the work being conscious, its authority is thereby lessened" (Meyerson 1908/1930,?421).

Bachelard was among the first critics of these forms of continuism, developing a position that aimed to account for the discontinuities that characterize scientific activity, from both historical and epistemological points of view. Jean Gayon has aptly reconstructed how the notion of epistemological rupture took on, from the start, the meaning of a theoretical and epistemological caesura (Gayon 2003, 80). Indeed, as noted in the previous section, Bachelard created the notion of epistemological rupture to foreground the specific difference between scientific experience and the

[14] On the reception of Duhem's work in the Anglophone post-positivist context, see Laugier (2015).

naive phenomenological experience of common sense. The object of science is, in fact, constructed by using the tools with which rationality is equipped, and therefore, scientific activity takes on the character of a struggle against what Bachelard defined as first-hand phenomenography. Through a constant dialectic between the abstract and the concrete, the scientific object comes into being. By "epistemological rupture," Bachelard meant the act that establishes the epistemic regime within which the various sciences operate. In this sense, the notion of epistemological rupture was in clear opposition to Meyersonian epistemology. On this point, for example, Canguilhem declared that "the epistemologist who is aware of discontinuities is by no means contemptuous of the epistemology of continuity, although he can be ironic about philosophers who believe in nothing else. Bachelard understood Duhem but found Émile Meyerson hard to tolerate" (Canguilhem 1977/1988,?16).

Undeniably, however, Bachelard's epistemology was also based on a form of historical discontinuism. The epistemological ruptures that Bachelard talked about are not located in an abstract and timeless space. The acts of founding and opening epistemic regimes occurred repeatedly throughout history, creating a discontinuous movement in the evolution of scientific thought. Bachelard clearly argued that the evolution of science was marked by profound "intellectual mutations" and a "revolutionary character" (Bachelard 1934/1984, 171–177). As an attentive connoisseur of contemporary science, Bachelard never hesitated to describe the innovations of Einsteinian physics, quantum mechanics, and the emergence of non-Euclidean geometries as authentic revolutions, in complete discontinuity with the past: "Thus, even in the historical evolution of a particular problem, real ruptures and abrupt mutations appear, which ruin the thesis of epistemological continuity" (Bachelard 1927, 263). Einstein's proposed description of the physical world caused a clear temporal and cognitive discontinuity with previous Newtonian theories. These discontinuities were, therefore, at the same time, related both to the epistemic dimension—the transition from common sense to science—and to historical dimension.

The key to understanding Bachelard's conception of historical time is contained in *Intuition of the Instant* (1932) and *The Dialectic of Duration* (1936). I describe these two texts as the "hinge" between the two sides of Bachelard's philosophy (the epistemological and the poetic[15]). Particularly in *The Dialectic of Duration*, Bachelard proposed a philosophy of temporality regulated by an intrinsic dialectic

[15] Bachelard devoted an entire body of work to aesthetic issues, in particular poetry, including *The Psychoanalysis of Fire* (1938), *Lautréamont* (1939), *Water and Dreams* (1942), *Air and Dreams* (1943), *The Poetics of Space* (1957), and *The Poetics of Reverie* (1960). According to Lecourt's famous formulation, in the context of Bachelard criticism, the existence of two Bachelards has frequently been recognized: one of the daytime (so to speak "*du jour*"), engaged in epistemological reflections, and a nocturnal one (so to speak "*de la nuit*"), dedicated instead to questions of imagination, poetry, and aesthetics (Lecourt 1974). In more recent literature, this duplicity is being increasingly questioned. On this point, see, for example, the recent conference titled "Bachelard, 'le jour et la nuit': oltre l'opposizione tra razionalità e immaginario," organized at the Italian Institute for Philosophical Studies in Naples by Massimiliano Biscuso, Fabrizio Palombi, and Gaspare Polizzi (Biscuso et al. 2020a, b).

that developed through the concept of rhythmanalysis.[16] On the basis of his reflections on the historical-epistemological field, Bachelard opposed any form of alleged continuity in the description of historical and psychological-individual temporality. In *Intuition of the Instant* and *The Dialectic of Duration*, however, his polemical references are no longer to Meyerson and Duhem—although they are probably implicitly invoked—but rather to Henri Bergson's philosophy of time. Though he accepted almost everything Bergson proposed regarding the psychology and philosophy of time, Bachelard could not agree with the continuity thesis (Bachelard 1936a/2000, 28–29). For Bergson, "time's true reality is its duration. The instant is but an abstraction, devoid of reality. It is imposed from without, by the intellect which can understand becoming only by apprehending motion" (Bachelard 1932/2013,?13).

To oppose Bergson's positions, Bachelard employed the theses of various writers who are considered minor today, such as Gaston Roupnel, Eugène Dupréel, and Lucio Alberto Pinheiro dos Santos. Through these references, Bachelard constructed a metaphysics of time based on the valorization of the reality of the instant. He proposed a theory of duration in resolutely constructivist terms, consistent with his epistemological principles (D'Aurizio and Palombi 2022). Completely inverting Bergson's thesis, Bachelard firmly believed that "*continuity is not given but made*" (Bachelard 1936a/2000, 19). Continuities "have to be constructed. They have to be maintained. Consequently, we do not in the end see the continuity of duration as an immediate datum but as a problem" (Bachelard 1936a/2000, 29). For Bachelard, indeed, time could not be conceived as an immutable and a priori structure within which events are situated. On the basis of Einsteinian physics and quantum mechanics, he attempted to build a materialist and dialectical epistemology and metaphysics of time in which the relationship between instants is what generates temporalization.

From this perspective, Bachelard derived a conception of time entirely focused on the present, or on the instant understood as the only reality of time. For Bachelard, time was a mathematical operation: "duration is but a number whose unit is the instant" (Bachelard 1932/2013, 22). Change—which implicitly refers to the unfolding of history—would not be possible if not in a discrete and discontinuous temporal structure. For Bachelard, Bergson's continuity denied the possibility of recognizing "the creative value of becoming" (Bachelard 1936a/2000, 24), canceling the dialectical fluidity of time. On the contrary, for Bachelard, we must find the seeds of becoming in the instant itself. Present and past are thus desubstantialized from the point of view of temporal ontology, taking on value only as a psychological dimension derived from memory and future planning (which are also, however, located in the present moment): "time is a reality confined to the instant and suspended between two voids" and thus "cannot transport its being from one instant to another" (Bachelard 1932/2013,?6).

[16]This is a term coined by the Brazilian philosopher Lucio Alberto Pinheiro dos Santos in *Ritmanálise* (1931). Bachelard took up this term in *The Dialectic of Duration* and some subsequent works. This concept was then further developed by Henri Lefebvre in *Éléments de rythmanalyse*.

Keeping wave mechanics in mind, Bachelard built an epistemology of duration based on rhythms or systems of instants, synchronies, and diachronies. Through these elements, he brought out the intrinsically dialectical and discontinuous progression of the emergence of temporality through shocks, ruptures, and oppositions. Continuity, in this conception, is only apparent, the result of a process of simplifying complex processes, or simply a form imposed on phenomena.

Another key figure in the emergence of discontinuism was Koyré, who interpreted the path of science as an effort that brought about "a veritable 'mutation' in human thought" (Koyré 1939/1978, 1). On this point, Koyré explicitly recognized his debt to *The New Scientific Spirit* in the formulation of the concept of intellectual mutation:[17] "The apparent continuity in the development of physics, from the Middle Ages to modern times (a continuity that Caverni and Duhem strongly emphasized) is illusory. [...] a well-prepared revolution remains, in any case, a revolution" (Koyré 1966, 152–153). Koyré was the best-known supporter of the notion of "scientific revolution," from which he constructed a discontinuous history of scientific thought. Predictably, some underlying theoretical positions were involved in using a concept such as scientific revolution.

Indeed, it implied a particular attitude regarding the epistemology and methodology of historiographical research. Whereas the continuist's historiographical work consisted of increasingly in-depth and meticulous research into sources that would allow lines of filiation, links, and connections to be traced between apparently unrelated historical events, the discontinuist organized historiographical material, trying to bring out new elements of certain events and criteria for demarcating historical phases. These different research objectives, therefore, determined different methodological postures.

Like Bachelard, Koyré also took a position against Duhem's historiographical and epistemological theses. A characteristic and recurring thesis in Koyré's studies—including *Galileo Studies* (1939), *From the Closed World to the Infinite Universe* (1957), *La révolution astronomique. Copernic, Kepler, Borelli* (1961), and the posthumously published *Newtonian Studies* (1965)—was the identification of a moment of clear historical discontinuity, with respect to medieval and ancient science, in the astronomical controversy that occurred between the sixteenth and seventeenth centuries. The abandonment of ancient physics, through the so-called Copernican revolution and the related opposition to Aristotelian-Ptolemaic theory were, for Koyré, the beating heart of the Scientific Revolution and modern science. The terrain of physics and cosmology was therefore the privileged battlefield, and in his extensive production, Koyré often targeted Duhem and his continuity theses:

> "If we had to assign a date to the birth of modern science," said Pierre Duhem, "we would undoubtedly choose 1277, when the Bishop of Paris solemnly proclaimed that multiple worlds could exist and that the set of celestial spheres could, without contradiction, be

[17] This expression was later replaced with that of epistemological rupture, starting from the publication of Bachelard's *Le rationalisme appliqué*.

> animated by a rectilinear movement." This is a curious assertion which places the Bishop's proclamation of two absurdities at the origin of modern science. (Koyré 1961,?33)

In addition to his criticism of Duhem, Koyré also addressed Crombie's historiographical model in a long essay titled "Les origines de la science moderne. Une nouvelle interpretation" (Koyré 1966, 48–72). Crombie's thesis aimed to show how the emergence of an attitude favoring experimental science—widely considered to be unique to modern science—could actually be dated back to the medieval era (Koyré 1966, 60). According to this interpretation, the methodological turning point, which requires that theories be correlated with observations, was originated by Western philosophers of the thirteenth century. While appreciating his detailed historiographical reconstructions, Koyré firmly contested Crombie's continuist conclusions.[18]

Koyré's opposition to this historiographical current was not based on the collection of historiographical data but rather on their interpretation, in terms of what could be defined as science. In this sense, the identification of historical discontinuity that supported the idea of a scientific revolution was closely related to an ontological problem:

> Yet, the problem of the origins of modern science and its relationship with the science of the Middle Ages remains a disputed question. The supporters of continuous evolution, as well as those of a revolution, all remain in their positions and seem incapable of convincing each other. This, in my opinion, is not so much because they disagree on the facts, but because they disagree on the very essence of what modern science is and, consequently, on the relative importance of certain fundamental characteristics. Furthermore, what for some is a difference of degree, for others is a difference of kind. (Koyré 1966,?48)

According to Koyré, the Scientific Revolution was characterized by at least two fundamental aspects: (1) the destruction of the cosmos (from the Greek *kósmos*, or world, understood as a finite, ordered, and harmonious system) and the appearance of the infinite and/or indefinite universe and (2) the geometrization of space, or the replacement of Aristotelian space, structured according to natural places, with an infinite and homogeneous Euclidean space (Koyré 1957, 8). According to his famous formulation, this second aspect corresponded to the identification of an epochal transition "from the world of approximation to the universe of precision" (Koyré 1961, 313, 2000). The underlying epistemological thesis maintained that only through mathematizing nature can a change in the structure of the forms of knowledge that would study it be set in motion. In this sense, from a disciplinary point of view, the process consisted in the consolidation of the defining differences and points of rupture—for example, between alchemy and chemistry, astrology and astronomy, and so forth.

As an heir to the historical-philosophical attitude of authors such as Tannery, Brunschvicg, and Meyerson, Koyré's objective was to reconstruct a philosophical history of science, tracking the efforts that human beings had made in their search for

[18] On the mutual respect between the two authors, see Crombie's autobiographical essay (Crombie 1987).

truth, in their "*itinerarium mentis in veritatem*" (Jorland 1981, 72). For this reason—and contrary to other historiographical currents, such as Marxism or externalism—Koyré supported an intellectual vision of scientific activity and a consequent internalist historiographical approach to the history of science.[19] He considered science to be a purely intellectual commitment, explicitly refusing to recognize any causal role for the process through which modern science emerged and consolidated with the practical-material and social spheres. For internalism, historical discontinuities and scientific transformations regard only the intellectual processes that are internal to scientific activity itself.

Unlike Bachelard, Koyré did not break so abruptly with Meyerson's philosophy.[20] Koyré shared Meyerson's idea that the history of science was the privileged place and means through which to study how thought functions (see Castelli Gattinara 1996, 113). But whereas Meyerson had analyzed the history of thought as an expression of its unitary structure (of a psychological type), Koyré—drawing inspiration from a phenomenological tradition—was convinced that this unity was to be found in the ontology underlying the comprehensibility of reality (see Jorland 1981, 95–98).

This primacy of the intellectual over the technical, practical, and material dimensions was a common ground between Duhem and Koyré. They both supported this position through a criticism of what, in their opinion, was an unwarranted overestimation of the centrality of empirical and practical experience in the historiographical description of the emergence of modern science.[21] According to the historiographical interpretation proposed by Duhem, none of the key principles of modern physics, such as the principle of inertia or the law of mixed proportions, could be derived from empirical observation. Furthermore, following the principles of the Duhem–Quine thesis, individual theoretical propositions could not be verified or invalidated through experimentation. Furthermore, by virtue of theoretical continuity, conceiving of an experiment that would allow a theory to be refuted in its entirety would not be possible. The interpretation proposed by Koyré, from *Galileo Studies* through his posthumous works, reflected the idea that not only did empirical experience fail to play a major role in the emergence of the Scientific Revolution but also it was often an obstacle to it:

> There has also been much discussion of the role of observation and experiment, of the birth of an "experimental attitude." It is certainly true that the experimental character of classical

[19] On the reconstruction of Koyré's opposition to externalism in the history of science, as proposed by authors such as Boris Hessen, Edgar Zilsel, Henryk Grossmann, Franz Borkenau, and George Norman Clark, see Ienna and Rispoli (2017, 2019) and Ienna (2022). I will return to this point in Sect. 3.1 of Chap. 3.

[20] Koyré had been Meyerson's student and dedicated his first book on the history of science, *Galileo Studies*, to him. Even beyond these formal and institutional aspects, in Koyré's subsequent research, Meyerson was always cited as a point of reference. On this, see also Fichant and Pécheux (1969,? 81).

[21] For a survey of the similarities and differences between Koyré's thought and Duhem's, see Clavelin (1987).

> studies is one of its most typical features. In fact, however, it is easy to misunderstand just what is involved here. The only role in the birth of classical physics played by observation, in the sense of *simple* observation, the observation of common sense, was that of an obstacle. The physics of the Parisian nominalists—even that of Aristotle—was often much more akin to such observation than was that of Galileo. As for experimentation—the methodical interrogation of nature—it presupposes both the language in which its questions are to be posed and a terminology which makes it possible to interpret nature's replies. But if it is in a mathematical (or more precisely, geometrical) language that nature is interrogated by classical physics then this language, or to put it more accurately, the decision to use it—a decision which corresponds to a change of metaphysical attitude—cannot itself be the product of the experiment which is conditioned by it. (Koyré 1939/1978,?2)

As an heir of Husserlian phenomenology, Koyré identified the mathematization of nature as the key turning point of modern science—that is, the transition from imprecise qualitative analysis to quantitative precision. Both for Edmund Husserl in *The Crisis of European Sciences and Transcendental Phenomenology* (1936) and for Koyré, Galileo embodied the figure of the ideal hero of this transformative process: "The name Galileo Galilei is inextricably linked to the Scientific Revolution of the sixteenth century [...] a revolution that implied a radical intellectual 'mutation' of which modern physical science is both the expression and the result" (Koyré 1966,?147).

The essential point of Koyré's position was to deny not the existence or relevance of experimental practices but rather their interpretation in intellectual terms. In a manner similar in many respects to Bachelard's concept of epistemological rupture, Koyré outlined a difference between the concept of common experience and *experimentum*:

> Galileo knows that the experiment [*expérience*]—or if I can be permitted to use the Latin word *experimentum* to contrast with common experience [*expérience commune*], with the experience that is only observation—is prepared, it is a question posed to nature, a question posed in a very special language, in geometric and mathematical language; he knows that it is not enough to observe what is, what presents itself normally and naturally to the eyes, that one must know how to formulate the question and that one must also know how to decipher and understand the answer, that is, to apply the strict laws of measurement and mathematical interpretation to the *experimentum*. (Koyré 1966, 46–47)

In Koyré's interpretation, the experimental dimension of modern Galilean science had nothing to do with the empirical experience of common sense. Experiments, unlike brute experience, are entirely constructed through mathematical language and are theory-laden. On this point, Koyré's contribution was both historiographical and epistemological, as I will illustrate with some examples. In the essay "Galilée et l'expérience de Pise. À propos d'une légende" (in Koyré 1966) as well as in *Galileo Studies*, Koyré showed how some of Galileo's most famous experiments were exclusively mental. What Koyré called the "legend" of Galileo's public experiments, with the dropping of heavy bodies from the Tower of Pisa, is well known in the history of science. In relation to this event, Koyré proposed a historiographical reconstruction highlighting how this experiment never actually took place and that the formulation of the law of falling bodies was entirely based on a thought experiment and mathematical calculation (Koyré 1939, 1966, 192–201). Likewise, in "Du monde de l' 'à-peu-près' à l'univers de la précision" (1950), Koyré deconstructed the myth of Galileo's discovery of the isochrony of the pendulum, which according to legend was traced back to the scientist's direct observation of the oscillatory movement of the chandelier of the cathedral of Pisa—which, however, was placed there only after he had left the city (Koyré 1950,?328).

Neither experiments nor scientific instruments are neutral; rather, they are built and made usable on the basis of the scientific theories that inform them. On this basis, Koyré's argument downplayed the contribution of technical-artisanal practices to the emergence of modern science. In this sense, he clearly stated that "In reality, the glasses-maker was in no way an *optician*: he was a *craftsman*" (Koyré 1950, 320), by virtue of the following facts: The optician constructs scientific knowledge on a mathematical basis, whereas the craftsman does not. From this clearly emerges a conception of technical practices as unreflective and incapable of posing concrete problems, solutions, or results of a scientific nature: "The technical thinking of common sense does not depend on scientific thinking, even if from this it can absorb various elements by incorporating them into common sense" (Koyré 1950,?314).

Within Koyré's interpretative horizon, even the same technical devices undergo a radical ontological transformation if they are theory-laden. Thus, the epistemological and ontological difference between a tool and a scientific instrument emerges. In this regard, Koyré mentions the story—perhaps mythical—of the completely accidental invention of the first telescope by the son of a Dutch glasses maker who, playing with his father's lenses, placed them face to face for the first time after observing that, in this way, he could extend his gaze. For Koyré, this is in no way sufficient for the invention of a scientific instrument. The glasses made by crafters were not yet optical instruments but only tools, "that is, something—which as ancient thought had clearly seen—prolongs and strengthens the action of our limbs, our sensory organs; [it is] something that belongs to the world of common sense but which cannot help us to overcome it" (Koyré 1950,?321).

From an epistemological point of view, Galileo's telescope had specific characteristics that, for Koyré, resulted from its being built on the basis of a theory. It became a tool capable of demonstrating something, of providing evidence to support this theory, only by virtue of presupposing the heliocentric theory: "The monocular was not a telescope. Transforming it into one was precisely the merit of Galileo" (Koyré 1950, 320). A theory-laden technological instrument

> is a tool, that is, a creation of scientific thought, or, better yet, the conscious realization of a theory. It is certain that, once a theoretical object has been created, it can become a practical object, an object of everyday use. It is also certain that practical considerations can inspire theoretical thinking [...]. But it is not the use of an object that determines its nature: it is the structure. (Koyré 1950,?325)

Some commentators, through in-depth historiographical and philosophical investigations, have noted how both Bachelard and Koyré, in some respects, maintained links with the continuity tradition (Castelli Gattinara 1996, 83).[22] Certainly this aspect is visible in Canguilhem's historiographical attitude toward this debate.

From an epistemological point of view, Canguilhem maintained an open and constant dialogue with Bachelard's thought. As we have seen, Canguilhem tended to emphasize how the ultimate meaning of the idea of epistemological rupture could be attributed to the epistemic dimension of the demarcation between science and common sense. Indeed, in this sense, Canguilhem argued that Bachelard elaborated the notion of epistemological rupture in disagreement with Meyerson.

For Canguilhem, that only the peculiar conformation of knowledge called science is a producer of truth, according to his well-known expression that "there is no truth but scientific truth" [*il n'y a de vérité que scientifique*], was no coincidence. In the wake of Bachelardian epistemology, Canguilhem extended the idea of a strong discontinuity between common sense and science, arguing that "There is no scientific knowledge without, on the one hand, very elaborate mathematical theories and, on the other hand, the manipulation of progressively more complex instruments. I would gladly say that there is no such thing as common knowledge [*connaissance vulgaire*]." From this perspective, therefore, talking about forms of knowledge that are not scientific would be a contradiction because "Knowledge that is not scientific is not knowledge. *True knowledge* is a pleonasm, as is *scientific knowledge* and *science and truth*; all these terms mean the same thing" (Badiou and Canguilhem 1965).[23]

From the proposition "there is no truth but scientific truth," Canguilhem deduced that other forms of knowledge, including philosophical knowledge, could not produce forms of truth. For Canguilhem, the task of FHE—and of philosophy in general—was to develop a meta-discourse on forms of scientific knowledge. The meaning of science is not contained in science itself but rather is a philosophical domain of investigation. For its part, a philosophically informed history of science must not be reduced to the analysis of what is considered science in a given historical moment. Instead, it must examine the intersection of this science with other forms of

[22] The image of an FHE centered on historical discontinuities has been emphasized and amplified by readings deriving from Bachelardo-Canguilhemism since the end of the 1960s. Although the internalist historiographical model used by Koyré was in clear opposition to the Marxist historiography of authors such as Hessen, Grossmann, Zilsel, and John Desmond Bernal, the Althusserians remained greatly influenced by his interpretative proposal precisely because it centered on the idea of "Scientific Revolution" and historical discontinuity (Fichant and Pécheux 1969). In the same way, Foucault and Bourdieu frequently associated these two forms of discontinuism, integrating the overlapping ideas of the two authors (Bourdieu et al. 1968; Foucault 1985). The tendency to unify the intersection and overlap of Bachelard's theses with those of the historiographical model proposed by Koyré is certainly a legacy of Canguilhem's teaching.

[23] This perspective was developed in a famous television dialogue coordinated by Badiou in which—in addition to Canguilhem—Foucault, Jean Hyppolite, Dina Dreyfus, and Paul Ricoeur also participated. This dialogue can be found in *Dits et écrits* (Foucault 1994, 448–464), and a video record has also been preserved (Badiou et al. 1965). This dialogue was preceded by Badiou's interview with Canguilhem (Badiou and Canguilhem 1965).

cultural production by analyzing the dynamics of how concepts migrate from one context to another.

In the context of classical epistemology, historicity and objectivity have traditionally been understood as contradictory to each other. On the contrary, as Braunstein rightly emphasized, for Canguilhem, precisely the historicity of scientific knowledge was a criterion for defining the border separating science from nonscience (Braunstein 2002a, 933). The essential element of Canguilhem's approach was, indeed, his firm opposition to neo-positivist and antihistoricist theories of scientific knowledge. Proof of the scientific nature of knowledge is demonstrated by the fact that

> A science that is conceived as having no history, that is, a science in which there is no rejection of the specific conditions of objectivity at a given moment and the substitution of more impartially defined conditions of objectivity, is not a science. (Canguilhem 1970,?235)

The category of objectivity tends to change over time as scientific knowledge is transformed. For this reason, a discipline that does not recalibrate its criteria of objectivity in light of advances in knowledge cannot be considered scientific. This is, for example, the case of "false sciences," such as astrology, which do not really have a history (Braunstein 2002a,?933).

In this context, Canguilhem formulated the concept of "being in the truth" [*être dans le vrai*][24] to analyze the porous historical layers within which the emergence of truth can be glimpsed. In his presentation at the 1964 conference "Galilée: la signification de l'oeuvre et la leçon de l'homme"—on the occasion of Galileo's 400th birthday—Canguilhem argued that "being in the truth does not mean always uttering the truth" (Canguilhem 1994a, 46). Referring to the analyses proposed by Koyré, Canguilhem thus stressed the substantial diachrony of Galileo's work with respect to the criteria of truth accepted by science in his time, which were based on Aristotelianism, as well as its anticipatory function.

In Canguilhem's interpretation, Galileo was the one who blew up the old criteria of truth, allowing their redefinition, and thus introduced an element of diachrony into the course of the history of science. Within the controversy relating to the two chief world systems, Galileo was right to support the Copernican positions even though he had made some errors—such as rejecting Kepler's replacement of the traditional circular conception of planetary orbits with an elliptical model (Han 1998, 136)—or did not yet have sufficient evidence of heliocentrism, such as the measurement of the parallaxes of fixed stars, which arrived in exhaustive form only in the nineteenth century (Canguilhem 1994a, 45–46):

[24] Compare the two interpretations of Canguilhem's notion of "being in the truth" found in Foucault's *The Order of Discourse* (Foucault 1979, 27–29; see Han 1998, 134–141) and Balibar's essay "Being in the True? Science and Truth in the Philosophy of Georges Canguilhem" (in Balibar 1994/2016). Foucault's interpretation seems to go in the direction that I would define as contextualist, whereas Balibar's tends toward a reading that links this concept to "scientific ideology."

> If, by experience, we mean the usual, pragmatic experience, Aristotelian physics accords better with it than Galilean physics; if, by experience, we mean the experimentation carried out as a function of a hypothetical explanation, none of Galileo's experiments (and today we know that he did much less than he was given credit for when his portrait was painted using Bacon as a model) succeeded in confirming the anticipations of calculus, none managed to convince scientists [*savants*] who were as un-Aristotelian as him. (Canguilhem 1994a,?45)

In this sense, Galileo was not uttering the truth—neither in the sense of respecting the culturally shared probatory standards of the science of his time nor with respect to the status of transhistorical truths. However, at the same time, he was uttering the truth because he would have determined a radical transformation of those forms of knowledge by virtue of the anticipatory function of his theory. "Being in the truth," then, did not mean uttering the truth but rather having glimpsed the truth, even before it was fully recognizable as such within a specific sphere of cultural production. This epistemological position formed the theoretical basis for an interpretation of the controversy between continuism and discontinuism.

From a historiographical point of view, Canguilhem had adopted a more complex and more nuanced methodological position than others historical epistemologists did in the way that it characterized the discontinuist methodology. The history of traditional science was typically based on studying the biographies of authoritative scientists or analyzing the formation and consolidation processes of scientific theories. Canguilhem's perspective, instead, articulated a completely original historiographical model, which, rather than looking directly at the work of scientists or scientific theories, emphasized the analysis of the intersections between theories and the culturally shared sphere of conceptual production in its historical evolution.

In this sense, it was a question of focusing on the processes through which scientific concepts emerged, were transformed, circulated, and rectified—key results in the history of scientific thought—across different disciplinary-cultural contexts. This model aimed to build a network of "conceptual filiations" for which it was not so much a question of creating a history of intellectual influences or inspirations but rather of epistemological research into the conditions of possibility and the formation dynamics of scientific concepts by observing them in action. As Taylan has noted, it was a question of reversing the direction of the analysis proposed by traditional historiographical models:

> not starting from authors taken in isolation but from a logical space, from the epistemological conditions necessary for the formulation of this or that concept. Instead of asking whether one author influenced another, we should ask which theoretical configuration was capable of accommodating this or that concept. (Taylan 2018,?60)

In this sense, Canguilhem wrote:

> we think that in matters of the history of science the laws of logic must not be put aside in favor of the laws of history. Thus, before putting the succession of theories in order, according to the logic of their adequacy and the homogeneity of their inspiration, we first need to ensure—in the presence of a given theory, where we try to identify this or that implicit or explicit concept—that we form an idea which is not lacking in concern for internal coherence. Without this, we would arrive at this paradox: that logic is everywhere

> except in the thinking of scientists, and that we can have a logic of the succession of theories that is, in itself, indifferent to logic. (Canguilhem 1955,?5)

Canguilhem applied this methodology to fundamental concepts of the history of the life sciences. I have already mentioned his historical reconstruction of the concept of "reflex" between the seventeenth and eighteenth centuries (Canguilhem 1955) as well as that of "milieu" (Canguilhem 1952/2008), to which can be added analyses of the concepts of regulation, organization, natural selection, development, cell, and others (Canguilhem 1994a, 1952/2008, 1977). In all this research, the cornerstone of Canguilhem's historiographical methodology was a careful and reflective analysis of the "conceptual cleavage [*fracture*]"[25] that occurred throughout the history of scientific thought within the intricate network of conceptual filiations (Canguilhem 1977/1988, 14). While maintaining a fundamentally discontinuous approach, Canguilhem underlined that we must not give in to the simplistic notion that a scientific idea

> begins with a flash of insight, a work of genius. Frequently the effects of that flash are said to be all-embracing, affecting the whole of a scientist's work. But the reality is different. Even within one man's work we often find a series of fundamental or partial insights. (Canguilhem 1977/1988,?15)

In this important passage for Canguilhem's conception of history, he cited Koyré's *Galileo Studies* to exemplify a good practice of researching "decisive" discontinuities for the history of scientific thought. In this sense, for Canguilhem, "a theory is woven of many strands, some of which may be quite new while others are borrowed from older fabrics" (Canguilhem 1977/1988, 15):

> The epistemology of rupture is appropriate to a period of accelerated change in science, a period in which change is measured in years or even months. By contrast, the epistemology of continuity has a natural affinity with periods in which knowledge is just awakening. But the epistemologist who is aware of ruptures is by no means contemptuous of the epistemology of continuity, although he can be ironic about philosophers who believe in nothing else. (Canguilhem 1977/1988, 16; translation modified[26])

The program of Canguilhem's FHE consisted in untangling this complex web of conceptual filiations to identify its authentic moments of "cleavage." The concept, as Taylan rightly pointed out,

> is a strategic element because it allows us to place ourselves, at the same time, on the terrain of filiation and transformation, of the continuous and the discontinuous, since it is the indicator of the contents that are maintained, expanded, or eliminated, that are transformed or abandoned. (Taylan 2018,?61)

[25] In this case, Canguilhem chose the term *fracture*, inspired by Cavaillès, which was interpreted and used in light of the notion of Bachelard's epistemological rupture (Canguilhem 1977/1988, 14 n. 30).

[26] The published translation used the term *discontinuity* for the French *rupture*, which is more typically translated as "rupture" in this context.

In this sense, Canguilhem pointed out that, for example, Galileo's case teaches us how scientific revolutions "did not sweep away tradition in one fell swoop" Canguilhem 1977/1988,?15).

These problems lead us to the subject of the next section, which examines another key aspect of the historiographical posture adopted within FHE, namely presentism and the idea of historical recurrence.

2.3 Presentism and Historical Recurrence

As emerged above, FHE is a field of research that has dedicated ample space to epistemological reflection on historical methodology. However, historical discontinuism was linked to other historiographical positions that created an additional common feature within this tradition. This section shows how the relationship between past and present was conceived in the French debates being analyzed and the implications of this conception for historiographical research practices.

During the twentieth century, debates developed around this issue in various cultural contexts. Each of these had its own peculiarities and specificities, underlying different polemical references and theoretical-political positions. In defining the terms of the debate, the reflections produced in the Anglophone and French (as well as Italian[27]) contexts had a particularly profound impact, even though they emerged independently from each other.

In the British context, this debate became the controversy between Whiggism and anti-Whiggism. A key contribution that, among other things, gave its name to the intellectual controversy was the 1931 volume *The Whig Interpretation of History* by the Cambridge historian of science, Herbert Butterfield. Butterfield proposed a strong criticism of the forms of historical anachronism deriving from a prospective view of history that started from the present: "The study of the past with one eye, so

[27] Even before these positions were consolidated, reflections on the relationship between present and past were underway in Italy. Omodeo (2018, 222, 2019, 32–37, 2020, 30–31) has emphasized how Croce in *Teoria e storia della historiografia* had already proposed a form of "presentism" according to the formula "every true history is contemporary history" (Croce 2007/1921, 12). In this sense, "History is living chronicle, chronicle is dead history [...] history is principally an act of thought, chronicle an act of will. Every history becomes chronicle when it is no longer thought, but only recorded in abstract words, which were once upon a time concrete and expressive" (Croce 2007/1921, 19). Gramsci echoed this form of historical presentism, politicizing its dimension by arguing the following: "Croce's concept of history: is it not purely bookish and erudite? Only the identification of history and politics rids history of this peculiar characteristic. If politicians are historians (not only in the sense of creating history, but in the sense that, by operating in the present they interpret the past), historians are also politicians and in this sense (which moreover is also present in Croce), history is always contemporary history, i.e. politics, but Croce cannot arrive at this necessary conclusion, exactly because it leads to the identification of history and politics, and thus of ideology and philosophy" (Gramsci 1977/1995, 383). On the Italian tradition in historical epistemology, see the appendix to this volume.

to speak, upon the present is the source of all sins and sophistries in history, starting with the simplest of them, the anachronism" (Butterfield 1931, 31–32). Adopting explicitly political terminology,[28] Butterfield started by analyzing some paradoxes arising within British historiography:

> What is discussed is the tendency in many historians to write on the side of Protestants and Whigs, to praise revolutions provided they have been successful, to emphasise certain principles of progress in the past and to produce a story which is the ratification if not the glorification of the present. This whig version of the course of history is associated with certain methods of historical organisation and inference—certain fallacies to which all history is liable, unless it be historical research. (Butterfield 1931,?v)

Butterfield's argument was not initially addressed to historians of science—even though the debate developed largely in this context—but, more generally, to all historians who based their historical research on value judgments and moral judgments about the past. Butterfield identified ideal types of this Whig approach to history in political historian Lord Acton, who based his historiographical perspective on the "moral function of history" (Butterfield 1931, 109 et seq.), and conservative historian Henry Hallam (Butterfield 1931,?4).

In the field of the history of science, the Whig approach translated into a tendency to see present science as superior to past science and, consequently, to construct historical narratives interpreting past stages of the evolution of scientific thought as part of an unfolding of constant progress—considered by some to be inevitable—toward the results obtained by contemporary science. The result was a model that glorified historical actors whose work seemed to lead to what, in the present, is considered "scientific truth," alongside a correlative historical marginalization of those who seemed to have fallen "into error." Butterfield's work was, in the first instance, a criticism of such positivist narratives. This approach was, for example, structurally implicit in Duhem's continuist historiographical model, which aimed to trace the medieval precursors of modern science. According to Butterfield, adopting such a model inevitably led to interpretative fallacies:

> Through this system of immediate reference to the present-day, historical personages can easily and irresistibly be classed into the men who furthered progress and the men who tried to hinder it; so that a handy rule of thumb exists by which the historian can select and reject, and can make his points of emphasis. On this system the historian is bound to construe his function as demanding him to be vigilant for likenesses between past and present, instead of being vigilant for unlikenesses; so that he will find it easy to say that he has seen the present in the past, he will imagine that he has discovered a "root" or an "anticipation" of the 20th century, when in reality he is in a world of different connotations altogether, and he has merely stumbled upon what could be shown to be a misleading analogy. (Butterfield 1931, 11–12)

[28]The Whigs were a British liberal political party active between the end of the 1700s and the mid-1800s. Their political orientation favored constitutional monarchy and advocated for the aristocracy, landowners, and wealthy middle classes. From a religious point of view, Whigs represented more evangelical and dissenting faiths within Protestantism, such as Presbyterianism. The Whig party was politically opposed to the Tory party, which, instead, supported absolute monarchy and favored the Anglican church.

According to Butterfield, the historian's aim should be to look at their own historical data by adopting the same point of view as someone from the period being analyzed: "Real historical understanding is not achieved by the subordination of the past to the present but rather by our making the past our present and attempting to see life with the eyes of another century than our own" (Butterfield 1931, 16). Only in this way, without having future effects as a point of reference, can we reveal the intricate network of relationships from which the facts emerge. The historian's task is not to understand—to borrow Hall's example—who discovered modern atomic theory but rather to understand how the formulation of modern atomic theory was possible, by trying to identify with the mentality of the era in which it took shape (Hall 1983,?48).

This objective, however, must be achieved through an increasingly specialized and detailed analysis of the period under analysis: "It is only by [...] looking at some point in history through the microscope that we can really visualise the complicated movements that lie behind any historical change" (Butterfield 1931, 21). The historian can explain an event like the French Revolution only "by discovering exactly what it was that occurred," describing its hidden elements always in "greater detail" and progressively giving more "definite concreteness" to "what really did take place" (Butterfield 1931, 72). This historiographical ideal was based on the idea that long-term narratives must be avoided—because these would be nothing more than forms of "abbreviation"—as must, above all, those types of studies that attempt to highlight recursive forms and structural analogies in the course of history—because this would flatten the multifaceted agency of individuals (Hall 1983, 50–52). From this perspective, the sum of the historical facts emerging from specialist studies is what can shape broader narratives.

However, as was rightly pointed out by Adrian Wilson and Timothy G. Ashplant, in *The Origins of Modern Science 1300–1800*, the text that made Butterfield famous in the field of the history of science, many statements could frequently be accused of Whiggism, such as "The theory of the impetus did not solve all problems, however, and proved to be only the half-way house to the modern view" (Butterfield 1957, 13). However, Butterfield emphasized in various points how his approach strove to maintain the antipresentist historical posture proposed in *The Whig Interpretation of History*: "the subject has not been turned into genuine history [...] if we construct our story of science by drawing lines straight from one great figure to another" (Butterfield 1957, viii). What should we deduce from this contradiction—that, given the facts, the game was inevitably more complicated for Butterfield or that the historiographical model that he had in mind was more complex (Wilson and Ashplant 1988, 4–5)?

The Whig Interpretation of History was, indeed, a text that emphasized what history should not be and what the historian should not do. However, the *pars destruenens* was not matched with a *pars construens* that was adequately structured and capable of providing concrete, positive examples of what a non-Whig history would be. The historian's profession is based on making choices, cutting, or, even better, foregrounding certain aspects of the primary material rather than others. In fact, constructing a narrative by placing every event or historical fact on the same level would be impossible. A problematic question therefore arises: What selection principle should the historian adopt? Butterfield was aware of this problem, so much so that he recognized the necessity of "a creative act of the historical imagination" or "imaginative sympathy" on the part of historians (Butterfield 1931, 91–92):

> if history could be told in all its complexity and detail, it would provide us with something as chaotic and baffling as life itself; but because it can be condensed, there is nothing that cannot be made to seem simple, and the chaos acquires form by virtue of what we choose to omit. (Butterfield 1931,?97)

Indeed, Butterfield did not hesitate to criticize historians who "do nothing more than transcribe their elaborate card-indexes" and do not seek to provide any form of mediation "between the subject-matter and the reader" (Butterfield 1931, 90–91): "It is by no means the historian's duty to whittle himself down to a mere transparency, and simply to transcribe information with colourless, passionless impartiality" (Butterfield 1931, 91). On the contrary, only through the historian's complex work of mediation does the past become understandable. Thus, a regulated form of presentism seems to be acceptable for Butterfield:

> By imaginative sympathy he makes the past intelligible to the present. He translates its conditioning circumstances into terms which we today can understand. It is in this sense that history must always be written from the point of view of the present. It is in this sense that every age will have to write its history over again. (Butterfield 1931,?92)

The historian is therefore guided both in research and in the composition of narratives by "instinct and sympathy"; they are anything but indifferent and passionless. Their objective, however, is not to investigate the past and force it into rigid analogies with the present but rather to make the past comprehensible and fully intelligible to contemporary readers. What marks the boundary between the Whiggish and anti-Whiggish models is that the former tends to describe "only one side of the historical story" (Butterfield 1931, 94), whereas the latter is based on the historian's total immersion in the mentality of the era in question in order to bring out its complexity.

As Nick Jardine pointed out, the reception of *The Whig Interpretation of History* had a somewhat paradoxical fate. In the years following its publication, historians and historians of science made little explicit reference to it (Jardine 2003, 125). Despite this, some of its theses against "present-based" historiography largely entered the shared *doxa* of these disciplines. While Butterfield's positions were radicalized by authors such as Robin C. Collingwood in *The Idea of History* (Braunstein 2008, 99), his criticism of presentism was immediately absorbed—sometimes without explicit reference to his work—by key authors such as Crombie, who made a long reference to this issue in the introduction to *Augustine to Galileo* (further extended in the second edition of 1961; see Abadía 2009, 58). Even an author as influential as Kuhn repeatedly expressed his support for antipresentist theses in various essays, including "The History of Science" and "The Relations between History and the History of Science," collected in *The Essential Tension* (Kuhn 1977), and "Possible Worlds in History of Science" in *The Road Since Structure* (Kuhn 2000, 58–89).

From the second half of the 1960s and throughout the 1970s, anti-Whiggism not only became the dominant historiographical approach but was also virtually the only imaginable model in the Anglophone cultural context. At the level of international debates, the establishment of the Science and Technology Studies (STS) movement led to the widespread diffusion of the anti-Whiggish approach. This approach largely permeated the sociology of scientific knowledge and, in particular, the so-called Strong Program, which was based, according to David Bloor's formulation, on the principles of impartiality and symmetry (Bloor 1976). This program significantly influenced the cultural branch of the historiography of science, represented by works such as *Leviathan and the Air-Pump: Hobbes, Boyle, and the Experimental Life*, by Steven Shapin and Simon Schaffer, who made anti-Whiggism a flag of their historiographical model for studying past scientific controversies (Shapin and Schaffer 1985, xviii). Similarly, Bruno Latour often served as a spokesperson for criticism of the FHE model, specifically in regard to its presentist or "anti-anti-Whiggist" tendency (Latour and Bowker 1987). Apart from a few reviews published shortly after its publication (Jardine 2003, 125), not until the 1980s did *The Whig Interpretation of History* become the subject of in-depth analysis, thanks to the first critical evaluations by historian of science Alfred Rupert Hall (1983) and political historian Geoffrey R. Elton (1984), who paved the way for broader discussions.[29]

More recently, presentism has started to arouse interest again thanks to the most recent programs in the field of contemporary historical epistemology.[30] As Omodeo (2019) and Hasok Chang (2021) have particularly emphasized, the demand for a presentist approach is related to the politicization of historical epistemology. On the basis of this *nouvelle vague*, the approach developed within FHE has been correlatively rediscovered and valorized (Loison 2016; Simons 2019a). In the literature, the FHE approach is typically interpreted as being in strong tension and contrast with the anti-Whiggism put forth in Anglophone debates (Latour and Bowker 1987; Braunstein 2008).

In the French context, in fact, this debate has taken on a completely different and, in some ways, opposite form to the British one. The previous section showed how the conception of temporality developed by Bachelard in *Intuition of the Instant* and *The Dialectic of Duration* was strongly based on the present time and on the concept of the instant. An echo of this approach was thus present in the relationship between past and present underlying Bachelard's epistemological perspective. In a famous speech given in 1951 at the Palais de la Découverte in Paris, titled "L'actualité de l'histoire des sciences," and in the first chapter of *L'Activité rationaliste de la physique contemporaine*,[31] Bachelard focused on the specificity of the history of

[29] See, for example, the discussions in Ashplant and Wilson (1988), Wilson and Ashplant (1988), Pickstone (1995), Jardine (2000, 2003), Spoerhase (2008), and Abadía (2008, 2009, 2011).

[30] On this point, see in particular Chang (2009, 2021), Loison (2016), Omodeo (2019, 32–37), Simons (2019a), and Vagelli et al. (2021).

[31] These are two particularly relevant texts because they are some of the rare occasions in which Bachelard directly addressed the issue of methodology in the history of science. These texts enumerate the historical-epistemological principles put forward by Bachelard in his other works.

science compared to other disciplines, such as political history, art history, the history of philosophy, and so forth. Unlike these, the history of science "cannot in any way be a history like the others," because it "evolves in a manifest progression" (Bachelard 1972b, 148). In the context of political history, however, what constitutes progress and regress is susceptible to debate and conflicting opinions, and in the history of art or philosophy, every moment of the narrative can claim degrees of transhistorical universality. Bachelard provocatively asserted the impossibility of writing a history of scientific regression or decadence.[32] Indeed, opening any book on the history of science—whether scholarly or popular—is enough to see how this history is a process of constantly increasing knowledge, never decreasing: "the axis [...] is clearly oriented toward a better understanding and a broader experience" (Bachelard 1972b, 149). The history of the decline of particular scientific theories is always accompanied by the emergence of new, more-explanatory theories. Furthermore, "If the history of science collected the errors made after the discovery of mathematical truth, it would just be a history of bad mathematics students and not the history of true mathematicians" (Bachelard 1972b, 150). For those rare cases in which an already-discovered truth, after a period of falling into disuse, is taken up again by later scientists, the task of the history of science is to show the process of this re-emergence, not its oblivion (Bachelard 1972b,?149).

Starting from these assumptions, Bachelard insisted on the need for a history of science centered on the problems of contemporary science. Contrary to what Butterfield claimed, for Bachelard, the history of science had to be a judged and evaluated history because only in this way could the historical meaning of the scientific enterprise itself emerge—that is, its substantial irreversibility. Indeed, when the historian of science deals with an "absolute growth" in knowledge, then the situation imposes a particular methodology:

> in complete opposition to recommendations that the historian should not judge, it is instead necessary to ask the historian of science to make value judgments. The history of science is, if nothing else, a fabric of implicit judgments on the value of scientific thoughts and discoveries. The historian of science who clearly explains the value of each new idea helps us to understand the history of science. In short, the history of science is essentially a judged history, judged in the details of its weave, with a meaning that must be relentlessly refined by truth values. The history of science cannot simply be recorded history. (Bachelard 1972b,?151)

The mere recording of events that occurred in history is not sufficient to produce a history of scientific reason. Historians who limit themselves to recording the events that occurred in the "scientific city" or to repeating/retracing the reports of the academies cannot produce narratives capable of bringing out the underlying epistemological dynamics. According to Bachelard's approach, the historian of science must account for how scientific progress occurs through the process of the selection, valorization, and promotion of those conceptions or past technical-experimental

[32] On the relations between Bachelard and the positivist history of science of authors such as Comte and Mach, see the reconstruction proposed by Pietro Redondi (1978, 164–197).

experiences that allow the development of more-explanatory and -inclusive theories of the totality of known phenomena. In *L'Activité rationaliste de la physique contemporaine*, Bachelard added:

> The history of empires and peoples rightly has as its ideal the objective account of the facts: it asks the historian not to judge, and if the historian imposes the values of his time on the determination of the values of the past, he is rightly accused of following "the myth of progress." However, here is an obvious difference: for scientific thought, progress is demonstrated, it is demonstrable, its demonstration is even an indispensable pedagogical element for the development of scientific culture. Progress, that is, as the very dynamic of scientific culture and the history of science, must precisely describe this dynamic. It must describe by judging, by valorizing, by preventing any possibility of a return to erroneous notions. The history of science can insist on the errors of the past only to reject them. (Bachelard 1951,?54)

Some key points for Bachelard's historical epistemology emerge from these passages. The first is the definition of the social function of the history of science. According to Bachelard, its task is to educate the community about the value of science. Anticipating the debate on the "Two Cultures" by many years (Snow 1959), Bachelard maintained that

> Above all, it is necessary to maintain interest in scientific thought, even if this is not easy in an era in which, with some levity, human errors for which it is not responsible are attributed to science. [...] Scientific culture must be integrated into general culture. (Bachelard 1972b,? 157)

Second, the concept of error plays an important role. If the anti-Whiggist approach requires that history not be written from the point of view of the "winners" of scientific controversies, a recurring point in Bachelard's epistemology was to show how science faces a constant process of overcoming epistemological obstacles. Epistemological history that has been evaluated and judged "is made [...] to help us become aware of the strength of certain barriers that the past of scientific thought had formed against irrationalism" (Bachelard 1951, 56). Knowledge of errors is what permits the refinement of scientific reason and thus its rectification: "In other words, if there is transmission of truth throughout the history of science, there is also always a certain permanence of error" (Bachelard 1972b,?158).

The third point concerns the criteria for value judgments on the history of science. For Bachelard, for the historian to identify, select, and highlight moments of history in which historical discontinuities arise that mark the path of progress, they cannot be ignorant of contemporary science. Only on the basis of such knowledge do ruptures become visible and meaningful:

> The historian of science, to judge the past well, must know the present; he must do his best to learn the science whose history he wants to write. In this way, the history of science, whether he likes it or not, is closely linked to the current affairs of science. (Bachelard 1972b,?151)

Value judgments are expressed starting from current scientific knowledge, yet they also take on ethical tones. Unlike philosophies centered on the myth of "facts," Bachelard's was an epistemology that, in this case, took on idealistic tones. As Redondi noted, this led him to formulate a "concept of value as the 'material ethics' of rational science" (Redondi 1978, 191). These rational values, however, change over time, and the history of science must be constantly rewritten and rejudged, starting from the present, in relation to the process through which they emerge. It is, therefore, not a teleological model precisely because of its constant revision of the selection "of the pertinent facts of the past and its narrative scenarios" (Gayon 2003, 108). For this reason, recurring history takes on, to use Gayon's expression, a "deliberately anachronistic and normative" character that is completely incompatible with both historicism and the "idea of an *a priori* law of progress" (Gayon 2003,? 108).

Thus, it is completely different from anti-Whiggist approaches, as well as from the "symmetry" proposed by the Strong Program of Edinburgh, which, for example, led to the consolidation of ethnographic approaches that valorized their total extraneousness to scientific knowledge as a methodological principle favoring the observation of the social dynamics taking place in the production of scientific facts.[33] Even though it seems like a paradoxical achievement, only through knowledge of contemporary science is the historian able to profoundly grasp the historicity of science itself: "Here, the consciousness of modernity and historicity are strictly proportional" (Bachelard 1972b, 151). Therefore, from the fabric of history, the historian of science must bring out the distinction between "error and truth, the inert and the active, the harmful and the productive," citing the Nietzschean motto that "You can interpret the past only on the basis of the highest power of the present" (Bachelard 1951, 54). For Bachelard, in fact, the history of science "must be remade often and reconsidered several times" (Bachelard 1972b, 153), in light of the current state of science and the progress that it is constantly making: "We must understand

[33]For example, the sociological program developed by Latour (in collaboration with Steve Woolgar) in *Laboratory Life* (Latour and Woolgar 1979) and continued in his subsequent works was based on this point. This book was based on ethnographic work carried out at the endocrinology laboratory directed by Nobel Prize winner Roger Guillemin at the Salk Institute in San Diego. In the French edition of the volume—which is different in some parts from the first English edition—Latour explained the objectives of the anthropology of science as follows: "I then asked myself what would have been said about scientific discourse if this had been studied with the same care with which ethnographers study pre-, para-, or extra-scientific cultures, societies, and discourses" (Latour and Woolgar 2006, 10–11). This text explicitly critiqued the Bachelardian epistemological model in the background. While many commonly maintain that "it is necessary to be specialists in order to talk about science" (Latour and Woolgar 2006, 22), Latour, on the contrary, supported a parallel with the anthropological method: "would the anthropologist who studies hunting practices not need to know anything? Could he who brilliantly studies witchcraft in the Bocage be crassly ignorant? Could it be said that to study the Bantu one must be born a Bantu? That to talk about magic, not only do you have to be a magician, but you shouldn't add any comments or explanations about the magical rites?" (Latour and Woolgar 2006, 23–24). If Bachelard's epistemology was based on the clear distinction between the scientific and prescientific spirit (between science and common sense), Latour aimed rather to defend a fluid frontier between these two poles (Latour 1987). He was, in fact, opposed to a historical epistemology whose historiographical model was structured, on the one hand, on discontinuism and, on the other, on the idea of a "sanctioned history," while his socio-anthropology of science aimed at a radicalization of the symmetric approach proposed by the Edinburgh Strong Program (Ienna 2019b, 253–270). On Latour and Michel Serres's relationship with historical epistemology, see Simons (2019b).

the importance of a historical dialectic *specific to scientific thought*. In short, it is necessary to continually formulate and reformulate the dialectic of the history prescribed [*périmée*] and the history sanctioned [*sanctionné*] by currently active science" (Bachelard 1951, 55). This dialectic between past and present creates the need for a process of disciplinary self-reflexivity such that "we believe there is a certain interest in following this history of science, this history of science that reflects on itself, this history that is always rethought and always restarted" (Bachelard 1972b, 155). The history of science, conceived as the history of the defeats inflicted on irrationalism by reason, is the history of an endless battle, and therefore, as Bachelard declared, it is "necessary for each generation of scientists to take stock of rationalism and continually review the balance of the history of science" (Bachelard 1951, 56). Regarding the need to constantly rehistoricize science, a common intersection between Butterfield and Bachelard can seemingly be traced—though from completely opposite points of view. For Butterfield, this rehistoricization is motivated by the need to make one's contemporaries understand past science; for Bachelard, the process of modernizing science requires a constant process of reanalyzing its history.

In opposition to Butterfield's model, the objective of Bachelard's epistemology was, as Bontems noted, to highlight the transhistorical laws that regulate the course of the history of science or, rather, to bring out those epistemological regularities that characterize the development of knowledge (Bontemps 2010, 29–69). It is therefore a history of science aimed at (and inseparable from) epistemological reflection. In this context, the concept of historical and epistemological "recurrence" assumes a crucially important role, which, together with the notions of obstacle and rupture, represents the third key element of the architecture of Bachelard's materialist dialectic.

As pointed out in the previous section, the concepts of obstacle and rupture stress both the historical discontinuity of knowledge and the epistemic discontinuity of knowledge. "Recurrence," on the other hand, is the dialectical element that allows these discontinuities to be put in relation to each other, showing their progressive character over the course of history. As Bontems indicated, through this concept, Bachelard aimed to show how

> the knowledge of an era is judged according to a further state of knowledge, which then rectifies it and clarifies its limits, but also that this more correct and deeper state of the system of knowledge tends to find the previous state as a simplified particular case. (Bontemps 2010,?223)

To this observation, Federica Bonicalzi added:

> history, conceived as the result of a dialectic of epistemological acts and obstacles [...] implies a judged history that consigns the negative of the past, that is, the epistemological obstacle, to the psychoanalysis of knowledge [...] and the positive of the past ("a sort of actual past" operating in the present) to scientific thought. (Bonicalzi 2007, 41–42)

Bachelard's reading of the relationship between Newtonian and Einsteinian physical theory is an exemplary case. In *The New Scientific Spirit*, as in other texts, Bachelard underlined the strong epistemological rupture that arose between these two theories, emphasizing how "It is a mistake [...] to regard Newtonian physics as the first approximation to Einsteinian physics" (Bachelard 1934/1984, 44). Yet, a relationship existed between these two in that Newtonian theory described a particular case of physical space regulated by Einsteinian laws:

> When the construction of Einsteinian rationalism was carried a little further, it was rightly seen that Newtonian physics had its place in expanded physics, a place which was defined with the greatest clarity by the fact that it was a simplification of the complete doctrine. (Bachelard 1972a,?109)

This relationship is precisely what Bachelard called "recurrence." While remaining in historical discontinuity, there is a corresponding form of continuity of rational connections: "The philosophy of the continuity of important concepts, rather than presenting a problem of historical connection, is a link whereby the rational progressively dominates the contingent" (Bachelard 1951,?56).

For Bachelard, recurrence marked the transition to a higher epistemic order, which included past scientific successes. These passages, indeed, contained the whole meaning of the dialectic of *The Philosophy of No*, which at the same time integrated what it denied. The figure of Newton cannot be interpreted as a precursor of Einstein (due to the epistemological rupture between their two theories), but at the same time, the Newtonian model is part of a history "sanctioned" by Einstein's theory. An event such as the development of the phlogiston theory, which was "prescribed because it was based on a fundamental error, on a contradiction of ponderal chemistry," presents a different kind of situation. In the case of ideas like the caloric theory, however, despite being incorrect in some respects, certain aspects of the theory have been sanctioned as valid results that opened up ideas that were exploited in subsequent science (Bachelard 1951, 55). In this sense, the recurrence-based, epistemologically informed history proposed by Bachelard brought into being "a well-ordered history that does not correspond to actual history at all. In this remade history, everything has value" (Bachelard 1949, 80). The rationalist value hierarchy put in place by contemporary science is what allows the history of science to be better understood than it was by those who actually experienced it (Bachelard 1972a, 109). For this reason, Bachelard maintained that the historian must be "very tactful in handling possible recurrences" when building a history of science that combines "the history of the development of facts with the history of the development of values" (Bachelard 1972b, 152–153).

As mentioned, Bachelard's objective was to describe the transhistorical invariants (found in the obstacles, ruptures, and recurrences, in the prescribed history and the sanctioned history, etc.) characterizing the process of knowledge acquisition. Obviously, this model cannot be assimilated into neo-positivism given how neo-positivism searches for logical invariants from an ahistorical point of view. As can be clearly seen, the Bachelardian model can neither be reduced to the forms of continuist positivism proposed by Duhem or Popper (because it is based on the concept of rupture) nor be compared to the model proposed for historical relativization that emerged from the pages of *The Structure of Scientific Revolutions*, as Kuhn posited an incommensurability between scientific paradigms (that Bachelard, instead, avoided through the concept of recurrence).

In the work of historicizing scientific thought, some FHE authors showed notable overlaps or intersections with emerging debates within French historiography, first with the foundation of the Centre Internationale de Synthèse Historique by Henri Berr and, later, with the historiographical movement that revolved around the *Annales* journal founded by Lucien Febvre and Marc Bloch.[34] A distinctive feature of the historiographical direction of the *Annales* was that it proposed a model that, instead of looking at event-based history or political transformations [*histoire événementielle*], concentrated on analyzing long-lasting historical temporalities [*longue durée*], investigating sources aimed at constructing a problem-oriented history [*histoire-problème*], sociocultural history [*histoire de l'imaginaire, des mentalés, des processes de civilization*, and *l'ethnohistoire*], or macroeconomic structures [*quantitative histoire*] and, in some cases, even foregrounding elements in the background of traditional narratives that seemed to be almost immobile in the course of history [*histoire immobile* or *histoire quasi immobile*].

In this context, Febvre and Bloch defended a presentist historiographical model by inviting scholars of their discipline "to be inspired by the problems posed in the present time, in which they lived, thought and wrote" (Diogo 2020, 78). For Febvre, the historian "does not preserve the past in his memory like the northern ice keeps thousands-of-years-old mammoths hibernating. The historian starts from the present—and it is through the present that the past is always known and interpreted" (Febvre 1992, 15). In *The Historian's Craft*, Bloch, for his part, stressed how only through constant contact with the present does the historian have the ability to unwind the film reel of events in the opposite direction to that in which it was shot, thus applying a regressive methodology (Bloch 1949, 50–57). This theoretical stance also clearly emerged in the first editorial of the *Annales* journal, which stated the following:

> While historians apply their good old tested methods to the documents of the past, more and more people are dedicating themselves, and not without a certain feverish spirit, to the study of contemporary societies and economies: two classes of workers made to understand each other but who, in everyday life, coexist without knowing each other. (Bloch and Febvre 1929,?1)

The editorial project of the *Annales* aimed precisely at resolving this tension. The historiographical model proposed by Bloch and Febvre presupposed that the historian lives in a constant back-and-forth movement between past and present in which each of the two poles has a heuristic function for the other (Hartog 2015, 110). The

[34] For a stimulating reconstruction of the intersections between FHE and these new historiographical models, see Castelli Gattinara (1996). For a general introduction to the debates that emerged within this current, see Burke (1990). For an analysis of the specific contribution made by the *Annales* to the history of science and technology, see Diogo (2020). On the difficult relationship between the history of mental life and the history of science, see Conry (1983). For a general analysis of the relationship of the *Annales* to the present day through its entire editorial history, see Les Annales (2020).

present guides the research of the past by enriching the questions that can be asked of the documents, and vice versa, the past illuminates the present by facilitating a more effective practice in the present. Therefore, for Febvre and Bloch, history must, in every era, be remade and reconstituted because "each historical moment builds its representation of the past according to what matters most" (Diogo 2020, 78). This tension implies a form of political epistemology of historical time: "This present is political in a very different sense, or even opposite to that of the 'political' history that has traditionally dominated contemporary history (in the disciplinary sense)" (Les 2020, 498). This reverberates within the methodology of this historiographical current: "The *longue durée*, however, had a social and political meaning: that of emphasizing a particular relationship with the past, of a historical discipline that became a counterpoint to the modernity of the *Les Trente Glorieuses*" (Les 2020,? 498).

Contrary to Bachelardian epistemology, however, the presentism proposed in the context of the *Annales* does not lead to forms of historical anachronism. These authors strove to avoid falling into the fetishism of facts like positivist historiography did, questioning how the historian constructs an event as such through the work of research. For this reason, Febvre developed a criticism of forms of historical narration based on the identification of precursors of subsequent eras. Particularly in *The Problem of Unbelief in the Sixteenth Century: The Religion of Rabelais* (Febvre 1942/1982), he aimed to deconstruct the anachronistic interpretation of the figure of Rabelais provided by Abel Lefranc. According to Lefranc, Rabelais was a precursor or forerunner of forms of modern rationalism by virtue of the frequent antireligious parodies contained in his work and from which Rabelais's alleged atheism could be deduced. Through a dense philological analysis, Febvre showed, on the contrary, how such *boutades* were completely in conformity with the Christian mentality of the sixteenth century and were already well appreciated in medieval times, even among church officials. From this initial problem, Febvre developed a broader historical investigation into the impossibility of atheism in the sixteenth century, showing how the conceptual apparatus [*outillage mental*] of the time prevented the existence of God from being questioned. In this sense, Febvre declared that "each civilization has its own mental tools" (Febvre 1942/1982, 150). As underlined by Maria Paula Diogo, the use of the idea of the "precursor"

> implied the idea of time as a chronological and linear order of events, autonomous in relation to individuals and phenomena. Moreover, it indicated that historical time was part of a deterministic course, followed a specific direction, and did not follow a random and contingent one. A precursor is someone able to escape his time and surpass it, announcing to his contemporaries what the future will be. (Diogo 2020,?81)

On the contrary, the perspective developed by the *Annales* on the basis of a criticism of event-based history [*histoire événementielle*] brought forth a radically different conception of temporality. For the authors associated with the journal, and in particular for Bloch, time was not an invariant structure in which historical events were inserted, but rather, individual and collective actions bring time into being—that is, create the sense of a time. For Braudel, "time moves at different speeds"

(Burke 1990, 39) depending on the elements that we decide to observe. The aim of his works on *La Méditerranée* was, indeed, to "distinguish, in the time of history, geographical time, social time, individual time" (Burke 1990, 42). According to this historiographic methodology, the historian's task is to foreground the intersection between these temporalities in the social totality.

The significance of the intellectual impact of the birth of the historical program of the new French historiography certainly did not go unnoticed by historical epistemologists, and culturally, it corroborated their presentist theoretical positions. As seen in the previous sections, Canguilhem developed an anti-neo-positivist model in which, on the one hand, separating epistemological reflection from historical reflection was unthinkable and, on the other, precisely the historicity of knowledge, in his system, was the basis of the boundary between science and nonscience.

For Canguilhem, error played a central role because, contrary to the positions of both eighteenth-century rationalists and nineteenth-century positivists, "Error is not a weakness but a strength, *rêverie* is not smoke but fire" (Canguilhem 1957, 10). As shown, for example, in his essay dedicated to the concept of reflex, Canguilhem deconstructed the historiographical prejudice that held that a concept considered scientific in the present moment could not have its origins in a theory considered to be nonscientific at that same moment. With echoes of Bachelard, Canguilhem maintained that the history of science is a work of constantly reinterpreting the past according to the current state of science. The act of historicization repeatedly redefines the boundaries between the "prescribed" and "sanctioned" theories. Writing the history of an irreversible process within the history of scientific thought progressively illuminates the gap between truth and error.

Canguilhem's position differed from Bachelard's in various aspects. Laurent Loison developed an interesting taxonomy of possible forms of presentism in epistemology, bringing out the relative specificities among them. In particular, he defined Canguilhem's view as "critical presentism," an approach in relative contrast to Bachelard's "normative presentism" (the use of the present as a norm for judging the past) (Loison 2016, 36). This critical presentism, incorporating some elements of normative presentism, aimed at the "elaboration of a mutual critical relationship between the scientific knowledge of different times. Its legitimacy derives from the recognition of the contingency and rationality of science" (Loison 2016, 36). Latour and Bowker also brought out this difference between Bachelard and Canguilhem's perspectives, underlining how whereas Bachelard's approach was markedly normative-presentist, Canguilhem's rather took the form of an "anti-anti-Whiggism" that did not correspond in any way to the affirmation of a Whiggish approach (Latour and Bowker 1987,?725).

As has already been widely noted, in Bachelard's historical epistemology, the axis of interest was more inclined toward epistemology than history, whereas, for Canguilhem, it was exactly the opposite, so much so that it was configured more as an "epistemological history." For this reason, avoiding falling into historiographical errors that could be sources of anachronism was of greater importance for Canguilhem.

In "The Object of the History of Sciences," Canguilhem made Bachelard's normative conception of the history of science his own. To describe the task of this discipline, Canguilhem proposed replacing the canonical image of the history of science as "the laboratory of epistemology"[35] with that of the "tribunal" or "school," or rather "an institution and a place where judgments are brought to bear on the past of knowledge, on the knowledge of the past" (Canguilhem 1994b/2005, 200). If the history of science is a tribunal, then it needs a judge, and this function is covered by epistemology: "It is the epistemologist who is called to furnish history with the principle of judgment by teaching it the most recent language spoken by some science" (Canguilhem 1994b/2005,?200).

As Chimisso (2003) noted, Metzger had already used this metaphor of the "tribunal" in 1935 during one of her interventions at the Centre de Synthèse Historique, titled "The Tribunal of History and the Theory of Scientific Knowledge" (in Metzger 1935, 181–190). In this text, Metzger focused on whether the "tribunal of history" had the epistemic legitimacy necessary to evaluate, judge, and "close the debates" within scientific controversies of the past or whether the historian was given the task of expressing a definitive judgment of "condemnation" or "acquittal" on past scientific disputes (Metzger 1935/2019, 181). Metzger responded to this question by warning that if one gave in to this temptation, then history would colonize territories that were inappropriate to it: "If it were possible to answer yes to this question [...] the tribunal of history would have rendered any speculative research unnecessary. The theory of scientific knowledge would no longer belong to the field of philosophy" (Metzger 1935/2019, 181). Consequently, history should deal with the following themes: "reflection on science, the argumentation made with the aid of science, and the dialectics of scientific discussion" (Metzger 1935/2019, 181). However, for Metzger, this did not imply that the historian's work is value-free and not subject to conditioning or position taking but rather implied that the work of historicization is potentially infinite. Indeed, a definitive judgment on the part of history would be impossible by definition: "since history, like any science, can progress, we can clearly see that the tribunal of history does not have absolute competence and that it is entirely unqualified to pass a verdict without appeal" (Metzger 1935/2019, 188). This was a concept that she later repeated in a review of a book by Enriques, in which she declared that the work of historicization is not "merely the gathering of documents" but that, on the contrary, the historian's task is to produce reconstructions and narratives "like the scholar works on the results produced in the laboratory" (Metzger 1934, 391). For this reason, the history of science is subjected to constant reinterpretation: "The tribunal of history has investigating judges, but their sentences are not accepted as final and absolute" (Metzger 1987, 391). On the other hand, in *Les concepts scientifiques* (1926), Metzger explained that because of sociological

[35]This is an expression used by Canguilhem in "The Role of Epistemology in Contemporary History of Science," in which he mentions this laboratory metaphor taken from the essay "The Origins of Classical Mechanics: From Aristotle to Newton," by Dutch science historian Eduard Jan Dijksterhuis (Canguilhem 1977/1988, 2–3).

influences, the analysis of the history of scientific thought must proceed through the analysis of scientific concepts in correlation with the cultural-ideological context within which they are formed, highlighting the establishment of the border between the prescientific and the scientific (Castellana 2009, 31–32).

The position developed by Canguilhem thus seems to be a gloss on and an integration of the perspectives developed by Metzger and Bachelard. Indeed, his historiographical model highlighted "both the contingency and rationality" of the scientific enterprise (Loison 2016, 36). Without denying a Bachelardian approach, in "Le rôle de l'épistémologie dans l'historiographie scientifique contemporaine," Canguilhem stressed how a certain dose of normativity is necessary in the process of historicizing science. How, for example, can we reconstruct the history of botany if not by starting from the actual state of the discipline today? If we do not investigate the past in a state of awareness of the contemporary discipline, we may fall into a crucial historiographical error. Following a strictly nominalist principle in the historical reconstruction of the evolution of a form of knowledge would entail a risk of losing focus on the circulation of the ideas and concepts in all the various branches that contributed to the development of the field—including those that may not have always been considered relevant to it. Thus, one cannot limit oneself to the image that contemporaries of a specific past era had of a certain field of research, because this would offer a limited image of the historical evolution of science and its conceptual production:

> A history of science free of epistemological contamination would inevitably reduce the state of a scientific discipline—plant physiology in the eighteenth century, say—to a summary of chronological and logical connections among various systems of propositions pertaining to various classes of problems and solutions. [...] Pure historians are interested only in what scientists thought they were doing and how they went about doing it. But a fundamental question must be asked: Does this science of the past constitute a past for the science of today? Taken in an absolute sense, "the past of a science" is a vulgar concept. The "past" is a catchall of retrospective inquiry. (Canguilhem 1977/1988,?3)

Only the current state of a discipline allows us, continuing backward in time, to follow all the networks of conceptual filiations that, as shown in various cases by Canguilhem, often undergo migration processes between different cultural contexts (Canguilhem 1977/1988, 4–5). An epistemologically informed history of science must, rather, reveal from the historical fabric how, starting from certain conditions of contingent possibility, arriving at the conquest of relatively solid truths or defining the transhistorical value of scientific knowledge was possible: "A science is a discourse governed by critical correction. If this discourse has a history whose course the historian believes he can reconstruct, it is because it *is* a history whose meaning the epistemologist must reactivate" (Canguilhem 1977/1988, 11). Through this double articulation, FHE distances itself from both naive positivism and forms of historiographical relativism.

Canguilhem took up Bachelard's concept of "recurrence" on various occasions, in which, while emphasizing its usefulness, he suggested a certain reflexivity and caution in using it. According to his interpretation, the fact that Bachelard shaped his epistemological reflection with case studies taken from physical-mathematical disciplines in mind facilitated his demonstrating epistemological recurrences.[36] However, if we move on to examine other disciplines—such as the life sciences—the discussion becomes considerably more complicated from a historiographical point of view and, for this reason, this methodology "can be not so much generalized as broadened. Yet it cannot be extended to other areas of the history of science without a good deal of reflection about the specific nature of the area to be studied" (Canguilhem 1977/1988, 14). For Canguilhem, this extension involved the redefinition of the terms of the debate about continuism and discontinuism highlighted previously.

The substantial difference with the anti-Whiggist models proposed in the Anglophone context consisted of the fact that, for an author like Canguilhem—subsequently taken up by Foucault on this point—the present state of science, as the result of a construction, is not a given but rather a "problem" that the history of science must confront (Braunstein 2016,?24).

Many have widely noted that Canguilhem was much more sensitive than Bachelard to the risks of anachronism deriving from the application of certain forms of presentism. Indeed, on this basis, he elaborated a critique of the use of the concept of precursor in the history of science. The relevance of such questions was not limited to the Anglophone context but rather extended to the Francophone debates as well.

For an author like Koyré, straddling the Francophone and Anglophone debates, the correct posture for historians of science was to try to put themselves in the perspective of the historical actors—that is, to trace the history of scientific knowledge as it appeared to its protagonists:[37]

> The history of scientific thought, as I understand it and strive to practice it, aims to grasp the path of this thought in the very movement of its creative activity. To this end, it is essential to contextualize the works studied in their intellectual and spiritual environment, to interpret them according to the mental habits, preferences, and aversions of their authors. (Koyré 1966,?4)

The objective was to bring out the mental habits of these actors by placing them in their intellectual and cultural contexts. In any case, this methodological stance took into account the cultural role played by the present in the historian's interpretation of the past. For Koyré, historical analysis turned out to be an infinite task, not only because it advances with time (that would be far too banal) but also because the

[36] In fact, in mathematical thought, which also constitutes the backbone of physical theories, "one speaks not of the normal but [of] what is normed" (Canguilhem 1977/1988, 13). In Bachelard's epistemological perspective, mathematical thought is not made up of mere formalism but instead has a real content of knowledge, allowing for identifying "theory with mathematics" (Canguilhem 1977/1988,?13).

[37] On the anthropological and social implications that this theoretical position had on historical epistemology for authors such as Meyerson, Metzger, and Koyré, see Sect. 3.1 of Chap. 3. In fact, some have argued that Metzger's works contain the first formulations of a criticism of the presentist approach—even prior to Butterfield's classic volume (Abadía 2008, 2009). This consideration could also be extended to Meyerson.

historian projects the interests and values of their time into history and, through these, puts into motion its reconstructions: "It is precisely for this reason that history renews itself, because nothing changes faster than the immutable past" (Koyré 1966,?354).

Canguilhem and Koyré—despite their theoretical differences—were united in proposing strong critiques of the use of the concept of precursor, which played a central role for the continuist position in the history of science. Indeed, according to Koyré, "Nothing has had a more pernicious influence on history than the notion of 'precursor.' To think of someone as a 'forerunner' of someone else is, surely, to deny oneself the ability to understand him" (Koyré 1935, 4). Emphasizing the existence of hypothetical precursors would interrupt the process of identifying and understanding a past act of scientific thought in its contemporaneity. For his part, Canguilhem, following Joseph T. Clark, emphasized the risks of being infected by the "virus of the precursor" by stating that "Strictly speaking, if precursors exist, the history of science loses all significance, since science itself would only apparently have a historical dimension" (Canguilhem 1994b/2005, 205). In both cases, what was at stake was the possibility of the discipline itself:

> The history of science must be as demanding and critical as science itself. By seeking to trace filiations without ruptures, all values, dreams, and plans, forebodings, and anticipations would be confused; precursors of everything would be found everywhere. (Canguilhem 1994a,?184)

These methodological aspects are fundamental to understanding the political significance of FHE, both as it was practiced by Canguilhem and Bachelard and as it was received by, among others, Althusser, Foucault, and Bourdieu. What was at stake was the deconstruction of the alleged neutrality of the historiographical enterprise and the openness to the possibility of a history judged according to values that form the basis—implicitly or explicitly—for a reflection on the determination of the future of science. The form of presentism that goes under the name of recurrent history is a historiographical thesis that is involved not only in the problematic relationship between present and past but also with the projection toward the future—in which its intrinsic political tension consists.[38] Bachelard's epistemology was perhaps the most explicit on this point: "In teaching a revolution of reason, one would multiply the reasons for spiritual revolutions" (Bachelard 1936b/1995, 187). As Bontems emphasized, for Bachelard, "the future of the scientific spirit already exists in the form of a horizon" within which the present of science moves. In this sense, "the anticipation of the future of science already belongs to its present" (Bontemps 2010, 36). In its dialectical nature, Bachelard's rationalism was an open rationalism:

[38] In the context of recent debates relating to the development of historical epistemology toward a political dimension, Omodeo emphasized that "Different visions of science are not just alternative forms of reflection depending on an indefinite plurality of equally valuable perspectives. Rather, they correspond to different visions of the future that are tightly connected to collective praxis" (Omodeo 2019, 5). On this point, see also Badino and Omodeo (2020) and Badino et al. (2022).

In historical epistemology, history serves to put the progress of a scientific lineage into perspective. Bachelard establishes filiations between the theories of the past; he celebrates the novelty and subtlety of present concepts; he imagines the future of science as an open horizon towards which to guide the greatest number of people. He spontaneously follows the recommendations of Aristotle's *Rhetoric*: when he deals with the past, he does so to judge; when he deals with the present, his speech takes on an epideictic form (he evaluates what is good and beautiful); finally, when he considers the future, he takes a political turn by defending progress. (Bontemps 2010,?33)

2.4 Epistemological Regionalism

A further way of reflecting on the fragmentation of reason involved developing a mode of analysis capable of addressing the plurality of scientific forms. Just as FHE foregrounded the presence of "temporal ruptures," it similarly emphasized—employing a spatial metaphor—the "regional" specificities in which rationality manifests itself. This approach marked the point when French epistemology took its strongest stance against neo-positivism and logical empiricism, whose epistemological position was based on a program of unifying science under a common logical language. As seen in the previous chapter, the meeting/clash between historical epistemology and this tradition took place mainly at two conferences in Paris, one in 1935 and another in 1937. Against the unity-of-science thesis, the new generation of French epistemologists was developing a philosophical program that focused on "underlining the diversity of the sciences or the plurality, the singularity, even the irreducibility of the different domains of science" (Jacob 1991,?287).

This difference in perspectives was rooted in French epistemological culture itself. The origins of this tendency to view the sciences as intrinsically plural were already visible in the nineteenth century in the way that Comte had conceived the theme of the classification of the sciences[39] (also called the hierarchy of the sciences

[39] The question of classifying sciences is a relatively canonical theme in epistemological reflection. Studying the organization of knowledge was one of the primary tasks of philosophy, starting from the classical era. The emergence of the expression "philosophy of science" itself—as well as the process of its institutionalization—was closely intertwined with the theme of classification. As we know, the first occurrences of this phrase happened simultaneously in France and England, and only in the modern era (Lecourt 2001, 13). In French, physicist André Marie Ampère inserted the concept of "*philosophie des sciences*" into the philosophical vocabulary with the 1834 publication of *Essai sur la philosophie des sciences, ou exposition analytique d'une classification naturelle des toutes le connaissances humaines*. That the expression was in the plural is no coincidence given that, for Ampère, the object of this particular subdiscipline was to classify the sciences in order to investigate their specificity (both with respect to each other and each of them with the faculties of the intellect). These were the same years in which Comte was working on his monumental *Cours de philosophie positive*, which would be published between 1830 and 1842. Even for Comte, the primary objective of the *philosophie des sciences* was to identify the principles for classifying human knowledge, their mutual concatenation, and their hierarchical order. In the English-speaking world, William Whewell was responsible for coining the phrase "philosophy of science" in *The Philosophy of Inductive Sciences, Founded Upon Their History* in 1840. Although the title contained the expression in the plural, it also appeared in the singular within the text. This tension

or encyclopedic scale) (Petit 1994). In the second lecture of the *Cours de philosophie positive*, Comte addressed this theme, from both a historical perspective and an epistemological perspective (Comte 1830/2009, 18–33). On the one hand, following the principles of the famous "law of the three stages," Comte had argued that every science gains access to the "hierarchy of the positive sciences" only when it emancipates itself from the theological-metaphysical stage. This claim gave rise to the possibility of imagining the organization of knowledge as the product of historical evolution. However, Comte had advanced a rational order of the sciences based on their generality and the simplicity of their object. In particular, he considered six fundamental sciences: mathematics, astronomy, physics, chemistry, biology, and social physics, or sociology (with the latter elected as the queen of all sciences). Beyond this brief summary of Comte's perspective, what is interesting is how he advanced certain key epistemological reflections regarding the nonreducibility of the sciences to one another. The following passage, taken from *System of Positive Polity*, is particularly on point:

> To a philosophic eye there is Materialism in the common tendency of mathematicians at the present day to absorb Geometry or Mechanics into the Calculus, as well as in the more evident encroachments of Mathematics upon Physics, of Physics upon Chemistry, of Chemistry, which is more frequent, upon Biology, or lastly in the common tendency of the best biologists to look upon Sociology as a mere corollary of their own science. In all these cases it is the same fundamental error; that is, an exaggerated use of deductive reasoning; and in all it is attended with the same result; that the higher studies are in constant danger of being disorganised by the indiscriminate application of the lower. (Comte 1851–1881/1851,?112)

This passage attracted the attention of authors such as Brunschvicg, who commented that "For the Cartesians, mathematics implied spirituality. Comte, on the contrary, founded a new spiritualism on the specificity, on the irreducibility, of the various scientific disciplines" (Brunschvicg 1912,?301).

This position was then developed by Bachelard with the so-called epistemological regionalism thesis. As mentioned in the previous chapter, Bachelard chose not to speak at the First Congress of Scientific Philosophy in 1935 or at the Congrès Descartes in 1937 precisely because he had already developed an epistemological position in clear opposition to the concept of the "unity of science" and "scientific philosophy" (see Chap. 1, Sect. 1.2). According to his perspective, however, scientific rationality took on different forms—irreducible to one another—depending on their main object of investigation and mode of reasoning. In fact, in some of his writings prior to 1937, Bachelard had "elaborated the programmatic lines of his 'historical' and pluralistic epistemology" (Castellana 2019, 51–52). The only form of the unity of science that Bachelard was willing to accept was the historicity of the scientific enterprise.

---

clearly highlights the author's dual objective: on the one hand, to classify the range of scientific knowledge, on the other, to seek, through historical investigation, the elements common to the multifaceted "discoveries" that characterized the Scientific Revolution and represented rationality.

The first time that Bachelard raised the problem of a pluralization of scientific knowledge (not yet called "regional") was in the article "Surrationalism," published in 1936 in the first and only issue of the surrealist journal *Inquistitions*. As is evident, Bachelard's proposal found a home in a very peculiar context with respect to the epistemological debate of the time, placing his work alongside rather literary-oriented articles by Tristan Tzara, Louis Aragon, Roger Callois, Jules Monnerot, and others:

> Rational pluralism touches domains metaphysically so different that one cannot hope to give them coherence by the simple syntheses of opposites. Yet is it really necessary to search for that static coherence which would correspond with a closed metaphysical system? Is there no place, in evolving reason, for a coherence in some way dynamic, which would determine the mobility even of the psychical? [...] Then closed rationalism yields to open rationalism. Reason, fortunately incomplete, no longer is able to slumber in a tradition. (Bachelard 1936b/1995,?189)

As certain attentive readers have pointed out, this "surrationalism" represents the beating heart of Bachelard's epistemology, especially in his relationship to mathematics.[40] Although "surrationalism" had already put forward the idea of rational pluralism and open rationalism, the text devoted specifically to the question of epistemological regionalism was *Le rationalisme appliqué*, published in 1949. In this regard, note the questions posed at the start of the seventh chapter, "The regional rationalisms":

> Is the idea of determining distinct regions in the rational organization of knowledge healthy? Isn't the whole philosophical tradition of rationalism, which is fixated on total unity, against it? And—a more serious objection—doesn't the idea of regionalizing rationalism go against all the efforts of contemporary epistemology to *found* science, to find the *foundation* of all science? (Bachelard 1949,?119)

Clearly, Bachelard's formulation of "epistemological regionalism" was shaped by the intellectual debates that emerged from the two congresses of 1935 and 1937. With this concept—in addition to highlighting the historical variations in the evolution of scientific thought—he aimed to highlight the peculiarities inherent in the different sciences. Obviously, Bachelard did not intend to answer the questions that he had raised, because the stated object of his research involved "examining the simple problem of the rational shaping of experience and, more precisely, the problem of the rational preparation of scientific experimentation" (Bachelard 1949, 119). Therefore, he asserted, the problem of the search for foundations was not in question, and the effectiveness of regional epistemologies, at an explanatory level, was left to the reader's judgment: "We will leave it to our reader to decide [...] whether the establishment of separate regions in a rationalist doctrine of knowledge

[40] See Castellana (1974, 2021b), Palombi (2017), Alunni (2018), and Ienna (2019a). On this point, see also the following section dedicated to the specific epistemological region of mathematics. On some developments in surrationalism, particularly with reference to Michel Serres, see Simons (2019b).

is philosophically useful, whether it represents an effective step forward in the current development of the scientific spirit" (Bachelard 1949,?119).

Furthermore, the notion of epistemological regionalism also polemically referred to Husserlian phenomenology,[41] which Bachelard had read attentively.[42] As is well known, in *Ideas Pertaining to a Pure Phenomenology and to a Phenomenological Philosophy*, Husserl used the term *regional ontologies* to indicate "those regions of reality that the various sciences try […] to theoretically delimit"[43] (Costa 2009, 91). For Husserl, through the implementation of eidetic and transcendental reduction, it was possible

> to delineate, for example, what distinguishes the "nature" region from the "ideal objects" region, or the field of application of the natural sciences from that of logic and pure mathematics, or the "culture" region, delimiting it from nature, as well as the field of sounds from that of colors. (Costa 2009,?91)

Bachelard inherited this metaphorical expression, transforming it into the concept of regional epistemology. As seen previously, however, Bachelard clearly distanced himself from the phenomenological approach by proposing the concept of phenomenotechnics. In fact, in *Le rationalisme appliqué*, Bachelard explained that "The regions of scientific knowledge are determined by reflection. They cannot be found delineated in an immediate phenomenology" (Bachelard 1949, 122). Here, Bachelard argued that the regions of common sense do not correspond to those of scientific-rational thought, relating the idea of region to that of "applied rationalism" or "technical materialism."

This shows how, for Bachelard, each epistemological region corresponded to a specific phenonenotechnic. In this sense, the autonomy of a region of knowledge is determined, first of all, by how it constructs its research objects. For example, over the course of their history, chemistry and physics developed "nuclei of apodicticity" on the basis of the construction of different evidential spaces (see Bachelard 1949,

[41] Husserl's idea of phenomenology as a rigorous science was just beginning to become widespread in France in those years, thanks to the endorsement of Koyré. It is well known that Koyré, Husserl's former student at Göttingen, sponsored the first translations of Husserl and organized a conference at the Sorbonne, the text of which, translated by a very young Emmanuel Lévinas, was published as *Cartesian Meditations*.

[42] For a broad survey of Bachelard's critical reaction to Husserl, see Barsotti (2002).

[43] In epistemological terms, notably, part of the inherent problem with regional ontologies was brought up by Husserl in the debate about the distinction between natural sciences and *Geisteswissenschaften* [human sciences], which was hotly debated in the German philosophical context (especially neo-Kantian) by figures such as Dilthey, Rickert, and Windelband. In this regard, Costa continued: "The reference to this possible way of dealing with the relationships between nature and spirit only served to highlight how the difference between these disciplinary sectors could be traced to a difference of a methodological and not ontological order. On the contrary, speaking of regional ontologies, Husserl wanted to specifically indicate that philosophical analysis must not start from existing sciences, questioning their methodological structure and thus determining what nature is, starting from what the sciences tell us is nature. Instead, it must ask itself: starting from what appearance of phenomena can such a science be established?" (Costa 2009,?92).

132). The criteria for evidence and for what can be recognized as a demonstration are not uniform for all sciences but instead vary according to the discipline. As seen in the previous sections, even scientific instruments are nothing more than theories materialized in objects (Bachelard 1934/1984, 13). The act of experimentation is an experience that intentionally modifies the phenomenon being examined; thus, it is directly fabricated by the tools and theories that inform it. The idea that represents the vital core of "technical materialism" or "applied rationalism" is that all theory is a practice and vice versa. Bachelard claimed, against all forms of empiricist reductionism, that the notion of epistemological rupture could lead to the discovery of the foundations of the sciences by looking at how rationality informs reality and not merely by reference to a "common knowledge of reality" [*"la connaissance vulgaire de la réalité"*] (Bachelard 1949, 120–121). In this sense, the foundations of different rational fields, such as those of arithmetic and physics, must be recognized as not superimposable (Bachelard 1949, 120–121).

But Bachelard's fragmentation of reason did not stop at the recognition of these disciplinary specificities, but rather, it continued toward a further fragmentation of rational activity through the identification of the specific axiomatics correlated with the multiplication of experiences that are used to measure rationality (Bachelard 1949, 121, 132). Bachelard, therefore, reversed the order in which a priori epistemological models—such as the neo-positivist one—had traditionally investigated rationality. Starting from the concrete practices of scientific activity, Bachelard built an epistemological model capable of tracing the varied and multiform applications—hence the title *Le rationalisme appliqué*—of rationality as a function of their specific subject matter. Constructing an object—which is the criterion for identifying an epistemological region—also brings with it the construction of a specific scientific "problem," which, to be resolved, requires the mobilization of particular methodologies that lead to the construction of regional axiomatics. For this reason, Bachelard described some exemplifying forms of specific rationality, such as "electrical rationalism" and "mechanical rationalism,"[44] whose creation was reconstructed on the basis of a careful historiographical investigation. What Bachelard wanted to highlight through this metaphor of regions was the way that rationality emerges from scientific activity itself, that it is not dropped down from the heights of an ideal world onto reality. For example, in Chap. 8, Bachelard followed the complex interchange between the theoretical horizon and the experimental practices that led to the emergence of "electrical rationalism." He chose this field because, apparently, "the history of electrical science is, at least in its early days, one of the least rational histories there is" (Bachelard 1949, 139). The objective was to show how, over a long process made up of obstacles and epistemological ruptures, we move, "little by little," freeing ourselves from "common concepts" and confused forms of knowledge and moving toward the rational formulation of "scientific concepts" (Bachelard 1949, 146–148). In the constitutive dialectic of this epistemological region,

[44]These are some examples of regions of rationalism that Bachelard used to demonstrate the cogency of his theoretical proposals. See Bachelard (1949, Chaps. 8 and?9).

Bachelard established a correspondence between theory and practice, between "experimental thinking" and "algebraic thinking" (Bachelard 1949, 156). Electrical rationalism, in this sense, could be reduced to neither a "logical space" nor a mere "chapter of natural history" (Bachelard 1949,?138).

For Bachelard, the investigation of rationality should pass through a historical-epistemological reconstruction of the specificities of the various regional rationalisms making up the scientific enterprise. However, once fragmented, reason is not forced into a progressive relativization and localization. From this multiplicity of axiomatic systems, something about rationalism in general can be discovered:

> Once rationalism has been fragmented to properly associate it with the matter that it informs, the phenomena that it regulates, and the phenomenotechnic that it establishes, we are led to pose the philosophical problem of the relationship of a general rationalism with the various regional rationalisms. (Bachelard 1949,?131)

Bachelard outlined two possible ways of describing the relationships between "regional rationalisms" and "general rationalism." The first way—that followed by the Vienna Circle and clearly rejected by Bachelard—consisted of defining rationality a priori in a way that was valid for all experiences, both present and future. Bachelard called this epistemological model "*fixiste*," in the sense that, with such a claim to universality, it was very close to the solipsistic solutions of idealism.

The second way was, instead, the one that was adopted by Bachelard himself and that took the name "integral or integrative rationalism" (Bachelard 1949, 132). This form of general epistemology was constructed a posteriori with respect to regional epistemologies by relating phenomena from well-defined types of experience. By integrating the epistemological results of each region, new axiomatic structures were built by extending the previous ones to account for a growing number of phenomena. The theory of epistemological regionalism could, therefore, be considered a contribution to what Fleck called "comparative epistemology,"[45] which first aims at the critical examination of the peculiarities of each individual axiomatic model and then reconnects all the specificities in an integrationist perspective:

> Integral rationalism must therefore be a dialectical rationalism that decides the structure in which thought must engage to inform an experience. It corresponds to a sort of central office of a factory that has found a rationalization. Therefore, the problem of defining a general rationalism that would encompass the common part of regional rationalisms no longer arises. In this way, only the minimal rationalism used in everyday life would be found. The structures would be erased.
>
> On the contrary, it is a matter of multiplying and refining structures, which from a rationalist point of view must be expressed as a structuring activity, as a determination of the possibility of multiple axiomatics to deal with the multiplication of experiences. (Bachelard 1949,?133)

Bachelard also returned to these questions in a 1950 speech at the Société Française de Philosophie titled "De la nature du rationalisme" (Bachelard 1950),

---

[45] This term was coined by Fleck in *Genesis and Development of a Scientific Fact* (1935). I believe that through an in-depth study, notable spaces of intersection and integration would emerge between the perspectives of Fleck and Bachelard.

in which he lauded the fact that the history of science was a history of progressive sectoral specializations[46] that simultaneously determined a specialization of reason. For this reason, "we are now dealing with the philosophy of 're,' 're,' 're,' 'restart,' 'renew,' 'reorganize'" (Bachelard 1950, 50), according to which rationalism appears to be something that is in continuous movement because it is always rectified. At the end of the speech, Bachelard was asked a series of questions that allowed him to clarify his epistemological position on the relationship between "regional" and "general" rationalism. In a long commentary, Jean Ullmo[47] raised objections to the use of the word *region* in the epistemological context as it created the specter of possible misunderstandings, proposing that Bachelard should, instead, use "specialized rationalism":

> But the word "regional" itself has the disadvantage of being opposed to "global," to "total," and of making people think that, in your mind, you are talking about "regional rationalism" at the same time as "regional rationalisms"—can you imagine the possibility of different co-existing rationalisms which would oppose each other? (Bachelard 1950,?64)

To Ullmo's objection, Bachelard responded:

> By considering the concept of regional rationalism, I was trying to show that, even in a small sector, the mind takes flight, the mind generalizes in an activity which seems particular, and I do not need to return to traditional rationalism. (Bachelard 1950,?66)

Canguilhem, in one of his many texts dedicated to providing commentary on and disseminating Bachelard's work, wrote:

> In the age of relativity and quanta, a rationalist philosophy cannot speak of reason as it did in the eighteenth century. That is not all: contemporary science now takes on a responsibility that the philosophy of the classical era attributed to itself: the questioning of the conditions of possibility of science and the critique of its foundations. There is no philosophical criticism except from the side of applied reason, and applications are only regional: the rational guarantees of objectivity are not the same in the sciences of electricity and mechanics. Rationalism in general does not exist, it can only be developed through integration. [...] Scientific phenomena are the historical product of a double effort, both theoretical and technical. (Canguilhem 2018,?625)

Let me summarize what has been explained so far. The epistemological theory of Bachelard's "open rationalism" can be synthesized in a scheme based on the intersection of two axes: a vertical axis that marks the opening of scientific reason in its historical dimension and a horizontal axis in which rationalism is divided into specific regions of specialization. Despite being caught up in this dual mechanism of?fragmentation, in Bachelard's opinion, reason is not lost in a progressive process of relativization but maintains its unity in view of the epistemological project of generalist integration, to be understood as immanent to scientific practice itself.

In the context of Canguilhem's studies, the regional question is not addressed in as direct and didactic a manner as it is in Bachelard's work but is largely present as a

---

[46] The same observation was also made in "Surrationalism" (see Bachelard 1936b/1995).

[47] Jean Ullmo (1906–1980) was a French historian and philosopher of science. His published works such as *La crise de la physique quantique* (1955) and *La Pensée scientifique moderne* (1957).

conceptual background in his historical-epistemological theorization. In particular, Canguilhem developed rational pluralism by following two trajectories: a philosophical one leading to the theme of vitalism[48] and a methodological-historiographical one leading to the theorization of a regionalization of the history of science.[49]

Starting with the philosophical path, Canguilhem's originality lay in his use of Bachelardian theoretical devices to demonstrate the specificity of the rational model underlying the problem of "life," which the "*philosophie biologique*" dealt with. In a significant article titled "Note sur la situation faite en France à la philosophie biologique," Canguilhem summarized his intellectual program clearly. Since Descartes in France, biological philosophy had been largely devalued, in that "life," in his philosophical system, was not granted any ontological originality. Furthermore, "as in Descartes, the philosophy of Alain, Brunschvicg, and Sartre did not recognize life as its own metaphysical object" (Canguilhem 1947, 324). Throughout his article,[50] Canguilhem noted a structural hostility to considering these issues without reductionism. Moreover, "life" had been widely perceived as a nuanced object by scientific reason: "In the eyes of reason, life is confused; in the eyes of reason, life is elusive" (Canguilhem 1947, 326). Using a beautiful analogy, he continued by saying that "reason is as regular as an accountant; life is anarchic like an artist" (Canguilhem 1947, 326; see also Wolfe 2024). Thus, Canguilhem's intellectual project began with rationalism's failure to problematize the specificity of this epistemological region.

The connection, which Canguilhem highlighted throughout his intellectual career, was between, on the one hand, "life" and "norm" and, on the other, "knowledge" and "life." Guillame le Blanc argued that "the philosophical analysis of life can only be done on the basis of the concept of norm," while "the concept of norm inevitably refers to the concept of life" (le Blanc 2008, 7). Life itself is the creation of norms, where vital norms—unlike social norms, which are organized in structural terms—are systematized spontaneously: every biological individual has their own intrinsic normativity. Canguilhem closed his article with an explicit statement regarding the sense in which he could be called a rationalist:

> If we understand reason not so much as the power to apperceive essential relationships embedded in the reality of things or of the mind, but as the power to establish normative relationships in the experience of life, then it is in this sense that we would like to call ourselves rationalists or, more precisely, that we can subscribe to Bachelard's beautiful formula in his book *Water and Dreams*: "Rationalist? we strive to become one." (Canguilhem 1947,?332)

Canguilhem's conception of the biological sciences focused on two main concerns. The first, in terms of epistemological regionalism, was to bring out the

[48] For an overview of the vitalist model adopted by Canguilhem, see in particular Wolfe (2015). For a detailed history of vitalism, see Wolfe (2019).

[49] Some studies have highlighted how Canguilhem's adoption of the "regional" metaphor in epistemology also had its basis in his work on social geography. In this regard, see Cammelli (2011, 525) and Braunstein (2013).

[50] Canguilhem mentions several times that in the German context, the theme of life had at least been taken into consideration.

specificity of the object "life," which is the object of these sciences. The second had to do with the relationship between science and technology. In this case, Canguilhem's privileged object of investigation was medical art, understood as the technical application of principles coming from various sciences. These concerns corresponded to two thematic planes related to vitalism and to the status attributed to the concept of norm.

In *La formation du concept de réflexe aux XVIIe et XVIIIe siècles* (Canguilhem 1955) and in the essays found in *Knowledge of Life* (Canguilhem 1952/2008), indeed, Canguilhem proposed a re-evaluation of vitalism—against the Cartesian mechanistic perspective—as the only perspective that took the specificity of life and living seriously. On the other hand, in *The Normal and the Pathological* (Canguilhem 1966/1991), he approached the science–technology relationship by historically reconstructing the relationships between the history of therapeutics and physiology, or what type of relationship "the living being" has with the "concept of life." Canguilhem responded to this question by analyzing how the concept of norm was necessary for creating the possibility of the science of "pathology," which needs to define what is "normal" and, consequently, what is "pathological." To establish medicine as a therapeutic practice, the desire to re-establish a "normal" state in the living being was necessary. Opposing the positivism of Adolphe Quetelet's *homme moyen*, which thematized "normality" in terms of the statistical average, Canguilhem once again espoused a "vitalist" position. In this sense, "life is in fact a normative activity"; the essential thing about normality is to be "normative"—that is, capable of establishing norms and capable of transforming them into one's actual life (Canguilhem 1966/1991, 126). This was the origin of medical techniques and the source of its primacy over the other sciences. Science, for Canguilhem, was always secondary to technology. Medical practice was what spurred scientific research through its failures.

The peculiarity of Canguilhem's thought therefore lay in its affirmation of the autonomy of "vital rationalism" (understood as a specific epistemological region) compared to other sciences, in particular physical-mathematical ones, which had reduced the dimension of life to mere mechanism.

Canguilhem's methodological position in the history of science was entirely consistent with his conception of "vital rationalism." Indeed, he renewed the typically French tendency to avoid using the term *history and philosophy of science* (in the singular), preferring rather the plural *history and philosophy of the sciences*.[51] Since his 1947 article, Canguilhem had supported the need to oppose a "history and geography of rationalities" to the research program "of the universality of reason" (le Blanc 2008, 200). Aiming to construct a regional history of the sciences, he started from the idea of representing the sciences as all on one level and irreducible to one another, in clear opposition to the ideas of the hierarchical classification of knowledge. Each "region" had "disciplinary frontiers" determined by their specific

[51] On this point, consult the appendix that Claude Debru devoted to this theme in his book *Georges Canguilhem, science et non-science* (Debru 2004, 89–99).

objects and methods. Rationalism, subjected to historical analysis, changed form while still managing to survive as the power to establish normative relationships and not as an observation of essential relationships included within the reality of things (also in this case, the epistemological vector moved from the rational toward the real).

If the object of science was constructed—in the phenomenological terms expounded by Bachelard—then the object of the history of science would be equally constructed. For Canguilhem, in fact, each region of the history of science had its own object (different from the object of the discipline in question) and its own methodology and means for studying it (Canguilhem 1994b/2005, 203–205):

> The object of the history of science has nothing in common with the object of science. The scientific object, constituted by methodical discourse, is second in relation to, even though not derived from, the initial natural object [...]. The history of science practices on these second, nonnatural, cultural objects, but does not derive from them, any more than these second objects derived from the first. The object of historical discourse is, in effect, the historicity of scientific discourse, inasmuch as this historicity represents the carrying out of an internally law-governed project, but one which is traversed by accidents, retarded or deflected by obstacles, interrupted by crises, i.e. moments of judgment and of truth. (Canguilhem 1994b/2005,?203)

Therefore, writing a history of physics or biology involves different stakes and different approaches to the material being historicized. In particular, the history of science involves paying attention not only to a certain group of sciences "but also to nonscience, to ideology, to political and social practice" (Canguilhem 1994b/2005, 204).[52] For Canguilhem, this also led to a consideration of the multiplicity of temporalities internal to the scientific enterprise itself: "The time of the advent of scientific truth, the time of verification, has a liquidity or a viscosity which is different for different disciplines in the same period of general history" (Canguilhem 1994b/2005,?204).

Pierre Jacob, despite being a detractor[53] (as an analytically inspired philosopher), underlined how epistemological regionalism represented a common trait within the French tradition, compiling a long list of epistemologists and related branches of study that were brought together in France under the influence of this type of approach. In a very critical sense, Jacob found the following reasons for the success of regionalism:

> Three phenomena make epistemological regionalism attractive. On the one hand, generalist philosophers of science, who sought a unitary model of scientific explanation, certainly

[52] I will return to the social dimension of Canguilhem's epistemology in Sect. 3.1 of Chap. 3.

[53] Jacob declared himself "distrustful" of the program of epistemological regionalism, preferring instead a "generalist approach" to epistemology. He wrote: "My thesis is that, by trying to dissolve rather than analyze the general concepts used in science, doctrinal regionalism places epistemology or the philosophy of science in a situation of pragmatic paradox which recalls that of the philosophers of mind (or psychology) who refer to materialism by eliminating it and proclaiming that neuroscience, which studies the nervous system without using the term 'belief' (or its translation into another language), demonstrates that beliefs do not exist" (Jacob 1991,?290).

> granted an excessive place to the physical sciences until the 1980s, to the detriment of other empirical sciences—life, cognitive, human, and social sciences. On the other hand, the generalist program in the philosophy of science has encountered two types of limitations. Finally, the generalist program highlights the reflexivity inherent in epistemology and, because it generates paradoxes, reflexivity is often considered an infirmity. [...] Let's examine the limits of the generalist program. Over the course of this century, the generalist philosophy of science has assigned itself two tasks: a) to discover a criterion for marking the border between scientific theories and non-scientific propositions; b) to build an inductive logic. It is widely admitted today, even by generalists themselves, that this program has not kept its promises. (Jacob 1991, 291–293)

Indeed, uncoincidentally, only following the post-Kuhnian turning point in the field of Anglophone epistemology—and with the correlative consolidation of the debate on the integration of the history and philosophy of science (HPS)[54]—did a form of epistemological pluralism develop that was based on the rejection of both metaphysical and methodological uniformity and the universalism of scientific knowledge. This began to gain ground through the debate on the so-called "nonunity" or "disunity of science" developed within the Stanford School (Galison and Stump 1996).[55]

As mentioned above, the spread of epistemological regionalism in France allowed the emergence of specific philosophies of science, such as those of mathematical sciences, life sciences, social sciences, logic, medicine, and so on. Each of these traditions is certainly deserving of study in their own right.[56] However, as far as my work is concerned, in the following section, I limit myself to examining how "mathematical epistemological regionalism" was thematized within the context of what I have called "the minor canon" of historical epistemology.

2.5 Militant Mathematism and the Creative Character of Mathematics

Among the exemplary forms of regionalism, the tradition of mathematical thought [*pensée des mathématiques*], which I have proposed as a minor canon of FHE, is worthy of a more specific focus. In sections two and five of the previous chapter, I described how, during the two Parisian conferences of 1935 and 1937, a relatively compact group took up a position against the logicism proposed by logical empiricism.

---

[54] For an introductory survey of this debate, see in particular the chapter "Il connubio fra storia e filosofia della scienza: la prospettiva HPS" in Badino et al. (2022, 107–112).

[55] In the same period, or shortly afterward, works were published that once again drew attention to epistemological regionalism (Dagognet 1996; Braunstein 2002a, 2013).

[56] For an introductory reconstruction of the main trends in the philosophy of physics, logic, mathematics, and biology in the French context, see Vinti (1977, 66–93). For a more detailed overview on this topic, see also Brenner and Gayon (2009).

What the intellectual trajectories of authors such as Brunschvicg, Bachelard, Cavaillès, Lautman, Gonseth, Desanti, Gilles Châtelet, and others had in common was the way they stressed the need to safeguard the specificity of mathematical thought. As seen in this chapter, Bachelard argued that "science does not have the philosophy it deserves," indicating with this expression a twist of epistemology toward the study of scientific practices that are in the process of being done [*en train de se faire*]. In Bachelard's view, scientific activity produced a philosophy itself. Consequently, the task of epistemology was to bring out this implicit philosophy from within science. In *The New Scientific Spirit*, in fact, he wrote:

> sooner or later, scientific thought will become the central subject of philosophical controversy; science will show philosophers how to replace intuitive, immediate systems of metaphysics with systems whose principles are debatable and subject to experimental validation. [...] Science in effect creates philosophy. (Bachelard 1934/1984, 2–3)

On the same basis, the expression mathematical thought, or mathematical philosophy [*philosophie mathématique*], also refers to the idea that in the implementation of mathematical practices is a reflective and creative form already in itself considered philosophical-epistemological. The general objective pursued by this "minor canon" was to study the evolution of the concepts and methods of meta-mathematical thought in order to bring out its autonomy in the production of truthful and objective processes—that is, the structure of mathematical rationalism. In this sense, the epistemological thesis of regionalism implied not only the autonomy of the various forms of rationality but also their consequent nonreducibility to logic. This was the initial terrain upon which the clash with neo-positivism or logical empiricism took place, which had, on the contrary, hypostatized mathematical-theoretical structures according to their static and timeless description. For authors such as Brunschvicg, Bachelard, Cavaillès, Lautman, Gonseth, Desanti, and Gilles Châtelet, the objective was rather to follow living mathematizing activity; they sought to capture moments of discontinuity and change in theoretical perspectives within the discipline either by observing how the search for structural analogies in mathematical problems and in the correlative processes of validating solutions and results had been transformed or by highlighting their intrinsic historicity. These authors based their claims for the independence of mathematics on the recognition of its creative capacity.

Although the evolution of French mathematical thought can be reconstructed starting from Cartesianism, passing through the Encyclopedists, and arriving once again at Comte's formulations of the concept of mathematical thought,[57] I believe, on the contrary, that the first turning point of this tradition can be found in Brunschvicg's reformulation of mathematical thought in *Les étapes de la philosophie mathématique* (1912) and *L'experience humaine et la causalité*

[57] I have already mentioned the Comtian roots of mathematical thought in Sect. 1.5 of the Chap. 1. However, I want to point out that even in Brunschvicg, there was a valorization of Comte's thought. Chapter 13 of *Les étapes de la philosophie mathématique* was, in fact, entirely dedicated to the analysis of Comte's perspective.

physique (1922), in which the beginnings of a post-conventionalist philosophy[58] (compared to the programs of Duhem and Poincaré) can be seen in the context of the French debates.

In *Les étapes de la philosophie mathématique*, Brunschvicg proposed a broad reconstruction of the evolution of mathematical thought from Pythagoras up to the problems posed by Cauchy, Galois, Riemann, Cantor, Dedekind, Du Bois-Reymond, Friedrich Frege, and Russell in the epistemological field (Castellana 2020, 13). In this work, as Desanti wrote in the preface to the 1972 reprint of this volume, the concept of mathematical philosophy took on a new meaning:

> The expression mathematical philosophy thus took on an original meaning. It was not a question of implementing, under the name of Philosophy, a speculative approach that completed what the mathematician had started. Saying something about "numbers," space, movement, a truth that mathematicians, caught up in the game of techniques could only glimpse, could only be an illusory undertaking. It is equally illusory to pretend to authoritatively define the status of first principles and therefore to identify the specific object of mathematical philosophy. Mathematical philosophy can only manifest itself in the constitutive movement of mathematics itself. This consists in identifying the intellectual operations that led to the establishment of norms appropriate to their objects. And this is why the philosopher who aspires to such a project must sit at the foot of [*se mettre à l'école*] mathematics, learn its language and enter into its content. The same movement would have designated and united, through the materials provided by empirical history, the reflective moments, the culminating points, in which an indissoluble bond between mathesis and philosophy had been woven. The slow emergence of forms of mathematical intelligibility has thus provided the interpretive scheme for the history of philosophy. (Desanti 1972, vi–vii)

In this regard, it was emphasized how, starting from *Les étapes de la philosophie mathématique*, a "surrationalist" orientation was created within the French philosophy of the second half of the twentieth century that implied the idea of a "creative power of mathematics" (Alunni 2015, 2). From this point of view, Brunschvicg argued for a "mathematizing idealism,"[59] for which "he proposed a philosophy of the concept, or more precisely of conceptual intuition (a fruitful oxymoron since Descartes) based on mathematics and describing the historical constitution of symbolisms in which the fundamental conceptual intuitions are somehow collected" (Badiou 2012,?10).

A further characteristic common to a number of the authors of this "minor canon" was the adoption of a mathematizing Spinozism, starting from Brunschvicg and continuing with the works of Cavaillès, Bachelard, and Desanti. The place that

[58]Castellana defined the period between Poincaré's death in 1912 and the publication of Bachelard's *Essai sur la connaissance approchée* in 1928 as the post-conventionalist moment (Castellana 2004b,?81).

[59]This expression used by Badiou refers to this entire tradition, starting from Brunschvicg. See Badiou (2006, 2012), as well as his more recent work, Badiou and Haéri (2015), in which he seems to depart more directly from Bachelard.

Brunschvicg gave to Spinoza[60] was very different from that reserved, for example, for Leibniz (who actively participated in mathematical debates). Spinoza, in fact, did not produce any mathematical work, but his rationalism provided the architecture for Brunschvicg's historical methodology (Peden 2014, 36). Spinoza's genius lay in the idea that "intuitive science is self-sufficient; it is the development of the internal dynamism that constitutes the nature of thought, the sign of spiritual automatism, to use an expression contained in the *Tractatus de Intellectus Emendatione*" (Brunschvicg 1912, 141–142). In this sense, "intuition is no longer an accident in the history of individual thought, a passing effort to maintain the distinct moments of reasoning under the simultaneity of the intellectual gaze" (Brunschvicg 1912, 141); similarly, "intelligence is an activity coextensive with human life; it is *judgment* and *will*" (Brunschvicg 1912, 142). In his reading of the history of mathematical thought, Brunschvicg thus tried to free himself from the contrast between historicist relativism and a logicist approach, which allowed him to propose an original vision based on "the Spinozist conception of truth."[61]

In the wake of the historical-critical method of inquiry developed in Germany and its Francophone reception, Brunschvicg also approached the analysis of the unresolved problems in contemporary science through an analysis of the epistemological transformations that, from a historical point of view, led to the crystallization of stable theoretical positions. Brunschvicg's main focus was on the relationships between physics and mathematics, in particular through "the attempt to put aside or at least weaken the idea that mathematics is 'the language of physics,' as 'the show still changes with the advent of mathematical physics'" (Castellana 2004b, 93). Brunschvicg's objective was to deconstruct the image of mathematics as the servant of experimental practices in physics, whose role was only to give shape, ex post facto, to an intuition that occurred in another epistemic field. On the contrary, he aimed to highlight how the transformations of contemporary physics had been linked to the emergence of new theoretical conformations in the mathematical field. These structures, in themselves, had generated ex ante the creative space within which the transformations of physics could develop:

> Planck's quanta could not find a place in the architecture whose framework and coherence was based on the system of classical mathematics. [...] A physics of the discontinuous is being developed (according to Langevin's expression), and this therefore requires new mathematical schemes. (Brunschvicg 1922, 397–398)

Mathematical thinking, therefore, created the space of possibilities within which physics worked, or the context within which modern physical theory took shape:

> Einstein's world is a world of numbers; these numbers presuppose neither an *a priori* truth as the condition of their formal expression, nor an intuitive image as a condition of their

[60] Brunschvicg was also the author of a famous text on Spinoza titled *Spinoza et ses contemporains* (Brunschvicg 1951). For a detailed analysis of Brunschvicg's mathematizing Spinozism in relation to Bachelard, see Alunni (2018).

[61] This expression is used as the title of one of the sections dedicated to Spinoza in *Les étapes de la philosophie mathématique*.

> physical meaning. Thus, it is necessary to understand that these figures are neither fictions nor abstractions; they correspond to coefficients that reality provides; mathematics intervenes only to highlight their objectivity, independent of the specific system of reference, and this succeeds by determining the *invariant* that allows passage from one system to another. Physics thus conceived is nothing more than *geometry* (or rather *cosmometry*), according to Descartes's formula, but in a non-static sense that goes beyond properly Cartesian methods. In fact, it is no longer a question of referring to a representation of space which is already complete and given [*déjà tout fait et tout donné*]. Space is not prior to measurement; it arises from the measurement that composes it into its parts, according to the procedures of Gauss and Riemann, that is, the space of the physicist is not the *a priori* representation that Descartes supposed and from which he implied the definition of matter; it is the fabric [*contexture*] of reality, gradually grasped by progression and extension in relation to the experimental coefficients. (Brunschvicg 1922, 429–430)

The scientific innovations of that period, such as relativity and quantum theory, clearly emphasized how, in the field of mathematical physics, calculus played a strategic role in the mathematical construction of the entities of theoretical physics. This gave rise to Brunschvicg's insistence on underlining "the role of *calculation*, the importance of *deduction*, and the idea that Einstein's physics was a 'mathematical construction'" (Alunni 2018,?42).

One of the main channels for the diffusion of this new mathematical philosophy was the previously mentioned *Revue de Métaphysique et de Morale*, which Brunschvicg founded in 1893, together with Élie Halévy and Xavier Léon,[62] and which quickly became one of the most important philosophical journals in France.[63] Despite its name,[64] this journal paid close attention to the sciences and the historical-critical approach to epistemology. In particular, *La Revue* granted the most space to the field of logical-mathematical philosophy, given the intense debates that, at the time, were generating non-Euclidean geometries and algebraic topology (Alunni 2018, 263; Castellana 2021a, 22). From the first editorial in its first issue, mathematics was clearly stated as representing "the older sister of philosophy" (Revue de Métaphysique et de Morale 1893, 3). Among the first collaborators were Maximilien Winter—who actively contributed to the creation and management of the journal (Alunni 2018, 259–287; Castellana 2020)—but also extremely important figures, such as Louis Couturat, Gaston Milhaud, Émile Boutroux, André Lalande, Émile Durkheim, Bergson, Alain, and Félix Ravaisson-Mollien.

In particular, Winter was a strong supporter of a historical-critical approach[65] to the study of mathematics, publishing *La méthode dans la philosophie des*

---

[62] These authors were particularly united by having been students of Alphonse Darlu at the Lycée Condorcet in Paris (Alunni 2018, 262–263).

[63] Collaterally, in 1901, the Société Française de Philosophie, founded by Léon and Lalande, began publishing the *Bulletin de la société française de philosophie* as well as the *Revue de Métaphysique et de Morale*, of which Léon himself was director from its founding until 1935.

[64] The name of the journal represented the desire of its founders to advance an explicitly post-positivist program (Castellana 2021a, 22) of defending metaphysics and morality from the "double threat of positivism and mysticism" (Alunni 2018,?263).

[65] I have already analyzed how Winter transposed and developed Mach's historical-critical method (see Sect. 1.2 of Chap. 1).

mathématiques in 1911. With this volume, in fact, he developed a dual theoretical perspective. On the one hand—going beyond Brunschvicg—he shifted attention away from exclusively studying the historical evolution of the mathematical sciences of the past—that is, to already-established mathematics—and toward analyzing mathematical work in its living state—that is, *en train de se faire* (Castellana 2020, 23–24). On the other hand, he set as the objective of his philosophical critique of mathematics the understanding of how the foundations of every science are formed, a process that is never complete and is constantly undergoing transformations (Winter 1911,?100).

Through Brunschvicg's teaching work at the Sorbonne and the debates that spread through journals such as the *Revue de Métaphysique et de Morale*, *Revue de synthèse*, and the *Revue Philosophique de la France et de l'Étranger*, mathematical thought was transmitted to a generation of scholars, including Bachelard, Cavaillès, and Lautman, who wrote their doctoral theses under Brunschvicg's guidance.

As I have already mentioned several times, during the consolidation phase of Bachelardo-Canguilhemism, Bachelard's thought was often presented, transmitted, and received as an epistemology of physics and chemistry, thus underestimating the fundamental importance that mathematics had played in its complex theoretical construction. Trying to relate Bachelard to the tradition of mathematical thought is thus of strategic value to presenting an alternative to the Bachelardo-Canguilhemian knot (and perhaps to contributing to unraveling it). This interpretation was largely based on the re-evaluation of the earliest of Bachelard's works, such as *Essai sur la connaissance approchée*, *La valeur inductive de la relativité*, *Physique et métaphysique*, the essays collected in *Études*, *The New Scientific Spirit*, and others,[66] in whose light even the more mature contributions, such as *Le Rationalisme appliqué* and *Le matérialisme rationnel*,[67] were reinterpreted.

Like Brunschvicg's, Bachelard's objective was to develop an epistemological reflection capable of providing an understanding of the evolution of the mathematical physics of his time. From the time of his doctoral theses written under the direction of Brunschvicg and Rey; in his later works, such as *La valeur inductive de la relativité*; and in the first essays contained in *Études*, Bachelard focused on the

[66] Bachelard prevented the republication of some of these texts during his lifetime, both in journals and in collections. Furthermore, having left precise instructions to his daughter, Suzanne Bachelard, and his philosophical heir Canguilhem, this editorial ban continued even after his death. *La valeur inductive de la relativité* was only republished in 2014, after its first edition in 1929. *Physique et métaphysique* is an extremely significant short text because it was the only work in which Bachelard directly dealt at length with Spinoza's philosophy and the problem of mathematical thought (Bachelard 1933a, b).

A significant part of Bachelard's unpublished works remains unknown to the academic community due to ongoing inheritance issues that have prevented the creation of an archive dedicated to Bachelard within French cultural and academic institutions. This Gaston Bachelard Archive should be housed in the Caphès Archives located within the École Normale Supérieure in Paris.

[67] Some of the contributions to this reinterpretation were made by Castellana (1974, 2004b, 2021b), Palombi (2017), Alunni (2018), and Ienna (2019a, 2021).

epistemological problems arising from microphysics and Einstein's theory of relativity. An essential point for Bachelard was to think about mathematics under the sign of a dialectical procedure that described the intrinsic creative activity of epistemic spaces. According to Bachelard, the task of mathematics is to organize reality.[68] In any case, the relativist physicist remains a realist, though of a very particular sort, because "he is not satisfied with discovering reality, he constructs it"; "in relativity, the real is demonstrated, not shown, and it is there that the inductive value of relativity must be placed, its force of inference, that is, of construction" (Alunni 2018,?48).

The concept of phenomenotechnics emerged in this context, understood as the ability of science to construct scientific phenomena contrary to the intuition of common sense (epistemological rupture). This term was introduced for the first time in "Noumène et microphysique" (1931–1932), one of the essays contained in *Études*. In the wake of the Spinozian reading of Brunschvicg,[69] Bachelard contested the idea of a noumenon unattainable to human knowledge; he therefore conceived of scientific activity as a noumenology capable of bringing out the structure of reality through an intrinsically mathematical phenomenology. The point of departure for Bachelard's analysis was the observation that the innovations of twentieth-century physics had reversed the traditional relationships that, in the nineteenth century, had underwritten the assumption of the empirically unified character of knowledge. Both empiricists and idealists arrived at concepts of unity, though from opposite trajectories. The empiricists had conceived of experience as uniform in that it was generated by sensations, whereas for the idealists, experience appeared to be unitary because it was impervious to reason. With the emergence of twentieth-century physics, however, a radical transformation took place, according to which "what is now hypothetical is *our* phenomenon; because our immediate grasp on reality has only the value of a confused, provisional, conventional datum" (Bachelard 1970, 13–14). So, what happens when science begins to be interested in what lies beyond—above and below, so to speak—sensory experience? The phenomenon, that of common experience, is transformed. It is "neither a judge nor a witness; it is an accused, and it is an accused who sooner or later will be guilty of a lie" (Bachelard 1970, 14). At this moment, mathematical thought comes into play through a complex phenomenotechnic revealing its real noumenal structuring:

[68] In this sense, in *Physique et métaphysique*, Bachelard approached Spinoza's thought by inserting a third dialectical pole, *natura constructa*, at the center of the classic pair *natura naturans/natura naturata*, by which he meant the phenomenotechnical power of mathematics to construct the real of mathematical physics (Bachelard 1933a). On Bachelard's mathematizing Spinozism, see Alunni (2016, 2018, 2021) and Ienna (2019a, 2021).

[69] Brunschvicg turned to Spinoza to argue that no limits to thought could be imposed by a "thing-in-itself," which was, therefore, considered an incoherent notion. "Since the thing in itself is never given in the world of knowledge," wrote Brunschvicg, following Spinoza, "the concept of thing-in-itself must not play any role, even a negative or a limiting one; *for there is no possible concept of a thing-in-itself*" (Peden 2014,?33).

> Obviously, the phenomenon initially beheld [*retenu*] can only be considered as a particular moment of a method. It will not take long to modify the method and, consequently, the phenomenon to obtain a fruitful sequence of experiences. When, then, will the series find authentic individuality? When it has been constructed mathematically. The mathematical formula gives it a form: it is by virtue of the mathematical relationship that the terms confused in the immediate phenomenon will be coordinated as a unit. However, the? mathematical relationships do not at all follow the connections that may appear in the primary observation. They follow the trace of a noumenal coordination, they act as the object of coordinated thought before being the object of experimental verification. (Bachelard 1970,?14)

This creates an epistemological rupture with common sense because the phenomenon no longer has any evidential value in itself. It is necessary to "sensitize it mathematically until the moment in which the mathematical reagents reveal the fundamental traits of reality by virtue of their noumenal coordination." In this way, the search for individualities, now located beyond phenomenal reality, will be conducted solely from a mathematical point of view. Contrary to nineteenth-century physics, it is no longer a question of translating the data collected through experience into mathematical language but "of expressing in the language of common experience a profound physical reality that possesses a mathematical sense before possessing a phenomenal meaning" (Bachelard 1970, 16). In this sense, "the *perfect atom* is more complex than the atom captured in its energetic reaction, at the moment in which it manifests the experimental trace of its existence. Our microphysical experience is always a mutilated mathematics" (Bachelard 1970,?17).

Bachelard conceived of this process as a "dialectic of creation" which—following Spinozian terminology—he defined as *natura constructa*, or artificial nature. This dialectic is structured according to an epistemological circuit that has its origin in mathematics, proceeds through physics, and then returns to a modified and extended mathematics, thus positing the possibility of mathematical physics (i.e., the field of research on the development of specific mathematical methods for physical problems) (Bachelard 1933a). In this sense, we can no longer speak of microphysics as "*a hypothesis between two experiences but rather an experience between two theorems*" (Bachelard 1970, 15). Epistemology must take charge of understanding how "the orderly proliferation of the abstract" occurs (Bachelard 1933a, 79). Therefore, a representation of human reason as a *natura construens* is what brings out, through mathematical reagents, a description of nature that is "much richer than all the structures discovered within reality. Man thus creates what idle nature has forgotten to produce" (Bachelard 1933a, 81): "The *a priori* frameworks of reason are enriched by their use and success. Reason, through physical science, asserts itself *a posteriori*" (Bachelard 1933a, 76). Regarding Bachelard's perspective, Jean-Claude Pariente rightly maintained that "reality is at the end, not at the origin, of knowledge. Knowledge, armed with mathematics is produced by experience, not induced from it" (Pariente 2015, 246). As Gaukroger argued, "mathematical theory must be more than a mere 'economic organization' whose rules are there just because they are useful or clear. Mathematical theory is inventive in its essence" (Gaukroger 1976, 218). For Gaukroger, in fact, Bachelardian epistemology is connected with ontological claims (Gaukroger 1976, 206–212). If a phenomenon is a tissue of relations, "the mathematical formalism of a theory is in some way synthetic and not merely classificatory. The object of the mathematical formalism is a system of relations" (Gaukroger 1976, 219). In this sense, the epistemological vector in Bachelard always started from the rational to arrive at the real, thus making it a realizing vector.

This conception of mathematics was not free of political overtones, as can easily be noticed in Bachelard's famous text "Surrationalism." Bachelard never made a secret of his socialist sympathies, and in many of his writings, Marxist terminology shined through. This aspect certainly united him with the surrealists, whose support of the Popular Front placed them in a common poetic-political-intellectual trend.[70] In this context, Bachelard, rather than expressing himself on poetic-aesthetic themes—which were hardly unusual for him—chose to publish a text that presented a true manifesto of his conception of scientific rationality (Bachelard 1936b/1995). In this short article, Bachelard described scientific reason as essentially defined by a dialectical movement that is leading from the inside (mathematical) toward the outside (experimental) and is subsequently synthesized into an a posteriori rationalism. Rationality, according to Bachelard, is polemical and aggressive, systematically divided and at war with itself:

> Hegelian dialectic confronts us with a dialectic *a priori*, a dialectic wherein the liberty of the mind is too unconditioned, too treacherous. [...] *It corresponds to one of those lifeless societies wherein one is free to do anything, but there is nothing to be done. Thus one is free to think, but there is nothing to be thought.* Much superior is the dialectic instituted at the level of particular ideas, *a posteriori*, after chance or history has brought forward an idea that, for that very reason, remains contingent. (Bachelard 1936b/1995,?187)

These were the same years in which French authors were developing their opposition to the program of logical empiricism, which was also an essential polemical reference point in the constitution of the surrationalist program:

> instead of realizing, surrealizing, the rational liberty which the mind had experienced in such precise and fragmentary dialectics, the logicians and formalists, on the contrary, unrealized, unpsychologized, the new spiritual conquest. Alas! after that labor of obstinate subrealism, the mind has not become more alert and more alive, but more weary and disenchanted. (Bachelard 1936b/1995,?187)

According to the combined reading of these last two quotes, for Bachelard, surrationalist epistemology clearly must be characterized by the observation of the dialectical movement internal to the dynamics of the historical development of science, without claiming to empty rationality of its material and psychological basis: "We have organized the scientific mind on simple foundations, historical foundations, forgetting that scientific history, like all history, is a recital of the misfortunes of reason, illusory struggles against illusions" (Bachelard 1936b/1995, 188). Surrationalism aimed to build an epistemology capable of "psychologically filling" and "putting into motion" the empty and abstract forms of rationality—

[70] Caillois was the one who involved Bachelard in this editorial project during the eighth International Congress of Philosophy in 1934 in Prague, during which Bachelard had presented his "Critique préliminaire du concept de frontière épistémologique" (Alunni 2018, 114–115).

however purified—created by logicians. Epistemology must bring out the dialectical pulsations[71] of scientific reason. The "rationalist commitment"[72] that drives Bachelard's surrationalism is, as Canguilhem suggested, "a permanent revolution" (Bachelard 1972c, 6). In subsequent writings, Bachelard further clarified that this epistemological posture was thoroughly materialist. However, it was not a philosophical materialism but a properly scientific materialism, as detailed in Sect. 2.1 of this chapter.

As has been mentioned, in these same years, Cavaillès was also continuing the intellectual impulse of Brunschvicg, who was the supervisor of Cavaillès's two doctoral theses, *Méthode axiomatique et formalisme* (main thesis) and *Remarques sur la formation de la théorie abstraite des ensembles* (complementary thesis) in 1937. *On Logic and the Theory of Science* (1947) and *Philosophie mathématique* (1962) were published posthumously. Cavaillès's theoretical objective was to imagine an alternative to transcendental analysis (with its foundation in consciousness), from Kant to Husserl, as well as to logicist-formalist analysis, from Leibniz to Carnap (Cortois 1994, 127). What these two opposed positions had in common was their tendency to dismember the unity of mathematical theory and practice into form and content. On the one hand, the form, emptied of its contents, was brought back to a mathesis universalis, or to a pure and universal syntax of a logical type. On the other hand, the material content of mathematics was instead reduced to physics (Cavaillès 1976/2021, 125). What disappeared in this process of radical abstraction was precisely the historical and dialectical character with which the rational activity of mathematics was constructed.

Cavaillès sought a way out of this theoretical impasse: on the one hand, through the philosophy of Spinoza and, on the other, in the so-called philosophy of the concept. What Canguilhem called his "militant mathematism" (Canguilhem 1977/1988, 16) emerged from their combination. According to Canguilhem, based on a letter that Cavaillès sent to Brunschvicg, Cavaillès's interest in Spinoza stemmed from his response to the phenomenology proposed by Husserl and his "exorbitant use of the cogito."[73] Bachelard also suggested another element, taken from another letter from Cavaillès, this time addressed to Lautman, in which he stated the following: "It is in relation to Husserl, a little against him, that I try to define myself" (Bachelard 1960, 24). In this sense, Canguilhem reminded us that Spinoza "represented the most radical attempt at a philosophy without cogito" (Canguilhem

[71] I use the concept of pulsations in the wake of recent philosophical-mathematical studies of Bachelardian orientation that have underlined the dialectical nature of mathematical thought (Guitart 2015; Alunni 2018; Zalamea 2018).

[72] *L'engagement rationaliste* is the title of a collection of Bachelard's essays edited by Canguilhem, at the beginning of which is the republished essay "Le surrationalisme" (Bachelard 1972c).

[73] This letter was cited by Gabrielle Ferrières (1950, 169) and then taken up by Canguilhem (1996, 27). In his interpretation, Canguilhem argued that Cavaillès's mobilization of Spinozian principles was a reaction to the way that Husserlian phenomenology used the cogito (in the version proposed in the *Krisis*). Thanks to Tran Duc Thao, Cavaillès had the opportunity to return to these themes.

1996, 27).[74] The last words of *On Logic and the Theory of Science*, already cited above, were emblematic in this regard and give us a better understanding of the organic link that united Cavaillès's thought to the tradition of historical epistemology: "It is not a philosophy of consciousness but a philosophy of the concept which can yield a doctrine of science. The generative necessity is not that of an activity, but of a dialectic" (Cavaillès 1976/2021, 136). In this sense, the search for the foundations of mathematical knowledge must not be entrusted to philosophies that claim to justify it through a return to the subjective pole or its reduction to logic. On the contrary, following Brunschvicg, the intrinsic creativity of mathematical knowledge and its specific criteria of rationality can be understood only through an internal analysis of the historical stages it has completed in its evolution and internal dialectic (see Jacumin 1967, 67). In *Philosophie mathématique*, in fact, Cavaillès claimed that it is not a question of tracing an "empirical history" but rather of highlighting the "dialectical development which ensures both their movement and the permanence of their validity" (Cavaillès 1994, 274). This statement should be read in conjunction with another passage from *On Logic and the Theory of Science*, in which the author outlined the model of historical evolution within science:

> Yet one of the essential problems for the doctrine of science is exactly that progress cannot be a mere increase in volume by juxtaposition, the prior subsisting with the new, but must be a perpetual revision of contents by way of deepening and erasure [*rature*]. What is after is more than what was before, not because it contains it or even because it extends it, but because it exits from it necessarily and bears within its content the singular mark, each time, of its superiority. (Cavaillès 1976/2021, 135–136)

It is therefore an epistemology that, recognizing the dynamic character of mathematical thought, acquires a historical dimension and does not claim to set the limits of this epistemological region a priori but seeks its foundations in the concrete dialectical-conceptual exchange within this discipline. From here, the need to historicize the Kantian notion of transcendental (Vinti 1977, 74; Angelini 2015, 199–200) emerged in order to account for how conceptual systems take shape:

> What marks history is the way the transcendental necessarily passes through its stages: the? obligation of this passage is recognized in a failure, the necessity of progress emerges in the indeterminacy of a discovery. The necessity appears after the fact. No analysis of the consciousness of actions allows one to make predictions, any more than it provides any stability. Intuitive is synonymous with effective (or effecting) consciousness, transcendental with constitutive (or constituent), relative to a given conceptual system. A closed system exists no more than an absolute transcendental: formal unity is the continuity of a series of acts, the concrete is the novelty of unification of a multiple still recognized conceptually as such. (Cavaillès 1994,?472)

Like Cavaillès, Lautman also developed an original historical-epistemological perspective on mathematics that was in complete contrast to the perspectives offered by neo-positivism. In 1937, he published his two theses (also written under the supervision of Brunschvicg), titled *Essai sur l'unité des sciences mathématiques*

---

[74] For further discussion of Cavaillès's Spinozism, see the reconstruction proposed by Peden (2014, 17–63).

dans leur développement actuel (main thesis) and *Essai sur les notions de structure et d'existence en mathématique* (complementary thesis). Shortly after, in 1939, *Nouvelles recherches sur la structure dialectique des mathématiques* was published, and his last book, *Symétrie et dissymétrie en mathématique et en physique*, came out posthumously.

Lautman started from a criticism of the logicist interpretation, arguing the impossibility of deriving mathematical notions starting from a small number of logical propositions. Instead, he advocated a qualitative reconstruction of the set of mathematical concepts (Vinti 1977, 73; Castellana 2019, 104). In this sense, a "rapprochement of metaphysics and mathematics" is what seemed to be necessary to recompose the synthetic unity of mathematics (Lautman 1939/2011, 197):

> Mathematical philosophy, as we conceive it, therefore consists not so much in retrieving a logical problem of classical metaphysics within a mathematical theory, than in grasping the structure of this theory globally in order to identify the logical problem that happens to be both defined and resolved by the very existence of this theory. (Lautman 2006/2011,?189)

Mathematics, as described by Lautman, is a "large living body" whose liveliness has generated exponential growth in the process of constituting each of its branches (Barot 2009, 28). These branches, despite their autonomous development, are characterized by structural affinities rooted in "a primordial unity," now lost, whose recomposition represents a "dialectical drama" that plays out between "a higher plane of being, an ontological plane of Plato's Ideas, and the historical plane of the scientific theories that are the incarnations, the realizations of such Ideas" (Barot 2009, 28). In this sense, Lautman argued that simultaneously carrying out both a historical and an epistemological reconstruction of mathematics—that is, an examination of this region of knowledge focused both on its structural elements and on its dynamic-evolutionary ones—was necessary:

> The structural conception and the dynamic conception of mathematics seem at first to be opposed: one tends in effect to consider a mathematical theory as a completed whole, independent of time; the other, on the contrary, does not separate it from the temporal stages of its elaboration. For the former, the theories are like entities qualitatively distinct from one another, whereas the latter sees in each an infinite power of expansion beyond its limits and connection with the others, by which the unity of the intellect is asserted. [...] we would however like to try to develop a conception of mathematical reality which combines the fixity of logical notions and the movement with which the theories live. (Lautman 2006/2011,?90)

Lautman's objective, in the notable expression of Zalamea,[75] was to reconstruct and account for the "creative dialectic of modern mathematics" (Zalamea 2006/2011, xxiii).

[75] Zalamea is a contemporary mathematician who has worked on constructing a new perspective on modern mathematics, starting from Francophone mathematical thought, in particular Cavaillès, Lautman, and Grothendieck (Zalamea 2006, 2016, 2018). The most important ideas are contained in his volume *Filosofia Sintetica de las Matematicas Contemporáneas*, recently translated into both English and French (Zalamea 2013, 2018).

Starting from this perspective, the clash between this new philosophy of mathematics with what Cavaillès called the "naive aspect" and "scandalous aggressivity" of the program of logical empiricism was inevitable (Cavaillès 1976/2021, 120). One of the texts that served as a manifesto of this new epistemological approach was the famous debate between Cavaillès and Lautman, titled "La pensée mathématique."[76] On February 4, 1939, the Société française de Philosophie invited these two authors to share the results of their recent doctoral theses at the Faculty of Letters of the Sorbonne in Paris. For both, the task of mathematical thought was to build an epistemology capable of explaining

> the qualitative changes, the transformations of the basic concepts of the mathematical edifice and of scientific rationality in general, not only from a formal point of view but according to a historical-critical methodology that takes into consideration the "internal dialectic of mathematical thought." (Castellana 2019,?98)

Against the logicist line represented by Frege, Russell, and Carnap, Cavaillès and Lautman rehabilitated Kurt Gödel's thought in an antianalytic key.[77] The program of reducing mathematics to logic had its roots in the so-called crisis of set theory, generated by the works of Russell and David Hilbert (Cavaillès 1994, 593). This intellectual debate had put forward the hypothesis that through axiomatization, the "epistemological" problem of the foundation of mathematics could be transformed into a problem that could be solved in a technical way (Cavaillès 1994, 597–598). Through the formulation of the incompleteness theorem, Gödel had shown that "it was impossible to insert mathematics into a single formal system; every system that includes arithmetic is necessarily incomplete" (Cavaillès 1994, 594, 598–599).

For Cavaillès and Lautman, the problem of both the foundation and the epistemological specificity of mathematics was to be found in the dynamic transformative activity and intrinsic creativity of mathematics itself. Indeed, Cavaillès maintained that "Mathematics is an evolution. All we can do is try to understand its history, that is, to place mathematics in the context of other intellectual activities, to try to find some characteristics of its becoming" (Cavaillès 1994, 600). The form of the historicity of mathematical thought that Cavaillès had in mind was characterized by two specific factors: Its becoming is, on the one hand, autonomous; on the other, it is completely unpredictable. Autonomy is related to the concatenations of "necessities" that the historian can observe in the diachronic transformation of mathematics (as he cannot escape history and is already aware of its trajectory) in such a way that the emergence of certain problems would correspond to the birth of a new concept (Cavaillès 1994, 600). Unpredictability is, instead, linked to the practice with which

[76]The report of this meeting was published only posthumously in 1946 in the *Bulletin de la Société Française de Philosophie*, accompanied by comments by Élie Cartan, Paul Lévy, Maurice Fréchet, Charles Ehresmann, Jean Hyppolite, Paul Schrecker, Claude Chabauty, and Paul Dubreil (Cavaillès and Lautman 1946). In French, this contribution can now be consulted in the volume *Oeuvres complètes de philosophie des sciences* (Cavaillès 1994, 593–630).

[77]For a general survey of the reception and uses of Gödel's thought in the French context, see Castellana (2021a).

the mathematician works. According to Cavaillès, how intuition will materialize cannot be predicted by virtue of the fact that, to use Kantian terms, mathematics is incapable of being based on analytical judgments (i.e., new mathematical concepts emerge as a function of posed problems and not due to the analysis of concepts that the mathematician already possesses). From this, Cavaillès derived the claim that mathematics is an experimental science (Cavaillès 1994, 601); that is, mathematical truths take form only through demonstrations.

However, some differences in perspective between these two authors emerged from the dialogue held at the Société française de Philosophie. In opening his talk, Lautman distanced himself from Cavaillès's perspectives on certain issues. Whereas Cavaillès had radically rejected any form of Platonism, Lautman instead highlighted how crucial focusing on the transitions between formal systemic principles and the concrete material realizations that emerged in mathematical practice was. As seen above, for Lautman, the structural analogies between the various branches of mathematics would allow us to infer the existence of mathematical entities beyond what could be demonstrated in practice. Georg Cantor's theorem in set theory, Évariste Galois's group theory, and Richard Dedekind's theory of algebraic groups played strategic roles because they were "abstract theories" that "studied [...] the possible ways of organizing elements whose nature is indifferent" (Cavaillès 1994, 606). As already noted, the question of conceiving the process therefore allows the passage from the formal to the material in a dialectical manner.

Cavaillès and Lautman's epistemological and political trajectories were largely intertwined, so much so that they were considered "spiritual brothers" (Castellana 2021a, 31). Both were, indeed, linked by a deep friendship and intense intellectual exchanges, and both lost their lives at a young age at the hands of the Germans during the French resistance. Also in this case, their approach to the philosophy of mathematics coincided with a committed political militancy on the left.

Notably, in those same years, the collective of mathematicians bearing the pseudonym Nicolas Bourbaki[78] was also active, made up at the time of its founding by authors such as Henri Cartan, Charles Chevally, Jean Dieudonné, Charles Ehresmann, and André Weil, who had much in common with the "surrationalist" trend in the epistemology of mathematics. Under this pseudonym, the collective published a series of works devoted to the foundations and various branches of mathematics, collected under the common title of *Éléments des mathématiques*. The original intention had been to try to bring order to the plurality intrinsic to mathematical thought by creating a hierarchical order within it on an axiomatic basis. The Bourbaki group particularly shared the idea that mathematical thought cannot be conceived in a monolithic and static way. Particularly significant was the definition of the mathematical universe proposed in "The Architecture of Mathematics":

> Thus, with these indispensable qualifications, we can become better aware of the internal life of mathematics, of its unity as well as of its diversity. It is like a big city, whose outlying

[78]This collective was formed in Besse-et-Saint-Anastaise in 1935, under the influence of André Weil (Alunni 2018,?267).

districts and suburbs encroach incessantly, and in a somewhat chaotic manner, on the surrounding country, while the center is rebuilt from time to time, each time in accordance with a more clearly conceived plan and a more majestic order, tearing down the old sections with their labyrinths of alleys, and projecting towards the periphery new avenues, more direct, broader and more commodious. (Bourbaki 1948/1950,?230)

The development of this historical-critical constellation in the epistemology of mathematical thought was also due, as seen in Chap. 1, to the international circulation of ideas.[79] Non-French authors such as Enriques and Gonseth were the main strategic allies of Cavaillès, Lautman, Bachelard, and others in the context of the two Parisian conferences of 1935 and 1937. I will not dwell on Enriques's thought here, because a section in the appendix to this volume is dedicated explicitly to Italian historical epistemology and focuses specifically on him. As for Gonseth, I limit myself here to pointing out his function as a strategic link in the construction of a transnational network committed to philosophical cooperation. Starting in 1938, the Swiss philosopher—with the support of the Institut International de la Coopération Intellectuelle—organized a series of conferences known as Entretiens de Zurich. The first meeting, which involved more than 30 scholars from various nations, was dedicated to "the foundations and methods of mathematical sciences." The objective of this meeting was to

provide a "platform" that could serve as a common basis for all current attempts at establishing the foundations of mathematics; to lay the foundations for an agreement on a doctrine of basic truths that could be recognized and accepted by all as legitimate and adequate to their needs. It was a doctrine in which the imagined relationships between the formal and the real, between the symbol and the thing symbolized, and more generally between the concrete and the abstract, would be made explicit, in order to satisfy all of us, mathematicians, logicians, and philosophers; in which the relationships we established, in fact, and even if we avoided talking about it, between intuitive judgments and methodical or formalized deduction, etc., would be clarified. All things upon which none of us could avoid having our own particular points of view and opinions, whether more or less systematic, bear the name of meta-mathematics, metalogic, or any other name, or have even remained in the background of reflection. (Gonseth 1941,?188)

Subsequent meetings focused on themes such as the idea of dialectics, experience and theory, the foundations and applications of calculating probability and statistics, and others, where Bachelard, Cavaillès, Lautman, Lebesgue, and Destouches participated, along with non-French authors such as Enriques, Łukasiewicz, Pauli, and Piaget (Castellana 2019, 47). On the basis of the network of cooperation and intellectual exchange initiated through these meetings, the debates in this field continued thanks to the foundation of the journal *Dialectica. Revue internationale de philosophie de la connaissance*, created by Gonseth in collaboration with Bachelard and Paul Isaac Brenays.

As I have pointed out previously, the establishment of a major canon of FHE through the affirmation of Bachelardo-Canguilhemism progressively obscured the

---

[79] I use the term *international circulation of ideas* here to refer to the model imagined by Bourdieu (2002), see also Ienna and Santoro (2016), Heilbron et al. (2018), Fleck et al. (2019), and Sapiro et al. (2020).

philosophical relevance of the debates around the epistemological aspect of mathematics. However, this does not mean that in the 1960s and 1970s the legacy of mathematical thought was not taken up by anyone. Desanti, for one, renewed its philosophical impetus by focusing on its intersection with markedly political themes. Desanti was a notable philosopher of mathematics with a Spinozist approach who was politically involved in the French far left. He attempted to find a connection between the mathematical Spinozism of Brunschvicg and Cavaillès and the phenomenology he had come into contact with through Maurice Merleau-Ponty and historical materialism, with the goal of founding a materialist epistemology (Castellana 1985, 69–118). Already in *Introduction à l'histoire de la philosophie*, Desanti was applying a Marxist methodology to the history of philosophy, dedicating himself to Spinoza as a case study. But the text that best demonstrated Desanti's belonging to the tradition of mathematical thought was *Les idéalités mathématiques. Recherches épistémologiques sur le développement de la théorie des fonctions de variables réelles* in 1968.[80] This work, taken from his doctoral thesis, sought to resolve the questions left unanswered by Cavaillès, starting from Spinozian thought and trying to explain mathematical creativity. As Castellana argued, Desanti's proposal moved toward a radicalization of Bachelard's epistemological regionalism. In this sense, epistemology must "address concrete scientific practice to avoid any totalizing philosophical discourse" (Castellana 1985, 70–71). Desanti's definition of the tasks of epistemology in this work was particularly interesting:

> The task of epistemology is to learn to traverse the entire thickness of this space, where the mathematician resides without ever having to use it. If such a path must still be called "philosophy," we can see that this philosophy can only be produced within an already established epistemological field: it consists in the knowledge of the mode of production and reproduction of the objects and statements specific to this field, in determining the structural laws that delimit the domain in which production practices are generated, maintained, and renewed. In this sense, it can be said that mathematics contains an implicit "philosophy": it requires knowledge of its specific mode of production [...]. It is the object of epistemology to recognize and design, through the field at the formalizable limit of mathematical objects, the sub-universes, the dense spaces whose connections constitute the evolving domain in which effective mathematics lives and is done. (Desanti 1968, vi–viii)

Following Bachelard's "applied rationalism," Desanti's epistemology also aimed to observe the mathematical sciences in action, *en train de se faire*, with the objective of reconstructing the epistemological space in which they moved and constructed their results—that is, of bringing out the epistemological stances implicit in scientific activity itself. To do this, Desanti explicitly dedicated himself to the "theory of real variable functions," identifying a particularly dense epistemological moment in the debates of the 1920s and 1930s. Without going into the details of his reconstruction, I must note for the purposes of this discussion how Desanti outlined the autonomous space of action within which mathematical rationality is constituted:

[80]The project of constructing a materialist epistemology would later be updated in *La philosophie silencieuse; ou critique des philosophies de la science* (Desanti 1975).

> Mathematics itself produces its soil, and there is no other soil for it than that which it has produced and incessantly reproduces [...]. There is no point in digging up the soil of mathematics in hope of discovering the original secret and mathematically silent subsoil upon which it was born [...]. We will never find ourselves faced with the event of the radical origin: here too the birth is unspeakable: it shows itself only in the product and from within. (Desanti 1968,?283)

Hence Desanti's famous statement that "there is no possible mathematical epistemology other than one within mathematics itself" (Desanti 1968, 286) meant that finding a foundation for the mathematical sciences is not possible except in mathematical practices themselves and in their ways of being practiced; that is, neither determining, a priori, the action space of mathematical creativity nor—as the neo-positivist model would have it—reducing this epistemological region to logic is possible. Furthermore, trying to circumvent the problem of foundations by researching the so-called origin problem is completely useless because one would always find oneself forced to give only ideal results. However, this does not mean that Desanti's epistemology was ahistorical; rather, it was based on a rejection of the progressive model of the evolution of science. According the following quote, in fact, one can see how Desanti fully embraced the Bachelardian sense of history as being crossed by ruptures and obstacles and in need of being judged and evaluated by the epistemologist:

> The epistemologist must deal with the history of "his" science, but his task is not to follow it, day by day, in its detail and particularities. The older "works" are the object of his analysis. He reworks them in his own way to highlight how this science was constituted and how its specific concepts are connected. (Desanti 1968,?7)

The task of epistemology is rather to work a posteriori with respect to the work of the scientist as it aims "to grasp the intrinsic productivity of mathematics and to eliminate, in Bachelardian style, all those inconveniences of various kinds that can hinder the true understanding of certain articulations of mathematical structures" (Castellana 1985, 76–77). For Desanti, the epistemologist and the mathematician move on different analytical-discursive levels. In the mathematician's daily life, they operate in what Desanti called a "flat space" in which the mechanisms that make their actions possible are taken for granted. The mathematician formulates theorems without calling into question how they bring objects—mathematical ideals—into being through those objects' relationships and interactions. The epistemologist, instead, proceeds in reverse because they must deconstruct the theorems to see how they were constructed and unravel the mathematical fabric to reveal the scheme from which it was conceived. In an article titled "En quoi l'histoire des mathématiques peut-elle être utile au mathématicien?" Desanti highlighted how this epistemological recognition was anything but an end in itself. In fact, he underlined how a historical epistemology of mathematics based on the foregoing principles was an essential tool for mathematical practice itself. In Desanti's view, the history of mathematics would, indeed, be reducible to neither the "simple collection of valid statements and their distribution according to a chronological chain," "the actualization of a purely linear development which would be pursued, so to speak, category by category, hypothesizing an epistemological development through which these concepts would emerge from each other," nor even "the simple reflexive reworking carried out by a subject on the basis of the norm of which he would be the only bearer" (Desanti 1974, 48). The epistemologically informed history of mathematics should, rather,

> derive the structure of the forms of mathematics to discover and liberate their mode of functioning, to test their possibility of affecting observable domains, to measure the constraints at the frontiers of mathematics and, where they exist, to highlight what is required to break through them. The objective of the history of mathematics should be *to show how mathesis works*, how it disappears. (Desanti 1974,?48)

Desanti held a central role in the French academic field. He exerted a strong influence on an entire generation of philosophers at the Ecole Normale Supérieure in Paris, including Jacques Derrida and, above all, Louis Althusser (to whom he passed on his interest in Spinoza).[81]

Alongside Desanti, Gilles Châtelet was also one of those authors who combined the development of mathematical thought with militant political commitment.[82] Starting from the end of the 1970s, he taught mathematics at the University of Vincennes, a lively political-scientific context within which he had already had the opportunity to participate starting from 1968, coming into contact with Foucault, Fèlix Guattari, and Gilles Deleuze (the latter had a particularly important role in the formation of Châtelet's thought). His most important epistemological work was *Figuring Space*[83] in 1993, which was rightly considered "the (provisional) conclusion of that great philosophical tradition which goes by the name of 'French epistemology'" (Cavazzini 2010, 11) and which, in various ways, went beyond the project of FHE. This volume was structured as a dialogue between three figures—a mathematician, a physicist, and a philosopher—whose work had been intertwined in a series of debates and controversies for more than two centuries. Châtelet's objective was to deconstruct the idea that mathematics is nothing more than a tool in the hands of what he called the "laborers of the exact sciences." The need to thoroughly investigate the age-old question of the relationship between the physical sciences

---

[81] In a private interview, Alunni recalled that Althusser, obsessed with the author of the *Ethics*, wanted to write a treatise entirely dedicated to Spinoza (a sort of myth within the Ecole Normale Supérieure), one which, of course, never saw the light of day. In the same vein, Desanti told him that Cavaillès, also obsessed with Spinoza, went into the garden of the École Normale, nailed a torn page of the *Ethics* to a tree, and riddled it with bullets fired from his revolver. Also, in his youth, Desanti, who was of Corsican origins, always had a gun with him, as was well known.

[82] In addition to his epistemological production, Châtelet established himself as a lucid political pamphleteer through the 1998 publication of his famous *To Live and Think Like Pigs. The Incitement of Envy and Boredom in Market Democracies*, in which, taking a clear anticapitalist stance, he addressed political theory and economic and social sciences (Châtelet 1998; on this point, see also Wolfe 1999). Other contributions of a political nature, but with various epistemological intersections, were collected posthumously by Catherine Paoletti in the volume *Les animaux malades du consensus* (Châtelet 2010b).

[83] More recently, Alunni and Catherine Paoletti edited a volume titled *L'enchantement du virtuel: mathématique, physique, philosophie*, which collected other epistemologically themed writings by Châtelet (2010a).

and the mathematical sciences, to which Brunschvicg had devoted much attention, was thus reactivated:

> What hand is holding the tool? What gestures are involved in this "functioning," which establishes a higher form of continuity between "structure" and "reality"? Is it simply a case of "applying" mathematics to physics or of awakening the physical [*physique*] in mathematics? Indeed, there is quite enough to exasperate even the most patient of understandings: the triumph of the operation is always accompanied by a sort of mystification, the "hand" seems to become increasingly visible and the "application" very quickly forgets the gestures that it mobilizes. (Châtelet 1993/2000,?1)

Placing himself fully in the tradition of Cavaillès, Lautman, Bachelard, and Desanti (and making full use of all of their theoretical mechanisms), Châtelet proposed an epistemology that did not pretend to reduce "axiomatic processes to a set of data and rules for the formation of formal statements," nor did he consider viable the sort of pragmatism that interprets mathematics as the instrument through which physicists operate. For Châtelet, mathematics was a form of thought capable of building the "conceptual horizon" of possible physical virtualities (Castellana 1985,?207).

This process of critical analysis also involved the definition of the status of philosophy and, in particular, of epistemology, which had been largely relegated to a marginal role in the investigation of the foundations. From this point of view, Châtelet criticized how the philosophy of science was often "too timid and stubborn, always seen as the unwanted guest at the banquet of the hard sciences" (Alunni 2018, 322). The philosopher, indeed, often found themself crushed by "two opposite temptations": on the one hand, limiting themself to observing what the sciences do only from afar—"having absolved himself of the responsibility of the patient examination of the concept"—and, on the other, becoming the "humble major domo of scientists, who are assumed to be in possession of the only real 'serious' truths," limiting themself in a decidedly more modest way to carrying out the "role of formal policing" (Châtelet 1993/2000, 2). For this reason, the philosopher "is rightly reproached for mixing up logic and mathematics, for reducing the latter to a formal game and confusing it with the administration of proofs by letting himself become intoxicated with the certainties of syntax" (Châtelet 1993/2000, 1). On the contrary, the philosopher must "wedge" themself into that liminal zone between the "metaphorical" and the "operational" (Châtelet 1993/2000,?2).

As Alunni emphasized, for Châtelet it was necessary to "capture the movements of thought" by focusing on "the dark and inarticulate areas" where the intrinsic creativity of mathematical activity can be brought out—for example, in the close observation of the "practice of surveyors and physicists, in the ways physicists 'talk with their hands,' and in the 'allusions' of surveyors" (Alunni 2018, 322). Epistemology must, therefore, account for mathematics's ability to "articulate" reality,[84] or

[84] Châtelet explicitly adhered to a Platonic model within the philosophy of mathematics. On this point, see Alunni (2018, 330–333). In this, he was linked to other authors, such as Lautman and Gödel (Castellana 2021a). In contrast, Cavaillès showed greater reticence regarding the Platonism of mathematical entities (Cavaillès 1994).

to determine the space of the knowable.[85] Analyzing scientific innovations of the twentieth century, Châtelet underlined how, in this period, the activity of mathematicians was attracted more "by the game of the operators themselves than by the substrate that they should be enquiring into":

> A geometrical language (Hilbert's "spaces") accompanies this play and tends to give it an autonomous contemplative pole and to forge the idea of a *functional figural* ("cone" of functions, spectrum, etc.). It is precisely this autonomy that was granted to the operators that allows them to be recognized as "observables" in quantum mechanics. Paradoxical result: it is the motion itself of the amplifying abstraction of mathematics that governs their incarnation as physical beings: *the more "abstract" mathematics is, the better it works in application.* Thus, to establish a new mathematical physics is to be capable of recognizing to what extent an effort of radical autonomy in mathematics necessarily involves the horizon of the virtualities of physics. (Châtelet 1993/2000, 5–6)

The physical-mathematical universe "involves thought otherwise than would the 'pure and simple' translation of a physical 'content' into a mathematical 'language'" (Châtelet 1993/2000, 6); mathematics has a creative autonomy, bringing into being the space within which physics operates epistemologically.

Although authors such as Brunschvicg, Cavaillès, Lautman, Bachelard, Gonseth, Desanti, and Gilles Châtelet followed philosophical itineraries with widely divergent epistemological positions, I have underlined some lines of continuity that distinguish a certain French way of reading their mathematical thinking. As I have shown, in the context of FHE, a philosophy of mathematics emerged that forcefully claimed the need to conceive mathematical rationality as autonomous and irreducible to other epistemological regions. This perspective has allowed us to highlight both the epistemological specificity of mathematics and the intrinsic creative capacity that sets its practices in motion.[86] This minor canon, for a long time considered a dead end in FHE, is starting to attract considerable interest again among both philosophers and mathematicians[87] by virtue of its ability to open up possible alternatives to the logicist and neo-positivist approaches that, for some time, have dominated the international scene.

---

[85] Even for Châtelet, the reference to "dialectics" became a cardinal point of the argument. On this, he wrote, "Such is our interpretation of dialectical immanence: dialectic is not the synthetic neutralization of two pre-existing and opposed terms, but the discovery of the articulation that opens out the dimension along which they will surge like 'sides'" (Châtelet 1993/2000,?13).

[86] I have limited myself to a general description of what I call the "minor canon" of FHE, which, for various reasons, could be further extended to other authors, such as Hermann Weyl, Gustave Juvet, Alexander Grothendieck, and many others, following what has been proposed for a number of years by Alunni and Castellana, who speak of a "surrationalist school" (Castellana 2004b, 2019, 2020, 2021a, 2021b; Alunni 2018). See also Zalamea (2019).

[87] I refer in particular to the work of Zalamea, who has re-evaluated these epistemological perspectives. See Zalamea (2006/2011, 2013).

References

Abadía, Oscar Moro. 2008. Beyond the Whig history interpretation of history: Lessons on 'Presentism' from Hélène Metzger. *Studies in History and Philosophy of Science Part A* 39 (2): 194–201. https://doi.org/10.1016/j.shpsa.2008.03.005.

———. 2009. Thinking about 'Presentism' from a Historian's perspective: Herbert Butterfield and Hélène Metzger. *History of Science* 47 (1): 55–77. https://doi.org/10.1177/007327530904700104.

———. 2011. Hermeneutical contributions to the history of science: Gadamer on 'Presentism'. *Studies in History and Philosophy of Science Part A* 42 (2): 372–380. https://doi.org/10.1016/j.shpsa.2010.12.003.

Alunni, Charles. 2015. Présentation. *Philosophie et Mathématique* 136 (1–2): 1–8.

———. 2016. Bachelard ancora e ancora. In *Metafisica della matematica*, ed. Gaston Bachelard, Charles Alunni, and Gerardo Ienna, 19–39. Rome: Castelvecchi.

———. 2018. *Spectres de Bachelard. Gaston Bachelard et l'école surrationaliste*. Paris: Hermann.

———. 2021. Préface. Gaston Bachelard, encore et encore. In *Métaphysique des mathématiques*, ed. Charles Alunni and Gerardo Ienna, 7–22. Paris: Hermann.

Alunni, Charles, and Éric Brian. 2002. La mémoire des gestes de science et ses enjeux. *Actes de la recherche en sciences sociales* 141–142 (1): 127–134.

Angelini, Andrea. 2015. Filosofia del concetto e soggettività. Jean Cavaillès tra fenomenologia e dialettica. *Discipline filosofiche* 25 (2): 197–215.

Ariew, Roger, and Peter Barker. 1992. Duhem and continuity in the history of science. *Revue Internationale de Philosophie* 46, 182 (3): 323–343.

Ashplant, T. G., and Adrian Wilson. 1988. Present-centred history and the problem of historical knowledge. *The Historical Journal* 31 (2): 253–274. https://doi.org/10.1017/S0018246X00012875.

Bachelard, Gaston. 1927. *Essai sur la connaissance approchée*. Paris: Vrin.

———. 1929. *La valeur inductive de la relativité*. Paris: Vrin.

———. 1932. *L'Intuition de l'instant. Étude sur la Siloë de Gaston Roupnel*. Paris: Stock. *Intuition of the instant*. Translated by Eileen Rizo-Patron. Evanston: Northwestern University Press, 2013.

———. 1933a. Physique et métaphysique. In *Septimana Spinozana. Acta conventus oecumenici in memoriam Benedicti de Spinoza*. The Hague: Curis Societatis Spinozanae Edita.

———. 1933b. *Métaphysique des mathématiques*, eds. Charles Alunni and Gerardo Ienna. Paris: Hermann, 2021. Original ed. Physique et métaphysique, in *Septimana Spinozana. Acta conventus oecumenici in memoriam Benedicti de Spinoza*. The Hague: Curis Societatis Spinozanae Edita, 1933.

———. 1934. *Le Nouvel Esprit scientifique*. Paris: Alcan. *The new scientific spirit*. Translated by Arthur Goldhammer. Boston: Beacon Press, 1984.

———. 1936a. *La Dialectique de la durée*. Paris: PUF. *The dialectic of duration*. Translated by Mary McAllester Jones. Geneva: Clinamen, 2000.

———. 1936b. Le surrationalisme. *Inquisitions* 1(1):1–6. Surrationalism. In *Surrealism*. Translated by Julien Levy, 186–189. New York: Black Sun Press, 1995.

———. 1938. *La Formation de l'esprit scientifique. Contribution à une psychanalyse de la connaissance objective*. Paris: Vrin. *The formation of the scientific mind: A contribution to a psychoanalysis of objective knowledge*. Translated by Mary McAllester Jones. Manchester: Clinamen Press, 2002.

———. 1940. *La Philosophie du non: essai d'une philosophie du nouvel esprit scientifique*. Paris: PUF. *The philosophy of no: A philosophy of the new scientific mind*. Translated by G. C. Waterston. New York: Orion Press, 1968.

———. 1949. *Le rationalisme appliqué*. Paris: PUF.

———. 1950. De la nature du rationalisme. In *L'engagement rationaliste*, ed. Georges Canguilhem. Paris: PUF. 1975.

———. 1951. *L'Activité rationaliste de la physique contemporaine*. Paris: PUF.
———. 1953. *Le Matérialisme rationnel*. Paris: PUF.
———. 1960. Préface à la deuxième édition. In *De la logique et de la théorie des sciences*, by Jean Cavaillès. Paris: Vrin. 1997.
———. 1970. *Études*. Paris: Vrin.
———. 1972a. Le nouvel esprit scientifique et la création des valeurs rationnelles. In *L'engagement rationaliste*. Paris: PUF.
———. 1972b. L'actualité de l'histoire des sciences. In *L'engagement rationaliste*. Paris: PUF.
———. 1972c. In *L'engagement rationaliste*, ed. Georges Canguilhem. Paris: PUF.
Badino, Massimiliano, Gerardo Ienna, and Pietro D. Omodeo. 2022. *Epistemologia Storica: Correnti, temi e problemi*. Rome: Carocci.
Badino, Massimiliano, and Pietro Daniel Omodeo. 2020. *Cultural hegemony in a scientific world: Gramscian concepts for the history of science*. Leiden: Brill.
Badiou, Alain. 2006. Logiques des mondes. *L'être et l'événement 2*. Paris: Seuil.
———. 2012. *L'aventure de la philosophie française: Depuis les années 1960*. 1st ed. Paris: La Fabrique Editions.
Badiou, Alain, and Georges Canguilhem, dir. 1965. *Philosophie et science*. https://www.youtube.com/watch?v=SKBNSabxhUA.
Badiou, Alain, Georges Canguilhem, Dina Dreyfus, Michel Foucault, and Paul Ricoeur, dir. 1965. *Philosophie et vérité*. https://www.youtube.com/watch?v=v3M0SJ2sJqg.
Badiou, Alain, and Gilles Haéri. 2015. *Eloge des mathématiques*. Paris: Flammarion.
Balibar, Étienne. 1994. Science et vérité dans la philosophie de Georges Canguilhem. In *Lieux et noms de la vérité*, 163-197. Le Château: Éditions de l'Aube. "Being in the True?" Science and truth in the philosophy of Georges Canguilhem. Translated by Andrew Fan. *Décalages* 2(2), 2016. https://scholar.oxy.edu/decalages/vol2/iss2/3.
Barot, Emmanuel. 2009. *Lautman*. Paris: Les Belles Lettres.
Barsotti, Bertrand. 2002. *Bachelard, critique de Husserl, aux racines de la fracture épistémologie/philosophie*. Paris: L'Harmattan.
Benoit, Audrey. 2019a. L'épistémologie historique en héritage. Althusser, Foucault et la fabrique conceptuelle de l'histoire. In *L'épistémologie historique. Histoire et méthodes*, ed. Jean-François Braunstein, Iván Moya Diez, and Matteo Vagelli, 103–116. Paris: Éditions de la Sorbonne.
———. 2019b. *Trouble dans la matière*. Paris: Éditions de la Sorbonne.
Biscuso, Massimiliano, Fabrizio Palombi, and Gaspare Polizzi. 2020a. *(1/2) - Bachelard, entre "le jour et la nuit": oltre l'opposizione tra razionalità e immaginario*. Naples: Accademia IISF. https://www.youtube.com/watch?v=_6ZkENr03TI.
———. 2020b. *(2/2) - Bachelard, entre "le jour et la nuit": oltre l'opposizione tra razionalità e immaginario*. Naples: Accademia IISF. https://www.youtube.com/watch?v=ME4rRYv0OQ4.
Bloch, Marc. 1949. *Apologie pour l'histoire ou métier d'historien*. Paris: Armand Colin. *The Historian's craft*. Translated by Peter Putnam. Manchester: Manchester University Press, 1992.
Bloch, Marc, and Lucien Febvre. 1929. Annales d'histoire économique et sociale 1(1):1–2.
Bloor, David. 1976. *Knowledge and Social Imagery*. Chicago: University of Chicago Press.
Bonicalzi, Francesca. 2007. *Leggere Bachelard: le ragioni del sapere*. Milan: Jaca Book.
Bontemps, Vincent. 2010. *Bachelard*. Paris: Les Belles Lettres.
Bourbaki, Nicholas. 1948. L'architecture des mathématique. In *Les grands courants de la pensée mathématique*, ed. François Le Lionnais, 35–47. Paris: Édition des Cahier du Sud. The architecture of mathematics. Translated by Arnold Dresden, *The American Mathematical Monthly*, 57(4), 1950:221–232.
Bourdieu, Pierre. 2002. Les conditions sociales de la circulation internationale des idées. *Actes de la recherche en sciences sociales* 145 (5): 3–8.
Bourdieu, Pierre, Jean-Claude Chamboredon, and Jean-Claude Passeron. 1968. *Le métier de sociologue*. Paris: Mouton Bordas. *The craft of sociology: Epistemological preliminaries*. Translated by Richard Nice, ed. Beate Krais. Berlin: de Gruyter, 1991.

Braunstein, Jean-François. 2002a. Bachelard, Canguilhem, Foucault. Le “style français” en épistémologie. In *Les philosophes et la science*, ed. Pierre Wagner, 920–963. Paris: Gallimard.

———. 2002b. La philosophie des sciences d’Auguste Comte. In *Les philosophes et la science*, ed. Pierre Wagner, 787–822. Paris: Gallimard.

———. 2008. *L’histoire des sciences. Méthodes, styles et controverses*. Paris: Vrin.

———. 2013. Die Geschichte des Regionsbegriffs in der Epistemologie. In *Situiertes Wissen und regionale Epistemologie: Zur Aktualität Georges Canguilhems und Donna J. Haraways*, eds. Astrid Deuber-Mankowsky and Christoph F. E. Holzhey, 35–49. Vienna: Turia + Kant. 10.37050/ci-07_02.

———. 2015a. Abel Rey et les débuts de l’Institut d’histoire des sciences et des techniques (1932-1940). In *L’épistémologie française, 1830-1970*, ed. Michel Bitbol and Jean Gayon, 163–180. Paris: Éditions Matériologiques.

———. 2015b. *La philosophie de la médecine d’Auguste Comte: Vaches carnivores, Vierge Mère et morts vivants*. Paris: PUF.

———. 2016. Foucault, Canguilhem et l’histoire des sciences humaines. *Archives de Philosophie* 79 (1): 13. https://doi.org/10.3917/aphi.791.0013.

Brenner, Anastasios. 2007. Le statut de l’épistémologie selon Meyerson. *Archives de Philosophie* 70 (3): 375–384.

Brenner, Anastasios, and Jean Gayon, eds. 2009. *French studies in the philosophy of science*. Dordrecht: Springer.

Brunschvicg, Léon. 1912. *Les étapes de la philosophie mathématique*. Paris: Alcan.

———. 1922. *L’experience humaine et la causalité physique*. Paris: PUF.

———. 1951. *Spinoza et ses contemporains*. Paris: PUF.

Burke, Peter. 1990. *The French historical revolution: The Annales School 1929-89*. Stanford: Stanford University Press.

Butterfield, Herbert. 1931. *The Whig interpretation of history*. New York: W. W. Norton & Company.

———. 1957. *The origins of modern science 1300-1800*. London: G. Bell and Sons Ltd.

Cammelli, Michele. 2011. Présentation. Le fascisme et les paysans (1935). In *Oeuvres Complètes, 1, 1926-1939*, by Georges Canguilhem, 515-534. Paris: Vrin.

Canguilhem, Georges. 1947. Note sur la situation faite en France à la philosophie biologique. *Revue de Métaphysique et de Morale* 52 (3/4): 322–332.

———. 1952. *La connaissance de la vie*. Paris: Vrin. *Knowledge of life*. Translated by Stefanos Geroulanos and Daniela Ginsburg, eds. Paola Marrati and Todd Meyers. New York: Fordham University Press, 2008.

———. 1955. *La formation du concept de réflexe aux XVIIe et XVIIIe siècles*. Paris: PUF.

———. 1957. Sur une épistémologie concordataire. Hommage à Bachelard. In *Hommage à Gaston Bachelard*, ed. Georges Bouligand et al., 4–12. Paris: PUF.

———. 1966. *Le normal et le pathologique*. Paris: PUF. *The normal and the pathological*. Translated by Carolyn R. Fawcett. New York: Zone Books, 1991.

———. 1970. Objectivité et historicité. In *Structuralisme et marxisme*, ed. Jean-Maria Auzias et al. Paris: Union Générale d’Éditions.

———. 1977. *Idéologie et rationalité dans l’histoire des sciences de la vie*. Paris: Vrin. *Ideology and rationality in the history of the life sciences*. Translated by Arthur Goldhammer. Cambridge, MA: MIT Press, 1988.

———. 1994a. *Études d’histoire et de philosophie des sciences concernant les vivants et la vie*. Paris: Vrin.

———. 1994b. L’objet de l’histoire des sciences. In *Études d’histoire et de philosophie des sciences concernant les vivants et la vie*, 9–24. Paris: Vrin. The object of the history of sciences. Translated by Mary Tiles. In *Continental Philosophy of Science*, ed. Gary Gutting, 198-206. Malden: Blackwell, 2005.

———. 1996. *Vie et mort de Jean Cavaillès*. Paris: Alliage.

———. 2018. Gaston Bachelard. In *Oeuvres complètes. Volume V, Histoire des sciences, épistémologie, commémorations (1966-1995)*, 623–629. Paris: Vrin.

Castellana, Mario. 1974. *Il surrazionalismo di Gaston Bachelard*. Naples: Glaux.

———. 1985. Epistemologia debole. In *Bachelard, Desanti, Raymond*. Verona: Bertani.

———. 2004a. Gaston Bachelard e il concetto di filosofia spontanea. In *Ri-cominciare. Percorsi e attualità dell'opera di Gaston Bachelard*, ed. Francesca Bonicalzi and Carlo Vinti. Milan: Jaca Book.

———. 2004b. *Razionalismi senza dogmi*. Soveria Mannelli: Rubettino.

———, ed. 2009. Introduzione. Storia della Scienza, Epistemologia ed Ermeneutica in Hélène Metzger. In *Il metodo filosofico nella storia delle scienze: testi 1914–1939*, by Hélène Metzger, 23–55. Manduria: Barbieri Selvaggi Editori.

———. 2019. *Federigo Enriques e la Nuova Epistemologia*. Lecce: Pensa MultiMedia.

———, ed. 2020. Il contributo di Maximilien Winter alla critique des sciences. In *Il metodo storico-critico per una nuova filosofia delle matematiche*, 9–85. Milan: Meltemi.

———. 2021a. *Il dibattito in area francofona sul pensiero matematico e Kurt Gödel*. Rome: Edizioni Studium.

———. 2021b. In *Il surrazionalismo di Gaston Bachelard con due saggi inediti*, ed. Pietro Console. Lecce: Milella.

Castelli Gattinara, Enrico. 1996. *Epistemologia e storia. Un pensiero all'apertura nella Francia fra le due guerre mondiali*. Milan: FrancoAngeli.

Cavaillès, Jean. 1976. *Sur la logique e la théorie de la science*. Paris: Vrin. *On logic and the theory of science*. Translated by Knox Peden and Robin Mackay. Falmouth: Urbanomic/Sequence Press, 2021.

———. 1994. *Oeuvres complètes de philosophie des sciences*. Paris: Hermann.

Cavaillès, Jean, and Albert Lautman. 1946. La pensée mathématique. *Bulletin de la Société Française de Philosophie* 40 (1): 1–39.

Cavazzini, Andrea. 2010. Gesti, forme, idee. Gilles Châtelet tra epistemologia e Naturphilosophie. In *Le poste in gioco del mobile*, by Gilles Châtelet, 11-36. Milan: Mimesis.

Chang, Hasok. 2009. We have never been Whiggish (About Phlogiston). *Centaurus* 51 (4): 239–264. https://doi.org/10.1111/j.1600-0498.2009.00150.x.

———. 2021. Presentist History for Pluralist Science. *Journal for General Philosophy of Science* 52 (1): 97–114. https://doi.org/10.1007/s10838-020-09512-8.

Châtelet, Gilles. 1993. *Les Enjeux du mobile. Mathématique, physique, philosophie*. Paris: Seuil. *Figuring space: Philosophy, mathematics and physics*. Translated by Robert Shore and Muriel Zagha. Dordrecht: Kluwer, 2000.

———. 1998. *Vivre et penser comme des porcs. De l'incitation à l'envie et à l'ennui dans les démocraties-marchés*. Paris: Exils. *To live and think like pigs: The incitement of envy and boredom in market democracies*. Translated by Robin Mackay. Falmouth: Urbanomic, 2014.

———. 2010a. In *L'enchantement du virtuel: mathématique, physique, philosophie*, ed. Charles Alunni and Catherine Paoletti. Paris: Éditions de la Rue D'Ulm.

———. 2010b. In *Les animaux malades du consensus*, ed. Catherine Paoletti. Paris: Lignes.

Chimisso, Cristina. 2003. The tribunal of philosophy and its norms: History and philosophy in Georges Canguilhem's historical epistemology. *Studies in History and Philosophy of Science Part C: Studies in History and Philosophy of Biological and Biomedical Sciences* 34 (2): 297–327. https://doi.org/10.1016/S1369-8486(03)00027-X.

———. 2019. *Hélène Metzger, historian and historiographer of the sciences*. London/New York: Routledge.

Clavelin, Maurice. 1987. Le débat Koyré Duhem, hier et aujourd'hui. *History and Technology* 4 (1–4): 13–35. https://doi.org/10.1080/07341518708581687.

Comte, Auguste. 1830. *Cours de philosophie positive. Tome 1*. Brussels: Rouen Frères Libraries-Éditeurs. *The positive philosophy of Auguste Comte*, vol. 1. Translated by Harriet Martineau. Cambridge: Cambridge University Press, 2009.

———. 1851–1881. *Système de politique positive ou Traité de sociologie, instituant la religion de l'humanité.* 1967 ed. Osnabrück: Otto Zeller. https://gallica.bnf.fr/ark:/12148/bpt6k5449v. *System of positive polity, or treatise on sociology, instituting the religion of humanity.* Translated by John Henry Bridges. Paris: Mathias, Carilian-Goeury, and Dalmont, 1851.

Conry, Yvette. 1983. Combats pour l'histoire des sciences: lettre ouverte aux historiens des mentalités. *Revue de synthèse* 111–112:363–406.

Cortois, Paul. 1994. Quelques aspects du programme épistémologique de Cavaillès. *Dialectica* 48 (2): 125–141.

Costa, Vincenzo. 2009. *Husserl.* Rome: Carocci.

Croce, Benedetto. 2007. *Filosofia come scienza dello spirito. Teoria e storia della storiografia*, eds. Edoardo Massimilia, Teodoro Tagliaferri, and Fulvio Tessitore. Naples: Bibliopolis. *Theory and history of historiography.* Translated by Douglas Ainslie. London: George G. Harrap, 1921.

Crombie, Alistair C. 1987. Alexandre Koyré and Great Britain: Galileo and Mersenne. *History and Technology* 4 (1–4): 81–92. https://doi.org/10.1080/07341518708581691.

Dagognet, François. 1996. A regional epistemology with possibilities for expansion. *Science in Context* 9 (1): 5–14.

D'Aurizio, Claudio, and Fabrizio Palombi. 2022. Il "ritmo delle idee" di Gaston Bachelard. *Note su La dialettica della durata. Iride* 35 (1): 149–158.

Debru, Claude. 2004. *Georges Canguilhem, science et non-science.* Paris: Presses de l'école normale supérieure.

Desanti, Jean-Toussaint. 1968. *Les idéalités mathématiques. Recherches épistémologiques sur le développement de la théorie des fonctions de variables réelles.* Paris: Seuil.

———. 1972. Préface. In *Les étapes de la philosophie mathématique*, by Léon Brunschvicg, i-ix. Paris: Blanchard.

———. 1974. En quoi l'histoire des mathématiques peut-elle être utile au mathématicien. *Raison présente* 31 (1): 41–49.

———. 1975. *La philosophie silencieuse; ou critique des philosophies de la science.* Paris: Seuil.

Diogo, Maria Paula. 2020. The perfect Pair: Bloch, Febvre, and the history of science and technology. *HoST - Journal of History of Science and Technology* 14 (2): 73–93. https://doi.org/10.2478/host-2020-0015.

Duhem, Pierre. 1905. *Les origines de la statique.* Paris: Hermann. *The origins of statics.* Translated by Grant F. Leneaux, Victor N. Vagliente, and Guy H. Wagener. Dordrecht: Kluwer, 1991.

———. 1906. *La théorie physique son objet et sa structure.* Paris: Chevalier & Rivière. *The aim and structure of physical theory.* Translated by Philip P. Weiner. Princeton: Princeton University Press, 1982.

Elton, Geoffrey R. 1984. Herbert butterfield and the study of history. *The Historical Journal* 27 (3): 729–743. https://doi.org/10.1017/S0018246X00018057.

Febvre, Lucien. 1942. *Le problème de l'incroyance au XVIe siècle. La religión de Rabelais.* Paris: Albin Michel. *The problem of unbelief in the sixteenth century: The religion of Rabelais.* Translated by Beatrice Gottlieb. Cambridge: Harvard University Press, 1982.

———. 1992. De 1892 à 1933. Examen de conscience d'une histoire et d'un historien. Leçon d'ouverture au Collège de France, 13 décembre 1933. In *Combats pour l'histoire*, 3–17. Paris: Armand Colin.

Ferrières, Gabrielle. 1950. *Jean Cavaillès, philosophe et combattant.* Paris: PUF.

Fichant, Michel, and Michel Pêcheux. 1969. *Sur l'histoire des sciences.* Paris: Maspero.

Fleck, Christian, Matthias Duller, and Victor Karády. 2019. *Shaping human science disciplines: Institutional developments in Europe and beyond.* Cham: Palgrave Macmillan.

Fleck, Ludwik. 1935. *Entstehung und Entwicklung einer wissenschaftlichen Tatsache.* Frankfurt am Main: Suhrkamp. *Genesis and development of a scientific fact.* Translated by Fred Bradley and Thaddeus J. Trenn, ed. Thaddeus J. Trenn and Robert K. Merton. Chicago: University of Chicago Press, 1979.

Foucault, Michel. 1979. *L'ordre du discours.* Paris: Gallimard.

———. 1985. La vie: L'experience et la science. *Revue de méthaphysique et de morale* 90 (January-March): 3–14. Life: Experience and Science. In *Aesthetics, method, and epistemology: Essential works of Foucault 1954-1984*, vol. 2. Translated by Robert Hurley, ed. James D. Faubion, 465–478. New York: New Press. 2003.

———. 1994. *Dits et écrits 1 1954-1969*. Paris: Gallimard.

Freudenthal, Gad, ed. 1989. *Études sur/Studies on Hélène Metzger*. Leiden: Brill.

Galison, Peter Louis, and David J. Stump. 1996. *The disunity of science: Boundaries, contexts, and power*. Stanford, CA: Stanford University Press.

Gaukroger, S. W. 1976. Bachelard and the problem of epistemological analysis. *Studies in History and Philosophy of Science Part A* 7 (3): 189–244.

Gayon, Jean. 2003. Bachelard et l'histoire des sciences. In *Bachelard et l'épistémologie française*, ed. Jean-Jacques Wunenburger, 51–113. Paris: PUF.

Gonseth, Ferdinand, ed. 1941. *Les entretiens de Zurich sur les fondements et la méthode des sciences mathématiques, 6-9 décembre 1938: exposés et discussions*. Zurich: S.A. Leeman fréres & cie.

Gramsci, Antonio. 1977. *Quaderni del carcere. 2 (VI-XI)*. Ed. Valentino Gerratana. Turin: Einaudi. *Further selections from the prison notebooks*. Translated by Derek Boothman. Minneapolis: University of Minnesota Press, 1995.

Grange, Juliette. 1996. Présentation. In *Philosophie des sciences*, by Auguste Comte. Paris: Gallimard.

Guitart, René. 2015. Bachelard et la pulsation mathématique. *Revue de synthèse* 136 (1): 33–74. https://doi.org/10.1007/s11873-014-0265-8.

Gutting, Gary. 2001. *French philosophy in the twentieth century*. Cambridge, MA: Cambridge University Press.

Hall, Alfred Rupert. 1983. On Whiggism. *History of Science* 21:45–59.

Han, Béatrice. 1998. *L'ontologie manquée de Michel Foucault: entre l'historique et le transcendantal*. Grenoble: Editions Jérôme Millon.

Hartog, François. 2015. Regimes of historicity: Presentism and experiences of time. Translated by Saskia Brown. New York: Columbia University Press.

Heilbron, Johan, Gustavo Sorá, and Thibaud Boncourt, eds. 2018. *The social and human sciences in global power relations*. Cham: Springer. https://doi.org/10.1007/978-3-319-73299-2.

Ienna, Gerardo. 2017. Materialismo tecnico. Fra fenomenotecnica ed epistemologie regionali. In *Il senso della tecnica*, ed. Saggisu Bachelard, P. Donatiello, F. Galofaro, and G. Ienna, 21–36. Bologna: Esculapio.

———. 2019a. *Natura constructa* et phénoménotechnique. Spinozisme et pensée des mathématiques chez Gaston Bachelard. In *L'épistémologie historique: Histoire et méthodes*, 43–58. Paris: Éditions de la Sorbonne.

———. 2019b. Science and Technology Studies. Socio-epistemologia storica delle negoziazioni disciplinari. Doctoral thesis. Bologna: Alma Mater Studiorum Università di Bologna.

———. 2021. Métaphysique des mathématiques. Bachelard et le spinozisme mathématisant. Translated by Charles Alunni. In *Métaphysique des mathématiques*, by Gaston Bachelard, eds. Charles Alunni and Gerardo Ienna, 23–48. Paris: Hermann.

———. 2022. The international and interdisciplinary circulation of Boris Hessen's Theses. In *Manuscripts and documents on the history of physics: A historical materialist textbook*, ed. Pietro D. Omodeo and Sean Winkler, 75–130. Venice: Verum Factum.

Ienna, Gerardo, and Giulia Rispoli. 2017. Boris Hessen al bivio tra scienza e ideologia. In *Le radici sociali ed economiche della meccanica di Newton*, ed. B. Hessen, 5–43. Rome: Castelvecchi.

———. 2019. Boris Hessen at the crossroads of science and ideology. From international circulation to the soviet context. *Societate Si Politica* 13 (1): 37–63.

Ienna, Gerardo, and Marco Santoro. 2016. Per una scienza delle relazioni internazionali in campo culturale. *Studi culturali* 13 (1): 61–69.

Jacob, Pierre. 1991. Il regionalismo epistemologico: una tendenza della filosofia contemporanea delle scienze in Francia. *Rivista di filosofia* 82 (2): 279–300.

Jacumin, Renato. 1967. *Jean Cavaillès: alla ricerca di una fondazione dell'operare matematico*. Udine: Dal Bianco.
Jardine, Nick. 2000. Uses and abuses of anachronism in the history of the sciences. *History of Science* 38 (3): 251–270. https://doi.org/10.1177/007327530003800301.
———. 2003. Whigs and stories: Herbert butterfield and the historiography of science. *History of Science* 41 (2): 125–140. https://doi.org/10.1177/007327530304100201.
Jorland, Gerard. 1981. *La science dans la philosophie. Les recherches épistémologiques d'Alexandre Koyré*. Paris: Gallimard.
Koyré, Alexandre. 1935. Introduction. In *Des révolutions des orbes célestes. (Du livre I, chapitres 1-11)*, by Nicolas Copernic, 1-24. Paris: Librairie Félix Alcan.
———. 1939. *Études galiléennes*. Paris: Hermann. *Galileo studies*. Translated by John Mepham. New York: Humanities Press, 1978.
———. 1950. Du monde de l' "à-peu-près" à l'univers de la precisión. In *Etudes d'histoire de la pensée philosophique*, 311–329. Paris: Armand Colin, 1961.
———. 1957. *From the closed world to the infinite universe*. Baltimore: Johns Hopkins University Press.
———. 1961. *Etudes d'histoire de la pensée philosophique*. Paris: Armand Colin.
———. 1966. *Études d'histoire de la pensée scientifique*. Paris: PUF.
———. 2000. Dal mondo del pressappoco all'universo della precisione. Translated by Paola Zambelli. Turin: Einaudi.
Kuhn, Thomas S. 1977. *The essential tension. Selected studies in scientific tradition and change*. University of Chicago Press.
———. 2000. *The road since structure. Philosophical essays, 1970-1993, with an autobiographical interview*. Chicago: University of Chicago Press.
Latour, Bruno. 1987. *Science in action: How to follow scientists and engineers through society*. Cambridge, MA: Harvard University Press.
Latour, Bruno, and Geoff Bowker. 1987. A booming discipline short of discipline: (Social) studies of science in France. *Social Studies of Science* 17 (4): 715–748.
Latour, Bruno, and Steve Woolgar. 1979. *Laboratory life*. 2nd ed. Princeton: Princeton University Press.
———. 2006. *La vie de laboratoire*. Paris: La Découverte.
Laugier, Sandra. 2015. Duhem, Meyerson et l'épistémologie américaine postpositiviste. In *L'épistémologie française, 1830-1970*, ed. Michel Bitbol and Jean Gayon, 67–90. Paris: Éditions Matériologiques.
Lautman, Albert. 1939. *Nouvelles recherches sur la structure dialectique des mathématiques*. Paris: Hermann. *New Research on the Dialectical Structure of Mathematics*. In *Mathematics, ideas and the physical real*. Translated by Simon B. Duffy, ed. Fernando Zalamea, 197-218. London/New York: Continuum, 2011.
———. 2006. *Les mathématiques, les idées et le réel physique*. Paris: Vrin. *Mathematics, ideas and the physical real*. Translated by Simon B. Duffy, ed. Fernando Zalamea. London/New York: Continuum, 2011.
le Blanc, Guillaume. 2008. *Canguilhem et les normes*. Paris: PUF.
Lecourt, Dominique. 1972. *Pour une critique de l'épistémologie: Bachelard, Canguilhem, Foucault*. Paris: Maspero. *For a Critique of Epistemology*. In *Marxism and epistemology: Bachelard, Canguilhem, Foucault*, Translated by Ben Brewster, 117-213. London: New Left Books, 1975.
———. 1974. *Bachelard ou le jour et la nuit*. Paris: Grasset.
———. 2001. *La Philosophie des sciences*. Paris: PUF.
———. 2008. *Georges Canguilhem*. Paris: PUF.
Les, Annales. 2020. Face au présent. Politique des temporalités. *Annales. Histoire, Sciences Sociales* 75 (3–4): 493–516. https://doi.org/10.1017/ahss.2020.127.
Limoges, Camille. 2012. L'épistémologie historique dans l'itinéraire intellectuel de Georges Canguilhem. In *Epistemology and history from bachelard and Canguilhem to today's history*

of science, ed. Henning Schmidgen et al., 53–66. Berlin: Max Planck Institute for the History of Science.

———. 2015. Introduction. Philosophie Biologique, histoire des sciences et interventions philosophiques. Georges Canguilhem 1940-1965. In *Oeuvres complètes. Tome IV*, ed. Georges Canguilhem, 7–48. Paris: Vrin.

Loison, Laurent. 2016. Forms of Presentism in the history of science. rethinking the project of historical epistemology. *Studies in History and Philosophy of Science Part A* 60:29–37. https://doi.org/10.1016/j.shpsa.2016.09.002.

Méthot, Pierre-Olivier. 2013. On the genealogy of concepts and experimental practices: Rethinking Georges Canguilhem's historical epistemology. *Studies in History and Philosophy of Science Part A* 44 (1): 112–123. https://doi.org/10.1016/j.shpsa.2012.06.002.

Metzger, Hélène. 1918. *La genèse de la science des cristaux*. Paris: Albert Blachard.

———. 1926. *Les concepts scientifiques*. Paris: Alcan.

———. 1934 *Review of Signification de l'histoire de la pensée scientifique*, by Federigo Enriques. Archeion, 16:390–391.

———. 1935. Tribunal de l'histoire et théorie de la connaissance scientifique. *Archeion* 17:1–14. The tribunal of history and the theory of scientific knowledge. Translated by Cristina Chimisso. In *Hélène Metzger, historian and historiographer of the sciences*, by Cristina Chimisso, 181–190. London/New York: Routledge, 2019.

———. 1938. In *Attraction universelle et religion naturelle chez quelques commentateurs anglais de Newton*, ed. Mario Castellana and Enrico Giannetto, 2021. Paris: Hermann.

———. 1987. La Méthode philosophique en histoire des sciences. Textes 1914-1939, ed. Gad Freudenthal. Paris: Fayard.

Meyerson, Émile. 1908. *Identité et realité*. Paris: Alcan. *Identity and reality*. Translated by Kate Lowenberg. London: George Allen & Unwin, Ltd. 1930.

Omodeo, Pietro Daniel. 2018. Soggettività, strutture, egemonie. Questioni politico-culturali in epistemologia storica. *Studi culturali* 15 (2): 211–234.

———. 2019. *Political epistemology. The problem of ideology in science studies*. Cham: Springer.

———. 2020. The struggle for objectivity: Gramsci's historical-political vistas on science against the background of Lenin's epistemology. *HoST - Journal of History of Science and Technology* 14 (2): 13–49.

Palombi, Fabrizio. 2017. *Elogio dell'astrazione. Gaston Bachelard e la filosofia della matematica*. Milan: Mimesis.

Pariente, Jean-Claude. 2015. Rationalisme et ontologie chez Gaston Bachelard. In *L'épistémologie française, 1830-1970*, ed. Michel Bitbol and Jean Gayon, 235–263. Paris: Éditions Matériologiques.

Peden, Knox. 2014. *Spinoza Contra phenomenology. French rationalism from Cavaillès to Deleuze*. Stanford, CA: Stanford University Press.

Petit, Annie. 1994. Genèse de la classification des sciences d'Auguste Comte. *Revue de synthèse* 115 (1–2): 71–102. https://doi.org/10.1007/BF03182484.

Pickstone, John V. 1995. Past and present knowledges in the practice of the history of science. *History of Science* 33 (2): 203–224. https://doi.org/10.1177/007327539503300203.

Redondi, Pietro. 1978. *Epistemologia e storia della scienza: le svolte teoriche da Duhem a Bachelard*. Milan: Feltrinelli.

Revue de Métaphysique et de Morale. 1893. Introduction. *Revue de Métaphysique et de Morale* 1 (1): 1-5.

Rey, Abel. 1909. Vers le positivisme absolu. *Revue philosophique de la France et de l'étranger* 67: 461–479.

Rheinberger, Hans-Jörg. 2005. Gaston Bachelard and the Notion of "Phenomenotechnique". *Perspectives on Science* 13 (3): 313–328.

———. 2010. *On historicizing epistemology*. Stanford, CA: Stanford University Press.

Sapiro, Gisèle, Marco Santoro, and Patrick Baert, eds. 2020. *Ideas on the move in the social sciences and humanities: The international circulation of paradigms and theorists*. Cham: Palgrave Macmillan.

Sfara, Emiliano. 2017. Il Sud e il Nord. I tre assiomi nella filosofia di Bachelard secondo Canguilhem. In *Il senso della tecnica. Saggi su Bachelard*, ed. Paola Donatiello, Francesco Galofaro, and Gerardo Ienna, 95–108. Bologna: Esculapio.

Shapin, Steven, and Simon Schaffer. 1985. Leviathan and the air-pump: Hobbes, boyle, and the experimental life. 2011 ed. Princeton, NJ: Princeton University Press.

Simons, Massimiliano. 2018. The Janus Head of Bachelard's *Phénoménotechnique*: From purification to proliferation and back. *European Journal for Philosophy of Science* 8:689–707.

———. 2019a. Obligation to Judge or Judging obligations: The integration of philosophy and science in francophone philosophy of science. In *The past, present, and future of integrated history and philosophy of science*, ed. Emily Herring, Kevin Matthew Jones, Konstantin S. Kiprijanov, and Laura M. Sellers, 13–130. New York: Routledge. https://doi.org/10.4324/9781351214827.

———. 2019b. Surrationalism after Bachelard: Michel Serres and *Le nouveau nouvel esprit scientifique*. *Parrhesia* 31:60–84.

Snow, Charles Percy. 1959. *The two cultures*. Cambridge, UK: Cambridge University Press.

Spoerhase, Carlos. 2008. Presentism and precursorship in intellectual history. *Culture, Theory and Critique* 49 (1): 49–72. https://doi.org/10.1080/14735780802024257.

Taylan, Ferhat. 2018. *Concepts et rationalités. Héritages de l'épistémologie historique, de Meyerson à Foucault*. Paris: Éditions Matériologiques.

Vagelli, Matteo, Laurent Loison, and Ivan Moya-Diez. 2021. Thinking crossroads: From scientific pluralism to pluralist history of science. *Journal for General Philosophy of Science* 52 (1): 87–95. https://doi.org/10.1007/s10838-020-09550-2.

Vinti, Carlo. 1977. *L'epistemologia francese contemporanea. Per un razionalismo aperto*. Rome: Città Nuova Editrice.

Wilson, Adrian, and Timoty G. Ashplant. 1988. Whig history and present-centred history. *The Historical Journal* 31 (1): 1–16. https://doi.org/10.1017/S0018246X00011961.

Winter, Maximilien. 1911. *La méthode dans la philosophie des mathématiques*. Paris: Alcan.

Wolfe, Charles T. 1999. Gilles Châtelet. Vivre et penser comme des porcs. *Chimères* 37:83–86.

———. 2015. Was Canguilhem a Biochauvinist? Goldstein, Canguilhem and the Project of Biophilosophy. In *Medicine and society, new perspectives in continental philosophy*, ed. Darian Meacham, 197–212. Dordrecht: Springer Netherlands. https://doi.org/10.1007/978-94-017-9870-9_12.

———. 2019. *La philosophie de la biologie avant la biologie*. Paris: Classiques Garnier.

———. 2024. A note on the situation of biological philosophy. *Revue internationale de philosophie* 307 (1): 95–110.

Zalamea, Fernando. 2006. Albert Lautman et la dialectique créatrice des mathématiques modernes. In *Les mathématiques, les idées et le réel physique*, 17–33. Vrin. Albert Lautman and the Creative Dialectic of Modern Mathematics. In *Mathematics, Ideas and the physical real*. Translated by Simon B. Duffy, ed. Fernando Zalamea, xxiii–xxxvii. New York, Continuum, 2011.

———. 2013. Synthetic philosophy of contemporary mathematics, Translated by Zachary Luke Fraser. Falmouth/New York: Urbanomic Sequence Press.

———. 2016. La resistenza sintetica di Lautman, Translated by Massimo De Pascale. In *La matematica come resistenza*, by Albert Lautman, ed. Mario Castellana, 87–90. Rome: Castelvecchi.

———. 2018. *Philosophie synthétique de la mathématique contemporaine*. Paris: Hermann.

———. 2019. *Grothendieck. Una guía a la obra matemática y filosófica*. Bogotà: Universidad Nacional–Editorial Nomos.

Chapter 3
The Socialization and Politicization of *Épistémologie Historique*

Abstract In addition to historical epistemology's focus on the fragmentation of scientific reason, both on the historical-diachronic level and in its regional pluralization, it foregrounds the intrinsically social and political dimensions of knowledge. Indeed, the existence of a third pole of mediation between the natural referent—that is, the research object—and the activity of the knowing subject is central to historical epistemology. However, what is controversial is how this mediation should be characterized. As the previous chapters show, historicizing scientific facts necessarily implies dealing with the cultural and social components that impact the processes of constituting and developing scientific knowledge. The first section reconstructs the reasons why authors who were affiliated with French historical epistemology (FHE) were been directly involved in the international debate on "internalism" and "externalism" (except for Koyré). I show how Meyerson, Metzger, Bachelard, and Canguilhem, while taking different positions, integrated social and anthropological perspectives into their theoretical and methodological systems. The second section shows how the legacy of Bachelard-Canguilhemism developed different directions with authors such as Althusser, Foucault, and Bourdieu. In particular, I describe how Canguilhem imparted a significant twist or bend in the development of this tradition by shifting the axis of interest from the physical-mathematical sciences toward the biological and life sciences. This process opened FHE to the human and social sciences, which had not been included previously. I also detail how, starting from common historical-epistemological references, Althusser, Foucault, and Bourdieu pursued three trajectories in their analyses of scientific knowledge, though they were marked by a common tendency to focus on the links between power and knowledge and between science and society.

In addition to historical epistemology's focus on the fragmentation of scientific reason, both on the historical-diachronic level and in its regional pluralization, it foregrounds the intrinsically social and political dimension of knowledge. Indeed, the existence of a third pole of mediation between the natural referent—that is, the research object—and the activity of the knowing subject is central to historical epistemology (Badino et al. 2022). However, what is controversial is how this mediation should be characterized. As can be observed from the previous sections,

G. Ienna, *Genesis and Development of French Historical Epistemology*, Studies in History and Philosophy of Science 64, https://doi.org/10.1007/978-3-031-88699-7_3

historicizing scientific facts necessarily implies dealing with the cultural and social components that impact the processes of constituting and developing scientific knowledge.

In the following, I demonstrate how authors belonging to the French historical epistemology (FHE) tradition addressed this problem. Starting in the 1930s, the international debate on the social aspects of scientific activity was marked by the controversy between "internalism" and "externalism." The only author from the French cultural context who directly and actively participated in this dispute was Koyré, who established himself, among other things, as one of the leading advocates of internalism. In the first section of this chapter, I show how, beyond this exception, the French cultural landscape remained rather indifferent to this debate for reasons intrinsically linked to the theoretical stance developed by FHE. Authors such as Meyerson, Metzger, Bachelard, and Canguilhem, despite taking different positions, integrated social and anthropological perspectives into their theoretical and methodological systems. Their epistemological positions, however, could hardly be traced back to either of the positions crystallized in the rhetorical opposition between "internalism" and "externalism" (as sketched out by internalist scholars).[1] French authors, in fact, developed a particular interest in the anthropological, social, and political aspects of knowledge without renouncing an internal analysis of the historical transformations of scientific content.[2] In this way, FHE inspired some of the theoretical and rhetorical strategies used in the 1970s by authors such as Kuhn to overcome this controversy. I also emphasize how the perspectives developed within this tradition were linked, in various cases, to the history of the social and human sciences (with particular reference to authors such as Comte, Durkheim, Lévy-Bruhl, and others).

In the second section, I show how the legacy of Bachelardo-Canguilhemism was developed in different directions by authors such as Althusser, Foucault, and Bourdieu. In particular, I describe how Canguilhem imparted a significant twist or bend in the development of this tradition by shifting the axis of interest from the physical-mathematical sciences toward the biological and life sciences. This process created the possibility of opening FHE up to the human and social sciences, which had not been included previously. I also detail how, starting from common historical-epistemological references, Althusser, Foucault, and Bourdieu pursued three trajectories in their analyses of scientific knowledge, though they were marked by a common tendency to focus on the links between power and knowledge, between science and society.

[1] Indeed, supporters of "internalism" provoked this debate by labeling as "externalist" the positions of various historians of science who, starting from a Marxist theoretical orientation, had underlined the need to analyze science as a social fact and thus pay attention to the social, economic, and ideological roots of knowledge.

[2] A brief look through Hessen's classic text, "The Social and Economic Roots of Newton's 'Principia,'" clearly shows that externalism does not imply an absence of internal content analysis, given the careful reading of Newton's *Principia* that he provided (Hessen 1931). See also the 2009 edition of Hessen's text included in Freudenthal and McLaughlin (2009).

3.1 Anthropological-Social Perspectives in Historical Epistemology

In this context, Comte's influence on French thought is as unavoidable as Hegel's on German culture. Already in 1819, the founder of positivism had written to his friend Valat: "My works are and will be of two orders: scientific and political" (Comte 1870, 99). Without taking this intrinsic duality of Comte's work into consideration, the subsequent developments of FHE and the French social sciences would be impossible to understand. Indeed, Comte's objective for the history of science—as well as its classification and organization—was to found a new science of social facts called "sociology." This new discipline, based on scientific evidence, would inform a "positive"—or scientific—policy that would allow society to be reorganized to eliminate its recurring crises.[3] From the first work that Comte published, titled *Plan des travaux scientifiques nécessaires pour réorganiser la société* (1822), this thematic nucleus was at the center of his thought. This initial intuition would then be further developed in the two monumental works that make up Comte's corpus: *Cours de philosophie positive* and the *System of Positive Polity*. Within this perspective, science would establish rational and effective social and political action.

Beyond the Comtian foundations of French epistemology, various twentieth-century authors developed epistemological positions capable of integrating anthropological-social elements. Before highlighting these aspects, however, I believe that I must briefly explore some contextual elements regarding certain debates that, at that moment, were prominent in other national contexts and without which understanding certain specificities of FHE would be difficult.

In the 1930s, the international controversy between "internalism" and "externalism" erupted at the Second International Congress of the History of Science held in London in 1931 in the wake of the presentations of the Russian delegation, led by Nikolai Bukharin (Bukharin 1971; Ienna and Rispoli 2019, 2021). The contributions of the Russian delegates focused on the application of the Marxist method to the history of science. Their main ambition was to highlight how social, economic, and political factors influenced the transformation processes of scientific knowledge. In particular, the text presented by Boris Hessen titled "The Social and Economic Roots of Newton's 'Principia'" was universally recognized as the "symbolic beginning" of this debate (Hessen 1931; Shapin 1992; Ienna 2022c).[4] In the same period, many authors, including Edgar Zilsel, Henryk Grossmann, Franz Borkenau, Joseph

[3] This perspective led to the consolidation of the positivist program in French sociology with authors such as Durkheim, Marcel Mauss, and Maurice Halbwachs. In particular, Durkheim based *The Rules of the Sociological Method* on the need to "consider social facts as things" to provide a scientific knowledge of society that would be capable of resolving its problems.

[4] Hessen's central argument was that the emergence of the innovations of Newtonian science was intrinsically linked to the material needs and cultural climate of seventeenth-century England. By analyzing the structure of the *Principia*, Hessen showed how the physics problems faced by Newton were connected to the needs of the emerging English bourgeois class (technical problems related to production, trade, maritime navigation, war, etc.) (Hessen 1931; Ienna 2022c).

Needham, and John Desmond Bernal, despite coming from quite different intellectual contexts, contributed to the development of an externalist approach to the history of science (Werskey 1988; Ienna 2022c). A common feature of these perspectives was the deconstruction of the tradition that treated the history of science as a celebration of great scientific minds. The rhetoric of the scientific genius sent by God or suddenly appearing out of nowhere was dismantled through the adoption of tools of sociological analysis capable of revealing the hidden dynamics of knowledge production. In addition, the communal basis of scientific activity was foregrounded, placing a strong emphasis on the collective dimension of scientific and technological enterprises. The Marxist tradition thus focused mainly on studying the social and economic conditions of the emergence of science and technology, while also giving some attention to their cultural aspects.

The internalists opposed the use of a Marxist interpretative framework for the history of science. In their view, scientific development was carried forward by exclusively intellectual factors, and they refused to accept the idea of the social, political, or economic conditioning of knowledge-production processes. In a famous passage in *Newtonian Studies*, Koyré argued for the total rejection of the entire externalist program:

> The new science, we are told sometimes, is the science of the craftsman and the engineer, of the working, enterprising, and calculating tradesman, in fact, the science of the rising bourgeois classes of modern society.
>
> There is certainly some truth in these descriptions and explanations: it is clear that the growth of modern science presupposes that of the cities, it is obvious that the development of firearms, especially of artillery, drew attention to problems of ballistics; that navigation, especially that to America and India, furthered the building of clocks, and so forth—yet I must confess that I am not satisfied with them. I do not see what the *scientia activa* has ever had to do with the development of the calculus, nor the rise of the bourgeoisie with that of the Copernican, or the Keplerian, astronomy. (Koyré 1965, 5–6)

From *Galileo Studies* to his posthumous works, Koyré supported the thesis that experiments not only never played a significant role in the birth of modern science but also were often an obstacle. On the contrary, he stressed the importance of thought experiments. Koyré thus interpreted the Scientific Revolution as a purely spiritual and intellectual transformation that was entirely internal to science (thus denying that this process of redefining knowledge could have been triggered by economic and social factors). Koyré's internalist thesis was accepted by an entire generation of historians of science, which included leading figures such as Bernard Cohen at Harvard, Alfred Rupert Hall in London, Herbert Butterfield at Cambridge, Alistair Crombie at Oxford, and Charles Gillispie at Princeton. Occupying such prominent positions, these authors made internalism the largely hegemonic approach in the history of science during the 1950s and 1960s. For this reason, externalism did not have an easy time in the institutionalization of the history of science as a discipline despite having stimulated the development of the historical sociology of science. During Koyré's so-called American period, his theses achieved great success and helped spread French epistemological culture in the United States.[5] The young Kuhn, in particular, was influenced by Koyré's classes and works, which then sparked his interest in other French authors, such as Brunschvicg, Bachelard, Meyerson, and Metzger.[6]

The debate over internalism and externalism continued internationally until at least the 1960s. In 1962, the publication of Kuhn's *The Structure of Scientific Revolutions* opened a new vision of the social dimension of science, leading post-positivist production in the English language toward the so-called historical turn in the philosophy of science[7] (Bird 2008; Stadler 2014). That Kuhn—the main architect of this historicist turn in Anglophone philosophy of science—was steeped in the French historiographical and philosophical culture that he had encountered, above all, in Koyré's classes in the United States was certainly no coincidence. The main theoretical objective of *The Structure of Scientific Revolutions* was to develop a general theory of discontinuism in the history of science while giving broad relevance to the social conditions determining "paradigm shifts." To do this, Kuhn's model—centered on the functional concepts of paradigm, normal science, and anomaly—incorporated sociological elements that he developed starting from a direct comparison with authors such as Fleck and from currents such as FHE and the Mertonian-style sociology of science (later rejecting relativist interpretations of

[5] Koyré—of Russian origin, then exiled first to Germany and subsequently to France—emigrated to the United States during the Second World War to escape the Nazi occupation. In the United States, he was among the founders of the École Libre des Hautes Études in New York in 1942, thanks to funding from the Rockefeller Foundation. Upon returning to France, he was appointed "directeur d'études" at the École Pratique des Hautes Études through the support of Fernand Braudel. Even after the war ended, Koyré continued to spend half the academic year in the United States. In 1955, he was appointed as a professor at the Institute for Advanced Study at Princeton University, taking the opportunity during his stays to give lectures and seminars at other universities, such as Harvard, Yale, the University of Chicago, the University of Wisconsin, and Johns Hopkins University. For further information on Koyré's trajectory, see Jorland (1981) and Zambelli (2016).

[6] Meyerson and Metzger were explicitly cited—along with Fleck—as key sources in the introduction to *The Structure of Scientific Revolutions*: "In particular I continued to study the writings of Alexandre Koyré and first encountered those of Émile Meyerson, Hélène Metzger, and Anneliese Maier. More clearly than most other recent scholars, this group has shown what it was like to think scientifically in a period when the canons of scientific thought were very different from those current today" (Kuhn 1962, v–vi). Despite these declarations, Kuhn then concealed his intellectual references, often downplaying the inspiration he received from authors coming from FHE (Kuhn 1977). Bachelard, in particular, was an exemplary case from this point of view. In a 1994 interview, Kuhn stated: "In preparation for my first trip to Paris, I remember I did read some Bachelard. But it was so close to my own thought that I did not feel I needed to read lots and lots more" (Kuhn 1994, 160). Koyré was the one who suggested the meeting in Paris between Kuhn and Bachelard by writing a letter of introduction for Kuhn. In another interview, Kuhn reported the moment of the meeting as follows: "A large burly man in an undershirt came to the door and invited me in; I said, 'My French is bad, may we talk English?' No, he made me talk French. Well, this all didn't last very long" (Kuhn 2000, 285). For further information on the interconnections between these two authors, see Castelão-Lawless (2004a, b) and Gil (2004).

[7] For this reason, Kuhn's work has been completely legitimized within the canon of international historical epistemology. See, for example, Rheinberger (2010) and Badino et al. (2022).

his epistemological theory[8]). On the basis of this framework, in a 1972 lecture titled "Mathematical versus Experimental Traditions in the Development of Physical Science," Kuhn attempted to overcome the polarizations of the debate between internalism and externalism, shifting attention to the theme of the unity and disunity of science:

> Virtually all the authors now regarded as internalists address themselves to the evolution of a single science or of a closely related set of scientific ideas; the externalists fall almost invariably into the group that has treated the sciences as one. But the labels "internalist" and "externalist" then no longer quite fit. Those who have concentrated primarily on individual sciences, e.g., Alexandre Koyré, have not hesitated to attribute a significant role in scientific development to extrascientific ideas. What they have resisted primarily is attention to socioeconomic and institutional factors as treated by such writers as B. Hessen, G. N. Clark, and R. K. Merton. But these nonintellectual factors have not always been much valued by those who took the sciences to be one. The "internalist-externalist debate" is thus frequently about issues different from the ones its name suggests, and the resulting confusion is sometimes damaging. (Kuhn 1977, 32)

Thanks to Kuhn, starting from the 1970s, the idea that the polemical clash between these methodological orientations in the history of science was based largely on mutual misunderstandings and on the intrinsically problematic nature of the categories "internal" or "external" to science progressively took hold.[9] This new

[8]I am referring in particular to how Kuhn's theses have been implemented within the sociology of scientific knowledge (SSK), especially with the so called Strong Program developed in Edinburgh and Bath. I agree with Hans-Joachim Dahms in maintaining that Kuhn's distancing from such perspectives was motivated by his proximity to Mertonian sociology of science (Dahms 2016, 119).

[9]The emergence of SSK in the Anglophone context of the 1970s and 1980s—and the subsequent cultural turn in the history of science (Omodeo 2020)—affected this controversy. Indeed, one of the fundamental objectives of SSK was to show how sociological tools could be used to study the content of science itself and not just the communitarian organization of scientific institutions (Barnes 1974; Bloor 1976; Mulkay 1979). With this rhetorical gesture, they believed they could demonstrate the existence of a fundamental inconsistency in the construction of the internal and external boundaries of science, and they thus considered the debate closed (Shapin 1992). One of the most significant works in this sense was *Leviathan and the Air-Pump* (1985), by Steven Shapin and Simon Schaffer. A retrospective essay written by the authors on the occasion of the second edition, in 2011, emphasized that overcoming the debate between internalism and externalism was one of the objectives of their work. In fact, they considered this debate largely unsatisfactory due to the intrinsically problematic nature of the "external" and "internal" categories themselves: "One incoherence concerned the placement of the boundary between what was deemed internal and what external to science. On what grounds were social and political things not accounted 'intellectual'? And how was it that the making and warranting of scientific knowledge was judged not 'social'? Did other *intellectual* practices—say religion and natural magic—count as external (since they were not considered to be 'scientific'), or was the external-internal boundary meant to mark the divide between what were taken as intellectual practices—presumably including things like religion and natural magic—and what were taken as nonintellectual—things like the production of material goods, the governance of nations, and the practices and patterns of quotidian social life. The authors wondered on what grounds historians allocated items to either side of the boundary. Did historical actors get to say what properly belonged, or did not belong, to their specific practices, or was the 'externalism-internalism debate' an expression of the various boundaries between science and nonscience recognized by twentieth-century historians?" (Shapin and Schaffer 1985, xv).

stance led the scientific community to consider the debate resolved and closed, at least for a while.

As I have mentioned, Koyré was the only author close to FHE who actively participated in this debate, in particular by expressing critical positions with respect to externalism. So why did other authors in this tradition choose not to participate directly in this debate (or were not considered part of it)? I believe that most of the theses developed by the architects of FHE could not be framed within the terms of the debate between internalists and externalists, and for this reason, this controversy did not receive wide consideration in France.[10] However, the lack of direct dialogue with international debates should not be considered as expressing a lack of interest in the social dimension of science on the part of FHE. In fact, authors such as Meyerson, Metzger, Bachelard, and Canguilhem had already proposed theoretical frameworks capable of focusing attention on socio-anthropological dimensions within the theory of knowledge, thus developing an epistemological position that would allow them to go beyond the internalism/externalism opposition (partly anticipating Kuhn's perspective). Attention to these issues was further extended in the 1970s thanks to the work of Althusser, Foucault, and Bourdieu. As I will detail below, these authors made marking out the connections between epistemology, politics, and society one of the distinctive features of FHE.

As noted previously, various theoretical positions developed within FHE also led to the treatment of anthropological and social issues. In particular, in dealing with the controversy between presentism and recurrent history and between Whiggism/anti-Whiggism, I limited myself to showing how common themes emerged between the community of historians and FHE in the early twentieth century. In particular, I emphasized how, within the so-called *Annales* school, the idea of reconstructing the "mentality" of a specific historical era was established as one of the distinctive characteristics of French historiography. At the same time, I described how, for Koyré, the perspective of the historical actors being studied must be identified with in order to propose a reconstruction of scientific knowledge as it appeared to its protagonists. This research topic formed the nerve center of an interdisciplinary dialogue between historians, anthropologists, and epistemologists that flourished in the period between the two wars thanks to institutions such as the Centre de Synthèse, the Institut d'Histoire des Sciences et des Techniques, and the Sorbonne, among others, and that thus bore the mark of mutual influences and collective elaboration.[11]

[10] For example, the first French translation of Hessen's famous speech at the 1931 London conference was only published in 2006 by Serge Guérout (Hessen 1931/2006). This text had already been informally translated and was available in an unpublished version from 1979 for users of the "science et société" collection of the Jussieu interuniversity library. A debate that partially overlaps with that on externalism and internalism can be found in the two-volume collection titled *À la lumiere du marxisme*, published between 1935 and 1937 (Baby et al. 1935; Cornu et al. 1937).

[11] On this point, see Castelli Gattinara (1996) and Chimisso (2001, 205–206; 2008). In addition to meetings and institutional relationships, starting from 1920, a group of scholars met weekly at

In particular, an interest in reconstructing the mentality of specific historical eras led FHE to confront not only historiographical issues in the strict sense—that is, the avoidance of anachronism—but also problems of an anthropological-cultural nature that were inherent in this type of research. Among these thinkers, Meyerson certainly played a central role. As seen previously, Meyerson was a supporter of epistemological continuity between common sense and scientific knowledge. Frédéric Fruteau de Laclos observed how this theoretical position allowed Meyerson to explore—through comparisons with psychology, logic, and metaphysics—"the fundamental anthropological principles of humanity" on which to elaborate "a general anthropology of knowledge" (Fruteau de Laclos 2009, 18–19). Meyerson's ambition, indeed, was to uncover "the general form of all reasoning" (Fruteau de Laclos 2009, 14) by integrating epistemology and the social sciences. Meyerson developed this project in *Du cheminement de la pensée* (1931) by engaging in a direct comparison with the most important representatives of French anthropology and sociology of the time, such as Durkheim and Lucien Lévy-Bruhl.[12] In particular, Meyerson took up Lévy-Bruhl's concept of mentality, a key tool of his intellectual project starting from *Les fonctions mentales dans les sociétés inférieures* of 1910 and later expanded in the famous *La mentalité primitive* of 1922 (Lévy-Bruhl 1910, 1922).

If a clear border between scientific knowledge and common sense cannot be located—as Bachelard would later argue—then the historicization of science requires an in-depth analysis of the cultural background within which scientific knowledge is situated. Such a project would involve a total immersion in the "mentality" of an era in order to bring out the ontological assumptions under which scientists operated. In a passage from *Explanation in the Sciences* (1921), in fact, Meyerson distanced himself from those historiographical models that tended to create anachronisms:[13]

> On the contrary, we must in a certain sense disregard our convictions and even our scientific knowledge insofar as possible in order to try to assume the same mental attitude toward the conceptions of the past that their contemporaries must have had. That is no doubt a precept of every historical reflection, but perhaps nowhere has it been more frequently disregarded than in the case of the history of science. (Meyerson 1921/1991, 530)

In this sense, as Taylan noted, Meyerson aimed to bring out the "subconscious of reasoning, or the metaphysical presuppositions that every knowledge of nature necessarily implies" by extending the scope of historical epistemology toward a

---

Meyerson's home in Paris with the aim of "discussing the impact of Durkheim's sociological and anthropological works, Lévy-Bruhl, and Boas on the culture of the time" (Castellana 2009, 30). These meetings were attended by, among others, Metzger, Brunschvicg, Koyré, Bachelard, and Enriques.

[12]On the role of Lévy-Bruhl in the epistemological debates of this period, see Chimisso (2008, 62–70).

[13]In relation to the historiographical precept of identifying with the mentality of the era being studied, Meyerson developed a historiographical model relatively close to the one subsequently proposed by Koyré (but in clear opposition to Bachelard's).

"more general history of the collective and historical ways groups identify reality" (Taylan 2018, 15). Contrary to the perspective developed by Brunschvicg, which was entirely oriented toward emphasizing the "evolutionary dynamism" of mathematical thought, in *Identity and Reality* (1908), Meyerson tried to grasp—under the sign of the concept of identity[14]—the existence of cultural invariants within human thought (Castelli Gattinara 1996, 96). In this theoretical framework, science was considered as only one cultural phenomenon among others through which one could reconstruct a general theory of intellect. In *Du cheminement de la pensée*, Meyerson explicitly stated: "Scientific research does not interest us for the results it produces, but for the reasoning that was implemented to obtain them; for us, science is only a set of intellectual operations, operations which are easier to grasp there than elsewhere" (Meyerson 1931, 66).[15] A similar position was also expressed in *Identity and Reality* in relation to an epistemological thesis regarding the social shaping of knowledge:

> what we are looking for is less the result than the method, the way by which we came. Now the scientist on this point does not differ from the ordinary man. He does not see himself in the act of reasoning. He therefore does not know directly the way by which he has come to such or such a conclusion; the motives which have influenced him in adopting it may be very different from those which he himself supposes. That is why it is advisable to control his assertions by turning, not to the individual thought, but to the collective thought, by searching for the genesis, the evolution, of these conceptions in history. (Meyerson 1908/1930, 10)

The relationship between subject and object is always mediated by sociohistorical elements, and therefore, our knowledge of the natural world is always based on metaphysical presuppositions that Meyerson considered unavoidable but, at the same time, worthy of being investigated: all knowledge "will contain elements that clearly belong, not to nature, but to ourselves, elements it will be totally impossible to eliminate" (Meyerson 1921/1991, 12).[16] On this basis, Meyerson considered the sciences as ontological systems. For example, thermodynamics establishes its own ontology. Despite having origins in common sense, a science dismantles its ontology by immediately establishing another: "science, while adopting common sense as its starting point, itself subsequently destroys this ontology. But—and this is the important point—it always destroys it in favor of a new ontology" (Meyerson

[14]Vinti underlined how the concept of identity should be considered not as "a state" but as "a process": "Identification—writes Meyerson—therefore means not only the act by which we recognize the identical where it exists, but also the act by which we restore to identity what previously seemed not to pertain to it" (Vinti 1977, 26).

[15]Notably, in Meyerson's introduction to *Identity and Reality*, he claimed that the best way to understand common sense is to start from an analysis of how scientific reason functions because the dynamics of the functioning of thought could emerge most clearly in this context (Meyerson 1908/1930, 5–6). On this point, see also Castelli Gattinara (1996, 99–100).

[16]As happened with Duhem's theses (see Chap. 2, Sect. 2.2), Meyerson's position was also taken up by authors such as Quine in the Anglophone post-positivist context to support the dependence of experiential statements on the ontological context of reference (Fruteau de Laclos 2009; Laugier 2015).

1921/1991, 27). The evolutionary process of knowledge, spurred by changes in scientific categories, corresponds to moments of change in the ontologies of scientific knowledge (Bensaude-Vincent 2008, 12–13; Laugier 2015, 74). However, contrary to the positions developed by Comte with the law of the three stages and by Bachelard with the theory of epistemological rupture, Meyerson considered identifying a clear line dividing the prescientific from the scientific to be impossible. Consequently, these transformations needed to be conceived in a more nuanced way. For Meyerson, changes in ontology were rather like progressive functional extensions to account for new phenomena (Fruteau de Laclos 2009, 244); therefore, such transformations are noticeable only in the long term as elements of the old ontologies remain active in the new ones (Taylan 2018, 19–20).

Metzger proposed a quite similar epistemological position. After achieving a degree in crystallography, she dedicated herself to the history of chemistry and developed relationships with authors such as Lalande, Meyerson, Brunschvicg, Rey, Sarton, and many others.[17] Notably, Lévy-Bruhl, who was her uncle,[18] played a decisive role in her education and training and was always "the first reader" of her texts.[19] Though she never obtained a stable academic position for her work as a historical epistemologist and historian of chemistry, she was part of an informal collaborative network that provided her with the opportunity to become one of the most interesting authors of that epistemological phase.[20] Metzger was one of the central figures at the Centre de Synthèse and the International Committee for the History of Science, as well as a collaborator with the Institut d'Histoire des Sciences et Techniques and the École Pratique des Hautes Études.

Like Meyerson, Metzger maintained that the historian of science needed to be immersed in the mentality of the period being studied, notably in a 1933 speech at the Centre de Synthèse Historique titled "L'historien des sciences doit-il se faire le contemporain des savants dont il parle?":

> if the historian does not want, or does not know how, to read the works of the past as the first readers of these works read them, he risks making misinterpretations, not only misinterpreting the studies done by our distant predecessors, but even altering our very vision of scientific progress itself. (Metzger 1987, 57)

This model, in addition to containing an implicit criticism of presentism,[21] was oriented toward the proposal of what has been defined as a "cultural anthropology of science" (Giannetto 2020, 7). This case also required translating into

[17] Some aspects of Metzger's contribution to historical epistemology have been discussed in the previous pages in reference to conceptual history and the debate on presentism in France.

[18] On the relationship of Metzger's perspective with that of Lévy-Bruhl, see Golinski (1987).

[19] On Metzger's complex and troubled relationship with Meyerson—who had a largely paternalistic attitude toward her—and on the situation of women in the philosophical-epistemological field of those years, see Chimisso and Freudenthal (2003).

[20] For further biographical information on Metzger, see Chimisso (2001, 2019), Castellana (2009), and Freudenthal (2009).

[21] On this point, see Abadía (2008, 2009).

historiographical terms the disciplinary attitude typical of ethnography and cultural anthropology in which interpreting other cultures through Western logical-cultural frameworks is considered methodologically incorrect. The anthropological objective underlying Metzger's historical-epistemological project was to bring out, through historical reconstruction, elements that could contribute to the advancement of a general theory of the functioning of human knowledge.

From Lévy-Bruhl, she drew, in particular, her interest in the investigation of what she called "spontaneous thought" (or, more rarely, "prelogical" or "prescientific" thought). Going beyond her uncle's anthropological model, Metzger, even before looking at the transformation dynamics of established scientific knowledge, was particularly interested in analyzing its genesis—that is, how a transition spontaneously took place from common-sense thinking to scientific thinking in a strict sense. As Castellana underlined, to do this, "it is not enough to identify the pre-logical within a nascent knowledge, but it is necessary to insert it into a very precise reconstruction, trying to bring it back to the basic *Weltanschauung*" (Castellana 2009, 31–32). This historical-epistemological model was, therefore, configured as a "history of the reception of scientific ideas, consistent with a conception of science understood as an enterprise that transcends mere personal or individual peculiarities" (Carrier 1988, 135).

Metzger began developing this perspective in her first work, titled *La genèse de la science des cristaux*, in 1918. This text was dedicated to reconstructing how crystallography had emerged as a new scientific discipline between the sixteenth and eighteenth centuries. She traced the genesis of this field in terms of its progressive autonomy from other forms of knowledge, in particular mineralogy, which led to its recognition as an "independent branch of physics" around the eighteenth century (Metzger 1918, 5–6, 15–16). To explain this development, Metzger focused on the contribution of "amateurs" and "laypeople" to the natural sciences in the seventeenth century as a cornerstone for the professional development of this area of investigation (Castellana 2009, 29; see also Chimisso 2019, chapter 6). Mezger maintained that to understand the genesis of a science such as crystallography, one must not be limited to studying its "definitive elaborations"—that is, the most complete, stable, and formalized theoretical formulations. On the contrary, one must pay attention to the slower phases through which this intellectual "construction site" was elaborated, through in-depth research into the "labor of a few modest workers, who, apparently without guidance" were those who brought "the materials [...] used to complete the building" (Metzger 1918, 15). In the same way, and this was one of Metzger's conclusions, scientific production has an essentially social character because it is inserted within the rhythms determined by the community:

> We have established that the history of crystal science, like the history of all sciences, is linked in broad terms to the general history of humanity; that only theories and works that develop as a whole along the path that society pushes them towards can achieve social success; that, on the contrary, opinions and methods that oppose or are on the margins of the current tendency of thought are definitively or temporarily destined for oblivion.
>
> The evolution of science is, ultimately, partially determined by its human conditions of development; it would neither be able to follow the rigid path of the pedagogical method nor

> the tortuous paths or contradictions to which the fantasies arising by chance in an isolated brain would lead them. (Metzger 1918, 224)

This focus on the transition from "spontaneous thought" to scientific conceptualization was further explored in Metzger's later works. In *Les concepts scientifiques* (1926), she shifted from studying the emergence of disciplinary systems to investigating how scientists use concepts, consciously or unconsciously, as tools for organizing their experiences of nature (Chimisso 2019, 30). For Metzger, in fact, scientific concepts emerged "from the need to discover the relationships between things, to explain their nature." These concepts are formed as "syntheses of instinctive hypotheses which aspire to be prolonged, completed, universalized, in short, transformed into a system." These concepts perform a precise cultural function in that they "can, in turn, be considered either as the backbone of the metaphysics that they support and corroborate or, on the contrary, as the dead residue of these same metaphysics, which have fallen into disuse" (Metzger 1926, 145). Through a simultaneously historical and epistemological approach, her goal was to account for the formation and transformation processes of conceptual groupings—which led to real "theoretical configurations"—in order to highlight the intrinsically "theoretical nature of observers" and their relationships to cultural systems. Also in this case, Metzger's objective was to look not at mature and already-established science but at the processes through which it took shape by navigating tortuous paths:

> It is a question of seeking—both in the genesis of the sciences and in the formation of this unanalyzed set of imprecise notions, vague statements, semi-obscure theories, prejudices, and intuitions which are the characteristics of the common sense of an era—the origin of new concepts, which can then be organized into an elegant framework through the mere play of thought. Furthermore, we examine this conceptualization when it is not yet in place, when it is still in the form of a question, as an aspiration to know, and not yet in the form of knowledge. (Metzger 1926, 1–2)

As one may easily imagine, this approach favored the creation of a methodological approach that dealt with the "theoretical frameworks" and "mentality" of a specific era with an anthropological spirit in order to avoid presentism. Through this approach, Metzger reconstructed the evolutionary dynamics of key concepts for the history of chemistry, such as that of "metal," whose meaning had evidently been transformed in the passage from the theoretical system of Paracelsian alchemy to Van Helmont's iatrochemistry (Castellana 2009, 33). In particular, Metzger advanced the idea of a theoretical "incompatibility" within the various conceptual systems that had some interesting parallels with the subsequent formulation of the notion of incommensurability in Kuhnian paradigms[22] (Castellana 2009, 35). For Metzger, the development of scientific knowledge did not take the form of a constant accumulation of knowledge that could compose an increasingly precise picture of reality but rather took the form of a succession of different—and often contrasting—ideas, conceptions, and visions of nature (Carrier 1988, 138). Each historical phase,

[22] For a comparison of Metzger and Kuhn's perspectives, see Melhado (1988) and Chimisso (2019, 143–148).

therefore, corresponds to specific "local *a priori*" concepts that determine specific forms of thought and, correlatively, different theoretical models (Chimisso 2019, 37). In this sense, "the *a priori* is not, and cannot be, always and everywhere equal to itself; or rather, there is not just one *a priori*, there are multiple *a priori* that are very different from each other and are sometimes heterogeneous and incompatible" (Metzger 1987, 83).

Anticipating Foucault,[23] Metzger advanced a form of historicization of the a priori, meaning with this term all the elements that contribute to generating the cultural-conceptual apparatus within which the scientist's own experience takes shape:

> the historian of science [...], during his research, has developed the conviction that certain forms of doctrine derive as much from experience and observation as from the mentality of the researcher; if we admit that things are like this, the *a priori* will not only represent the preconceived notions preceding the experience and on which the description of the experience is based; the *a priori* will also represent the fundamental tendencies that generate these notions; or rather, to use scholastic language: to the actual *a priori*, which is concretized through first notions, we will add the potential and latent *a priori* which, in contact with the experience of life (and not only with scientific experience), effectively take the form of the *a priori* in action. (Metzger 1987, 82–83)

As I have emphasized in the previous pages, Metzger's historiographical methodology cannot be reduced to or framed within either the internalist or externalist current explored above. On the contrary, her model of historical epistemology had already gone beyond this schematism. In a speech delivered at the Centre de Synthèse Historique in 1930 titled "Réflexions sur l'histoire sociale des sciences," Metzger explicitly argued that historians of science cannot avoid considering the wider sphere of influences external to science:

> And so, ultimately, there is no aspect of human history to which the historian of science is indifferent: political and social history, philosophical history, the history of literature, including the history of the theater which I have not yet told you about, the history of industry and commerce, we might add, as well as the history of art and the history of occultism, can all aid the researcher who is studying the writings of chemists. (Metzger 1987, 146)

In this sense, precisely the attention to the history of "mentality" led Metzger to free herself from the traditional historiographical model (aimed at retracing "the lives of scientists on a purely biographical level") to concentrate instead on the "formation of the very foundations of the theory, their modifications under the pressure of its internal logic, or under the pressure of external social or scientific influences" (Metzger 1923, 11). As Chimisso noted, starting from *La genèse de la science des cristaux*, Metzger combined her attention to the developmental dynamics within crystallography with an emphasis on the "social and cultural circumstances that made it possible for particular theories, and indeed disciplines, to develop and

[23] The trajectories through which historical epistemology was hybridized with elements of anthropology were the basis, as we will see in the next section, for Foucault's opening of the investigation into historical a priori.

succeed," especially in their emergent phase, in which, not yet being consolidated, they could not "develop according to an 'internal necessity'" (Chimisso 2019, 77). Likewise, her study of the influences of religion, theology, and philosophy on the formation of Newtonian science, developed in *Attraction universelle et religion naturelle chez quelques commentateurs anglais de Newton* (1938), can be considered as an effort in this direction—that is, in thinking about science in its cultural and social contexts (Metzger 1938).

Bachelard's proposed theoretical innovations[24] further opened the path toward the integration of sociological considerations within the FHE tradition.[25] In *Le rationalisme appliqué*, having fragmented scientific reason into various regional specificities and diachronic discontinuities, Bachelard located the foundation of dialectical rationality in the social dimension of science. To do this, he distanced himself from the model of Cartesian and subjectivist epistemology, advancing instead a "non-Cartesian" epistemology. For Bachelard, the Cartesian theory of knowledge was completely insufficient for describing the dynamics of how scientific knowledge was produced.[26] As he followed scientific activity in its constant industriousness, in the process of "being done," Bachelard learned that it was clearly not an undertaking carried out by solitary individuals. Scientific activity was a collective undertaking, and following this line of thought, he proposed that the Cartesian *cogito* could be overcome by a *cogitamus*.

As has been widely noted, Bachelard was a critical reader of phenomenology and constructed some of his most famous concepts, such as phenomenotechnics and epistemological regionalism, in a dialectical relationship with it. The same applies to the concept of corrationalism that Bachelard structured in dialogue with Husserl's thought regarding the concept of intersubjectivity.

Bachelard needed to overcome the presumption of determining subjective certainty a priori in order to pose the problem of scientific objectivity as both a

[24]Bachelard, along with Meyerson, Metzger, Koyré, and, subsequently, Canguilhem, stressed the strategic importance of studying the historical evolution of concepts. Already in his first work, *Essai sur la connaissance approchée*, Bachelard dedicated the entire second chapter to the "rectification of concepts" (Bachelard 1927), an idea that continued to be important in his subsequent works. I will not deal with this point here, because the social dimension of Bachelardian epistemology emerged from other types of reflection.

[25]I have explored these issues extensively elsewhere. See Ienna (2018, 2022a).

[26]The essential character of Bachelard's non-Cartesian epistemology consisted of the rejection of the applicability of the criterion of certainty, obtained through the initial and unconditional evidence of the cogito ergo sum, to science. The notion of "non-Cartesian epistemology" is much more complex than can be summarized here, so I will limit myself to marking its intersections with "corrationalism." For further information on the evolution and progressive definition of this concept, see Bontems (2013). As Castellana noted, Bachelard's *The Philosophy of No* was not the first moment of a negation of the negation of the Hegelian model but rather had the character of an already immediately positive negation, or a *dialectical generalization*: "the dialectic, or philosophy of no, identifies the various scientific constructions and reconciles them; in this way, all contemporary theories incorporate the previous ones" (Castellana 1974, 146; 2021, 119). Bachelard, in fact, argued that "Generalization by negation must include what it denies" (Bachelard 1940/1968, 117).

collective achievement and a historical achievement. In both Bachelard and Husserl, the possibility of intersubjectivity had its origins in the discovery of intrasubjective pluralization.[27] The rationalist *cogito* is animated by a rationalist "I" and a "you" who are in dialogue "abstractly" with each other within the subjective pole. Unlike Husserl, however, in Bachelard's case, it was an incorporation of a shared social norm within the scientific field. The "I" measures itself against a hypothetical alter ego, "you," who objects to it with the most refined rational arguments in order to monitor its actions. The mirror of a form of psychological-cultural control is what takes the name of "intellectual self-surveillance" (Bachelard 1949).

This doubling within the scientific subject is what allows it, at an argumentative level, to move to the level of analyzing "scientific intersubjectivity." The *cogitamus* "unites the rationalist *I* and *you* in the same thought, and consequently in a thinking coexistence. With this *cogitamus*, the *I* and the *you* are culturally bound to each other" (Bachelard 1949, 57). However, it is not a question of socialization, understood as the incorporation of forms of moral censorship,[28] but rather, as Bachelard claimed,

> Even if [knowledge] is acquired in the solitude of a subject, it bears the mark of emulation. A thought that has surveilled itself to acquire precision in its application faces the surveillance of others. It is the thought of an *I* that is ready to compete with a *you*. (Bachelard 1949, 63–64)

The "applied" character of Bachelard's investigation led him to emphasize how rationalism has a "polemical" and "dialectical" nature. The emergence of a problem forces rational subjects to meet and enter into conflict on a rational basis. It is the problem that sets their research in motion and puts them to the test, imposing, in some cases, a questioning of the very foundations of their knowledge:

> To become aware of their agreement, two rationalist minds do not need a complete identity; it is enough for them to establish each other in the role of objectively controlled thought. Controlled roles, functions that operate on a normalized object, are the best topics of *discursive agreement*. In other words, the rational *cogitamus* is less aware of a common *asset* than of a common *income*. It is an announcement of productive thought. It determines an obligation to think in agreement; in short, it is a common awareness of apodictic knowledge. (Bachelard 1949, 57)

---

[27] I have analyzed this subtle passage in detail in Ienna (2022a).

[28] On this delicate question, Bachelard disagreed with Freud: "when Freud tries to convince us of the need to socialize the instances of surveillance, we object that the socialization of the super-ego is carried out on too primitive bases, by identifying the super-ego with a social primitivism, undoubtedly well-qualified to explain neuroses but completely insufficient for a complete analysis of the instances of mixed surveillance and guidance. Particularly when we come to examine the super-ego of the scientific city in a search for scientific progress, we will see the anagogical values of surveillance in action" (Bachelard 1949, 70). Here, Bachelard's objective is, in my opinion, to separate the idea of "surveillance" from that of social (and moral) coercion, which are unjustified from a rational point of view. On this point, see Desanti (1984).

According to Bachelard, this dialectical structure of rationalism allows for the establishment of a "discursive agreement" that, however, does not merely consist of an agreement between two subjects in the order of empirical "mutual recognition." This discursive agreement rather has the characteristics of an "intellectual self-surveillance" that—through the dialectic between "controlling" and "controlled"—brings into being a normativity intrinsic to rationality. In this sense, knowledge has a social character even before the actual encounter of empirical subjects because the structure of reason and its working dynamics presuppose the existence of a rational otherness.[29] The empiricism of recognition—and possible mutual recognition—therefore occurs only ex post with respect to the corrationalism animated by Bachelard's *cogitamus*. Through this dynamic, "There is the consecration of a method, proof of efficiency of thought, socialization of truth" (Bachelard 1949, 58).

Having brought to light the intrinsically social nature of rationality, Bachelard's historical epistemology went further, not stopping at this first level of analysis. Indeed, Bachelard added some interesting elements regarding the authentically social-collective dimension of scientific knowledge. The organizational model on which Bachelard envisioned the "scientific city" was the scholastic one, precisely because it is driven by the dialectic between "teaching [*enseignant*] rationalism" and "taught [*enseigné*] rationalism," which, for this reason, represents the highest model of social life (Bachelard 1949, 19). In the scientific city, every scientist is an eternal student of their colleagues, always in need of dialectical comparison, laying the foundations for what Castellana called a "new scientific humanism" (Castellana 1974, 162–170; 2021, 130–135). Dialectically, through the analysis of "errors," a further step forward can be taken. Bachelard also recognized the possibility that two minds, in their collective work, might find themselves united in the same error (Bachelard 1949, 58). Starting from Nietzsche's *The Will to Power*, Bachelard maintained that "error *descends* towards beliefs, while truth *rises* towards evidence" (Bachelard 1949, 58):

> if we pose the problem of error at the level of *scientific* errors, it appears very clearly, or rather concretely, that *error* and *truth* are not symmetrical as a purely logical and formal philosophy would suggest. In the sciences, truths are grouped into a system, while errors are

[29] Indeed, Bachelard claimed, "In these conditions, it seems to us that the *cogito* of mutual obligation, in its simplest form, should be expressed like this: I think you will think what I have just thought, if I inform you of the fact of reason that forced me to think ahead of what I was thinking. This is the *cogito* of mandatory mutual induction. This rationalist *cogito*, moreover, does not properly belong to the order of mutual recognition. It is formed before the agreement of *I* and *you*, because it appears, in its first form, in the solitary subject, as a certainty of agreement with the rational other, once the preliminaries have been established. One can *force* recognition: since I recognize that what I have just thought is normal for normal thinking, I have the means to force you to think what I think. Indeed, you will think what I have already thought to the extent that I make you aware of the problem for which I have now found the solution. We will be united in the proof as soon as we have the guarantee of having clearly posed the same problem. Moreover, by recurrence, the solution to a problem determines a new clarity in its statement. The problem–solution relationship is an epistemological instance that dominates the empiricism of observation" (Bachelard 1949, 58).

> lost in a shapeless morass. In other words, truths are linked apodictically, whereas errors pile up haphazardly. (Bachelard 1949, 58–59)

Underlining the collective dimension of scientific thought implies the existence of a "scientific culture."[30] That culture is organized in a system justified by apodictic connections consisting of scientific truths. On the contrary, errors "pile up" in a "shapeless morass." The asymmetry between these two poles is evident:

> on the one hand, there are truths which are rationally coordinated and codified in books, provided with the guarantee of the scientific city, and on the other hand, there are some errors which linger in a few bad books, most often marked with detestable originality. (Bachelard 1949, 59)

Scientific reflection is, indeed, an "essentially cultural" reflection that is not limited to an effort by the subject: "It is therefore necessary to examine scientific notions to see this cooperation of the servants of proof" (Bachelard 1949, 34). Faced with the scientific object, a single Cartesian subject no longer exists, but rather, in its place is a rationally coordinated community characterized by a scientific culture:

> Man is no longer *alone* in front of the scientific object. We do not cultivate ourselves alone. We *no longer* cultivate ourselves alone. Culture poses a different problem than *knowledge*. Culture incorporates both a historicity which is not historicity of the subject, and a rectification of the historicity of knowledge of the subject. Culture normalizes its own history. (Bachelard 1949, 137)

For Bachelard, the problem in this case was not so much knowledge itself but rather the conditions for the emergence of scientific objectivity based on conditions of social coproduction concretized in the meeting/clash of two forms of historicity: that of the subject and that of culture. In this sense, he not only recognized the internal articulation of the scientific subject in the "*doublet* controlling and controlled" (Bachelard 1949, 67) but also highlighted another essential aspect of the "dialogic" and "dialectical" structure within "scientific culture," namely its internal coherence. In his view, scientific culture is animated by the coexistence of various forms of experimental thought that find unity in the progressively stronger coherence that they assume in theoretical thought, even though, at times, they may appear to contradict or compete with one another. In this perspective, scientific thought is never definitive but rather is in a constant state of reorganization or "rectification," and precisely the rationalist undertaking is what brings out this cohesion (Bachelard 1949, 40).

Bachelard maintained that the objectivity of scientific culture is guaranteed by its intersubjective foundation (Bachelard 1949, 52). Corrationalism takes on an authentically social dimension that Bachelard, using Marxist terminology, defined as "the union of proof workers [*travailleurs de la preuve*]" or "the union of minds in truth" (1949, 56, 59). This collectivity therefore makes up a scientific city animated by a hard-working community that contributes to the production of "rationally

[30] Indeed, Bachelard subsequently declared that "we have, in a previous work, greatly insisted on the need to include culture in the scientific city that guarantees it" (Bachelard 1953, 31).

coordinated truths" (Bachelard 1949, 59). According to Bachelard, this scientific city is, however, located on the margins of the social city and has to struggle—in accordance with the principles of *epistemological rupture*—against common sense to affirm the validity of rational thought. The distinction that he makes between the consensus of common experience and the consensus of scientific experience, which is at the basis of the emergence of the autonomy of scientific culture, is of fundamental importance here:

> When we put the *consensus* of common experience into contradiction with the *consensus* of scientific experience, as we will often have the opportunity to do, we must obviously stop taking the universal *consensus* as the basis of rationality. The blank universal, that which creates the theme of declarations such as "there is no science except of generalities," ends up losing all virtue of application. The *current scientific city* establishes itself as a psychological reality in the very proportion that it must react against the mentality built on *common consensus*.
>
> If the mind were formed *directly* in the scientific city, we could do without a psychoanalysis of psychologism and directly pose the principles, not of reason (a very useless position), but of the rational organization of scientific culture. But this is not the case, and the scientific city is established on the fringes of the social city, so it must struggle against psychology to create its non-psychologism. (Bachelard 1949, 22–23)

This passage illustrates the canonical configuration of Bachelard's epistemology. The scientific city is regulated internally by the *nomos* of the consensus of the experience produced by scientists. This is a key point as it foregrounds the social roots underlying the notion of epistemological rupture: "Following this path, we are led to consider the limited consensus of the society of scientists as a highly specialized consensus" (Bachelard 1949, 132). Bachelard assumed the collective dimension and coproduction of scientific truths while avoiding that his epistemology devolved into a relativizing sociologism. Opposing the use of the logic of "common-sense consensus" to found scientific reason does not, therefore, mean enclosing oneself in mere solipsism.

This position can be interpreted as an attempt to sensitize FHE to social issues that were clearly in opposition to Meyerson's model. Common knowledge is always subject to external influences, both social and historical, with which the scientific spirit makes an epistemological rupture. Being on the margins of the social city does not mean a (self-)marginalization by being placed in a liminal position. On the contrary, it represents the idea that the scientific city has society itself as its place of emergence, which is where the process of rationalization of knowledge (through epistemological rupture) begins. The passage marking the transformation of knowledge tout court into scientific knowledge consists of the elimination of impurities deriving from conditioning via common sense. Despite this, scientific knowledge remains an intrinsically sociohistorical product resulting from the dialectical agreement—always provisional—between "proof workers" whose product is always rectified and rectifiable.

Bachelard recognized the existence of various cultural and pedagogical levels that render the culture produced in the scientific city transcendent, in contrast to the primary culture (Bachelard 1949, 19). But how does this transcendence arise? According to Bachelard, it requires transforming individual psychological values into epistemological values through a process of rationalizing and aligning the propositions that make up scientific culture (Bachelard 1949, 48). This triggers a circular dynamic in the production of scientific knowledge. The corrationalism of the union of proof workers effectively manages to found a "cultural super-ego" of science that is incorporated as "intellectual self-surveillance"—that is, as a shared practice within the scientific city:

> A thoroughly intellectualized surveillance, based on a super-ego psychoanalyzed as a super-ego, will allow us to refine the psychic controls which alone give culture its true effectiveness. In other words, we must attempt to replace the super-ego of historical formation—contingent and arbitrary—with a coherent super-ego, a super-ego open to culture. This cultural super-ego must also be clearly separated from general social ties. This super-ego, which we accept as judge, must be judged by ourselves. (Bachelard 1949, 71)

Importantly, Bachelard's socialization of FHE was completely coherent and organic with the theory of epistemological regionalism. In particular, scientists unite in social cells that make up the scientific city. In this way, they can develop dialogic and dialectical reason on the basis of a process of social "diversification" (Bachelard 1953, 2). This process of "differentiation," however, is not intended to create spheres of epistemological incommunicability or incommensurability. According to Bachelard, scientific reason is clearly determined by a scientific city that organizes its specializations. What he had in mind was a pluralist form of epistemology that manages through processes of "intense socialization"—characterized by the dialogic dialectics described above—to find a harmonization and coherence of scientific reason. Just as rationalism is fragmented so that it can be appropriately associated with the matter that it shapes and the phenomenology that it establishes, so too does the corrationalist activity of science come to stand out in various cities. In this sense, Bachelard identified, for example, a mechanistic city, an electrical city, a physical city, a mathematical city, a chemical city, and others.

Bachelard's argument regarding the social dimension of knowledge had a circular structure. To summarize, I will quickly retrace the stages in reverse order. Within the scientific city, a scientific culture is institutionalized, one whose internal *nomos* is rooted in the "consensus" among the participants, shared epistemic values, and the good practice of "intellectual self-surveillance" in one's daily work used to function as a "scientific super-ego."[31] The community made up of "proof workers" creates a shared value system within the scientific city that determines the emergence of a

[31] However, notably, Bachelard's positions were quite distinct from Robert K. Merton's 1942 formulation of four rules for determining the normative structure of science (Merton 1973). As noted above, Bachelard's description of the *nomos* regulating community life in the scientific city was not proposed in a systematic way, but rather, it had to be made to emerge from the text. Despite this, Bachelard's thought also included intuitions developed in other disciplinary sectors, such as the sociology of science. However, what is most remarkable was the intellectual effort that he made in building an epistemology capable of holding together the following: (1) the dynamics of knowledge construction on the phenomenological-subjective pole; (2) the historical dimension of knowledge; (3) the role of scientific culture in the orientation of concrete practices; and (4) the emphasis on the epistemic properties emerging from society on the basis of the nonreducibility of the scientific city to a mere sum of Cartesian subjects.

scientific culture that is transcendent compared to the culture of common sense. In this order of discourse, subjective psychological acts are depsychologized and take on epistemological value. A series of coordinated epistemological values shapes a shared scientific culture within the scientific city. This culture, in turn, acts on the individual knowing subject. Scientific practice is oriented by the transcendental value that scientific culture assumes with respect to the (conscious and unconscious) work of the individual. In this sense, the scientific city cannot be reduced to the sum of the Cartesian subjects that compose it but rather has to do with a psychological reality shared among the members of the scientific community. This psychological reality operates as an "intersubjective super-ego," as a further intrinsic surveillance of the scientific city. The community structures of the scientific city—which guarantee the intersubjective value of knowledge—are incorporated into research practice. Precisely through "intellectual self-surveillance," Bachelard sustained an approach that did not lapse into sociological relativism, thus guaranteeing the objectivity of knowledge. He thus re-entered the intrasubjective pole of the *cogito* that, having incorporated the dialogic structures of the scientific city, discovers itself to be already plural, marking the transition to the *cogitamus*. In short, for Bachelard, an epistemology that fails to take into account the process of "intense socialization" at the base of the production of scientific knowledge, as well as its historicity, is only "an epistemological utopia, the utopia of the individualism of knowledge" (Bachelard 1953, 2).[32]

Another author who played a key role in opening FHE to social issues was certainly Canguilhem, who had been engaging with the social and human sciences since his training at the Centre de documentation sociale at the ENS, directed by Célestin Bouglé,[33] one of the founding figures of French sociology. In fact, under his supervision, Canguilhem worked on Comte and positivism for his Diplôme d'études supérieures. In this context, he came into contact with the French sociologists associated with the journal *Année sociologique*, edited by Durkheim (but also including social history, social geography, and other fields). During his years as a high school teacher—even before the beginning of his actual academic career—Canguilhem had never stopped being interested in issues related to politics, sociology, and the social sciences in general. Evidence of this can be found in both the issues he covered in his high school courses and his publications from that time, such as an analysis of the relationship between Marxist and positivist sociology, the problem of suicide, the theme of the milieu, and the epistemological status of the

[32] Once again, how critical Bachelard was of the neo-positivist program is clear. Obviously, Bachelard's expression also has strong similarities to Fleck's concept of *epistemologia imaginabilis*, which is mentioned at the beginning of this book and which was aimed at exactly the same polemical objective.

[33] Célestin Bouglé (1870–1940) was a philosopher and sociologist with a positivist orientation in the tradition of Comte. Along with Mauss, Henri Hubert, Maurice Halbwachs, and others, he was one of the main figures at the journal *Année Sociologique*, founded by Durkheim in 1898.

human and social sciences, among others.[34] Particularly relevant was the 1935 volume *Le fascisme et les paysans*, published anonymously through the Comité de Vigilance des Intellectuels Antifascistes, which was both a political pamphlet and a sociological investigation (Canguilhem 1935).

These interests and research activities in his younger years were crucial in the formation of Canguilhem's subsequent epistemological-intellectual profile. As has been said, his originality lay in having opened FHE to issues of life and to the analysis of the sciences and technologies related to living things, in contrast to the previous French tradition. Up to that point, French epistemology had been dedicated mainly to the exact sciences, such as mathematics, physics, chemistry, and so forth. This shift in focus spurred by Canguilhem led to the consideration of new elements.

The specificity of some of Canguilhem's research objects, such as the distinction between the "normal and pathological" and the relationship between "the living and its milieu" necessarily entailed a consideration of the social dimension of epistemology. The need to address the themes of "normal and pathological" emerged precisely from debates at the intersection of politics and the social sciences. As Cammelli has shown, in the French cultural context, starting from Comte, the conditions had been created for the development of a debate regarding observations of the interactions between scientific knowledge, power and government, and life (understood in both social and biological terms) (Cammelli 2007, 4). Comte had, indeed, seen the "central problem for positivist philosophy and government" in the definition of the distinction between the normal and the pathological. Subsequently, Durkheim devoted the entire third chapter of *The Rules of Sociological Method* to analyzing this problematic knot (Durkheim 1895), with the aim of delineating the therapeutic and preventive benefits that sociology, understood as a form of social medicine, could provide. For Durkheim, based on a form of medicalization of society, normal and pathological took on the value of hermeneutical categories through which one could analyze social order and disorder (see Canesso 2024, chapter 2). Canguilhem, despite being critical of these approaches as a function of his vitalist perspective, took a position in these debates in 1943 with the publication of the first version of his *Essay on Some Problems Concerning the Normal and the Pathological*. For this reason, Cammelli claims that

> outside the French context, a reflection specifically dedicated to the problem of the normal and the pathological like Canguilhem's may have appeared, at best, as limited to the narrow and obscure field of the epistemology of biology and medicine. In France, however, subjecting the question of the normal and the pathological to philosophical reflection meant calling into question an entire theoretical and political experience, touching its very core. (Cammelli 2007, 6)

---

[34] On this point, see the first volume of Canguilhem's complete works for the texts of this phase of his production (Canguilhem 2011). See also Sfara's (2018) volume for a survey of Canguilhem's unpublished manuscripts; Canesso's (2024) book and Cammelli's (2007) essay are both explicitly devoted to Canguilhem's relationship with sociology. Canguilhem's political vision is explored in Talcott (2019, 1–76) and in the volume edited by Gandolfi et al. (2024). For a reconstruction of his relationship with the history of the human and social sciences, see Braunstein (2016).

I have already stressed how conceptual history was a principal axis of Canguilhem's methodology, which was in a line of continuity with the analyses proposed by Meyerson and Metzger. Canguilhem also wanted to reconstruct the logical and intellectual spaces within which scientific concepts emerged and were transformed, paying attention to the various implications underlying the dynamics as concepts migrated from one sociocultural context of production to another. Attention to the social dimension of Canguilhem's epistemological science grew, starting from the 1950s and intensifying between the 1960s and 1970s, thanks in part to his exchanges with newly emerging lines of research in historical epistemology by thinkers such as Foucault and Althusser.

One of the significant traits that emerged in the essays collected in *The Knowledge of Life* (written between 1946 and 1952) was Canguilhem's tendency, using the methodology of conceptual history, to bring out the ideological, political, and social implications underlying scientific theories such as cell theory (Canguilhem 1952/2008, 25–56). In his 1955 essay "The Problem of Regulation in the Organism and in Society," in the collection *Writings on Medicine*, Canguilhem retraced how a frequent comparison [*assimilation*] of the "life of an organism" had been made to "the life of a society" and vice versa (Canguilhem 1955/2012, 67). Using conceptual history, he explained how the understanding of society and living organisms had been reciprocally influenced and characterized by a process of mutual theoretical implication. Implicitly from this, certain positions regarding the structure of society would correspond to as many positions about the nature of the organism. However, Canguilhem stressed the problems underlying the processes through which concepts migrated from one theoretical context to another. In his view, what was at stake was more than just a mere metaphor or explanatory analogy. Rather, it was a question of constituting the basis of a "political theory" and a "sociological theory" that "tends to subordinate the social to the biological" or to compare "society to an organism" (Canguilhem 1955/2012, 67–68; on this point, see also Canesso 2024; Gandolfi et al. 2024).

This philosophical perspective also had effects on Canguilhem's historiographical methodology. Canguilhem conducted much of his exploration of the relationship between science and society in the context of his university classes. A first important opening can be seen in the course "Le statut social de la science moderne,"[35] held between 1961 and 1962.[36] With these classes, Canguilhem aimed to describe "the

[35] By the expression "modern science," Canguilhem meant "the science that has been built since the beginning of the seventeenth century, by Galileo, Harvey, and Descartes" (original text by Canguilhem cited in Sfara 2018, 189).

[36] Here I note—especially to understand the effect of the reception of Bachelardo-Canguilhemism on subsequent FHE—that one copy of this course was typewritten by Balibar (one of Althusser's students) and ordered into chapters and sections. Another copy was, instead, prepared by Canguilhem himself. In a letter addressed to Canguilhem, Balibar wrote that two more copies of the typescript of the course would be sent to Althusser and Macherey (Sfara 2018, 188). All this demonstrates, once again, the back-and-forth dynamic between the Sorbonne and the École Normale.

historically changing set of relationships between science [...] and society (in a collegial, national, cosmopolitan sense)" by highlighting, for example, the socioinstitutional structure of modern science, the role played by academies and universities in the public valorization of the figure of the scientist, and so forth.[37]

In the general economy of the course, Canguilhem insisted that the history of science not be considered as separate from "general history." To support this position, he relied, once again, on Bachelard's thought, explicitly mentioning corrationalism and the idea of the scientific city as theoretical antidotes to excessively intellectualized positions in the history of science. If, for Duhem, using a "magnifying glass" implicitly meant adhering to certain theories in the field of optics, he "does not seem to think for a moment that using a magnifying glass means adhering to a society, its techniques, its economy."[38] The course therefore developed as an application—and a translation—of corrationalism to the history of science. Through his lectures, Canguilhem proposed a social history of the evolution of scientific knowledge that focused in particular on the modern era even though it started from the classical world and passed through the Middle Ages. According to Canguilhem, in this historical phase, the relationships between the birth of modern science and the wider sphere of social phenomena such as technological developments,[39] the emergence of new economic needs, religion, the modern state, and so forth came to be more clearly observed.[40] Thus, also in this case, Canguilhem's classes were a driving force for the diffusion of Bachelardo-Canguilhemism within French culture, which would largely influence the success of FHE as an intellectual program.

In "L'objet de l'histoire des sciences" (1966), Canguilhem incorporated the ideas of his Sorbonne course, arguing not only that the opposition between internalism and externalism—of which he seems to have had only indirect knowledge[41]—is fruitless but also that both positions of the debate seem founded on a misleading theoretical assumption: "the assimilation of the object of the history of science to the object of a science" (Canguilhem 1994, 14–16). Further, he claimed that

> The externalist conceives the history of science as an explanation of a cultural phenomenon through the conditioning of the global cultural environment and, consequently, assimilates it to a naturalistic sociology of institutions, completely neglecting the interpretation of a discourse that has a claim to truth. The internalist sees in the facts of the history of science, for example, those relating to simultaneous discoveries (infinitesimal calculus, conservation

[37] The original text by Canguilhem cited in Sfara (2018, 189).

[38] The original text by Canguilhem cited in Sfara (2018, 190).

[39] Uncoincidentally, Canguilhem, in the same period, held a seminar titled Science and Technology at the Institut d'histoire des sciences et des techniques (Sfara 2018, 203–211).

[40] For a complete summary of the course contents, see Sfara (2018, 187–203).

[41] In the passages in which Canguilhem addressed this debate, he did not directly cite firsthand externalist literature but rather referred to an article by Buchdahl titled "On the Presuppositions of the Historians of Science." On the other hand, for his criticisms of externalism, he limited himself to mentioning the works of Koyré. As argued in the previous pages, the debate in question had not yet arrived in France at that time.

> of energy), facts of which one can only make history without theory. Here, consequently, a fact relating to the history of science is treated as a fact of science, according to an epistemological position which privileges theory over empirical data. (Canguilhem 1994, 15)

As a counterpoint to these positions, in the following paragraph, Canguilhem referred to Metzger's positions on conceptual history, which he seemed to value as what I would define as an "excluded third" from the discussion. Through Metzger, Canguilhem claimed that the objective of the discipline was to generate "a discourse on other discourses" put forth by science (Canguilhem 1994, 16)[42] that, consequently, needed to be placed in context. This implied that the object of investigation of the history of science "is not only related to a group of sciences lacking intrinsic cohesion but also to non-science, to ideology, to political and social practice" (Canguilhem 1994, 18).

During the 1962–1963 academic year, Canguilhem had the opportunity to return to the ideas of his *Essay on Some Problems Concerning the Normal and the Pathological*, extending its scope. In fact, in this period, he developed the first reflections that would lead to the drafting of "New Reflections on the Normal and the Pathological (1963–1966)," which today makes up the second section of Canguilhem's masterpiece. Although his vitalist philosophical project remained unchanged, the main additions and extensions focused on the relationship between, on the one hand, social norms and vital norms—to which the chapter "From the Social to the Vital" was explicitly devoted—and, on the other, studies on the concepts of norm and normality in the psychiatric field (Canguilhem 1966/1991, 237–256). The idea of bringing epistemology to reflect on the epistemic status of medicine and the multifaceted dimensions and aspects of the concept of norm allowed Canguilhem to insert himself into the indefinite border area between what he called "the social" and "the vital." Rather than attributing theoretical, epistemological, or genetic prominence to one of these two dimensions, Canguilhem's objective was to show how they are structurally co-implicated (Duroux 1993; Cammelli 2007). Developing a vitalist perspective in which life corresponds to the ability to establish norms, Canguilhem placed the emphasis on the transformative and discontinuous dynamics of the living-milieu relationship, opposing static visions of society, and thus bringing him closer, at least in this regard, to a Marxist approach.[43]

---

[42] As I have pointed out, this is the theoretical position of the noncoincidence of the object of the history of science with the object of the science whose history is being made. The historian constructs the object according to research questions that are stimulated by the current state of science (Canguilhem 1994, 18).

[43] Canguilhem's relationship with Marxist theory is decidedly complicated and often ambiguous in his texts. However, the publication of the first volume of his complete works (Canguilhem 2011) has shown how he had engaged with the Marxist tradition several times early in his career (well before his works on scientific ideology) despite his never having defined himself as a Marxist or communist. For more on this point, see Mazauric (2019) and Cammelli (2007, 2011).

Starting from these theoretical assumptions, Canguilhem formulated further innovations. The specificity of his research object—that is, a conceptual history of the life sciences—necessarily forced him to simultaneously address aspects of the history of the social and human sciences[44] that had various intersections with medicine and biology. For example, consider the importance of a figure such as Quetelet, with his theory of the "average man," in the reconstruction of *The Normal and the Pathological* (Canguilhem 1966/1991, 154–162). His interest in the history of the human sciences continued during the 1950s. In this period, Canguilhem focused on the history of psychology, publishing an article titled "Qu'est-ce que la psychologie?" (Canguilhem 1958). This article provided the basis for his course at the Sorbonne in 1960–1961 titled "Histoire de la psychologie scientifique" (Braunstein 2016) and was, subsequently, further developed during 1967–1968 under the title "Science et idéologie dans la constitution de la psychologie" (Sfara 2018, 151). Indeed, uncoincidentally, when Canguilhem was appointed by the publisher Hachette in 1970 to direct a working group aimed at drafting an introductory manual on the history of science, he included disciplines such as sociology, economics, psychology, and linguistics alongside the more canonical exact sciences (Canguilhem 1970). Thus, Canguilhem was the first to legitimize the idea that a historical epistemology of the human and social sciences could take shape by shifting focus from the physical-mathematical sciences to the life sciences, preparing the ground for Foucault, Althusser, and Bourdieu, whose contributions began to emerge in the 1960s.

A further line of research emerged from his dialogue with Foucault and Althusser regarding the relationship between science and ideology. This topic was addressed first in university courses between the end of the 1960s and the beginning of the 1970s[45] and then in a series of writings collected in the volume *Ideology and Rationality in the History of the Life Sciences* in 1977. Canguilhem approached this problem with the aim of

> refurbishing without rejecting the lessons of a teacher whose books I read but whose lectures I was never able to attend. For whatever liberties my young colleagues may have taken with the teachings of Gaston Bachelard, their work was inspired by and built on his. (Canguilhem 1977/1988, ix)

[44] However, notably—as the first volume of Canguilhem's complete works testifies—his interest in the epistemological status of the human sciences had already emerged sporadically earlier in his career (Canguilhem 2011).

[45] Note the titles of the courses: "Science et idéologie dans la constitution de la psychologie" (1967–1968); "L'idéologie médicale au XIXe siècle" (1969–1970); "L'idéologie médicale au XIXe siècle: neurologie et psychiatrie" (1970–1971). On this point, see Canguilhem and Gohau (1978), Chimisso (2015), and Sfara (2018). Although in the preface to *Ideology and Rationality in the History of the Life Sciences* Canguilhem makes explicit reference to "the influence of work of Michel Foucault and Louis Althusser" (Canguilhem 1977/1988, ix), in reality, it was dedicated to clarifying his new conceptualization of scientific ideology; indeed, in the article "What Is a Scientific Ideology?" Canguilhem does not mention these two authors, preferring instead to start from a reading of Marx. For an analysis of the influence of Althusser and Foucault on Canguilhem's concept of scientific ideology, see Chimisso (2015) and Mazauric (2019).

For Canguilhem, confronting the topic of scientific ideology was a way to interact with the French intellectual climate at the end of the 1960s and, at the same time, resolve the theoretical tensions that the adoption of Bachelardian epistemology *à la lettre* had generated in the construction of historiographically solid narratives (Chimisso 2015). Indeed, Canguilhem wrote: "What is a scientific ideology? This is a question that arises, or so it seems to me, in the *practice* of the history of science, and its answer may be of importance for the *theory* of that subject" (1977/1988, 27). As seen previously, Bachelard's idea of "epistemological rupture [*rupture*]" represented a clear demarcation between science and common sense. Radicalizing Bachelard's perspective, Althusser and the Althusserians argued that precisely the "epistemological break [*coupure*]"[46] signals the existence of a qualitative, theoretical, and historical distinction between science and ideology. Science emerges when it frees itself from the false representations of the world stimulated by ideology; what is not scientific is ideological. Following this line, speaking of "scientific ideology" would be nothing more than an oxymoron because these two terms are in clear opposition to each other: Where there is science, there is no ideology, and vice versa (on this, see also the next section).[47] On the level of historiographical methodology, Althusser's thesis leads to various risks. Indeed, what would happen to the historicity of science following the moment in which the original epistemological break that generated the clear distinction between science and ideology happened? Taking this perspective literally, one would fall into the error of conceiving the history of science as free from ideological and, consequently, cultural and social conditioning.[48] Is simply affirming the existence of such an epistemological break therefore enough to "purify science from every ideological sin" (Canguilhem and Gohau 1978, 58)? Would we not risk, paradoxically, falling into an alternative form of Popper's epistemological positions?

Canguilhem, for his part, having explored the processes of conceptual migration between different contexts, was completely aware of this rigid demarcation. Looking carefully at the dynamics of the historical evolution of the sciences, they must necessarily be conceived in a more nuanced way. In this sense, at the end of the 1960s, Canguilhem began using the concept of scientific ideology to show how science and ideology have a much deeper relationship than the oppositional one proposed by Althusser.

---

[46] Even on a terminological level, Althusser transformed Bachelard's expression of "*rupture épistémologique*" into "*coupure épistémologique*" to strengthen its meaning. For a comparison of these two authors, see Cavazzini (2015).

[47] As detailed in the next section, Althusser did not deny that scientists or scientific practices are affected by ideology. This is the theme of *Philosophy and the Spontaneous Philosophy of the Scientists, and Other Essays*, in which he highlighted how scientific practices are tainted by ideologies deriving from the spontaneous philosophy of scientists. The task of philosophy, in this sense, is to "draw a line of demarcation between the ideological of the ideologies on the one hand, and the scientific of the sciences on the other" (Althusser 1967/1990, 83).

[48] See *Sur l'histoire des sciences* by Fichant and Pêcheux (1969) for the methods of applying this model within the Althusserian school.

For Canguilhem, the tribunal of the history of science must lead neither to a "purge" nor to an "execution": "The history of science is not the inverted progress of the sciences, that is to say, the putting into perspective of the stages that have been overcome and of which today's truth would be the vanishing point" (Canguilhem 1994, 14). On the contrary, his historiographical perspective provided space for forms of cultural production that were not strictly scientific. Canguilhem argued that "scientific ideology would be both an obstruction to and a condition of possibility of the constitution of science. The history of science would need to include a history of scientific ideologies, explicitly recognized as such" (Canguilhem 1977/1988, 32; translation modified). Having made a clear separation from Althusser's perspective, in a 1978 interview Canguilhem explicitly declared that he felt closer to Foucault's positions elaborated in *The Archaeology of Knowledge*:

> I agree with you that the rationalization of a sector of knowledge does not necessarily lead to the radical elimination of ideology. On this point, I share Michel Foucault's opinion: on the one hand, ideology is not exclusive to being scientific (the example of physiological science in twentieth-century medicine is apt), and on the other, a discourse that rectifies its mistakes or makes its formalizations more rigid does not necessarily break all ties with ideology. (Canguilhem and Gohau 1978, 59)

As Chimisso explained, Canguilhem used the term *scientific ideology* to indicate two types of relationships with scientific knowledge. The first can be defined as an ideology "that precedes science," or that broad series of discourses that represented a prelude to the emergence of authentically scientific theories (Chimisso 2015, 68). For Canguilhem, scientific ideology is not antiscience—that, in fact, would not have a history—because its prescientific stage could not be imagined (Canguilhem 1977/1988, 33). The example he gave was ancient atomism, which, despite not providing properly scientific conceptions, was already in the sphere of discursive production and so tended to become independent of religious or superstitious descriptions of the natural world. Democritus and Lucretius are thus legitimated to enter the history of science on the basis of their critical attitude toward the culture they had inherited and their tendency to use rationality. In this sense, Canguilhem argued that "The existence of scientific ideologies implies the parallel and prior existence of scientific discourses. Hence it also presupposes that a distinction has already been made between science and religion" (Canguilhem 1977/1988, 33).

The second form of scientific ideology, according to Chimisso, can be called parasitic. This presupposes that an epistemological rupture has already occurred because it uses methods and explanatory models produced in the scientific field, but in an improper manner. It is therefore a notion much closer to the Marxist notion of ideology as false consciousness. Among the examples given by Canguilhem in this case is Herbert Spencer's adoption of an improper extension of Darwin's model of evolution to explain how human societies function.

Canguilhem's adoption of the concept of scientific ideology was useful for resolving internal tensions in his historiographical methodology because it was aimed at constructing conceptual genealogies. These helped illustrate the process of epistemological modification crystallized in the formulation of scientific abstractions. Thus, adopting the notion of scientific ideology was a strategy aiming to avoid producing historical narratives with ideological flaws. If a scientist "fails to recognize and incorporate scientific ideology into his work, he runs the risk of producing nothing more than ideology himself, by which I mean in this instance a history that is a false consciousness of its object" (Canguilhem 1977/1988, 39). In those same years in France—as well as in Italy, the United Kingdom, and the United States[49]—a social movement with a 1968 spirit known as the "critique of science" had emerged, aimed at denouncing the non-neutrality of scientific knowledge (Quet 2013; Debailly 2015). It was a movement driven mainly by professional scientists in clear tension with the positions developed by Althusser and his school. In contrast, scientists belonging to the "critique of science" supported the need to expose how scientific practices and theories were informed by ideological stances. One of the most important events of this movement in France was the 1973 publication of the volume *[Auto] critique de la science*, edited by Alain Jaubert and Jean-Marc Lévy-Leblond (Jaubert and Lévy-Leblond 1973). Note that in the context of internal debates within what has been defined as "*épistémologie historique*" in the strict sense,[50] references to these scientific movements were largely omitted. Authors such as Canguilhem, Althusser, Foucault, and Bourdieu, despite participating in the French (and Parisian) academic-intellectual field of those years, neither mentioned nor directly addressed the contributions of the "critique of science."

In this section, I have highlighted how Meyerson, Metzger, Bachelard, and Canguilhem developed theoretical models capable of integrating and legitimizing research of a social-anthropological nature within the theory of *épistémologie historique*—an element that has been given little consideration in the secondary literature. In particular, I believe that paying attention to this issue allows a deeper understanding of the subsequent developments of this tradition. In this phase of development, Canguilhem's intellectual work and his teaching—and the correlative diffusion of Bachelardo-Canguilhemism—were absolutely crucial. Shifting FHE's focus from the exact sciences toward the life sciences and medicine led this intellectual tradition into previously unexplored theoretical territory. This transformation of the axis of interest took the form of a "twist" or "bend" within French epistemology. With Althusser, Foucault, and Bourdieu, who applied the principles of Bachelardo-Canguilhemism to new and diversified objects of investigation, research trajectories opened up that, despite their heterogeneity, led to an increasing interest in the social and human sciences. Furthermore, they constructed three theoretical perspectives on the functioning of systems of domination and their relationships to

---

[49] For a reconstruction of the debates on the non-neutrality of science in the United States, the United Kingdom, and Italy, see Werskey (2007); Guerraggio (2010); Baracca et al. (2017); Bharucha (2018); Schmalzer et al. (2018); and Ienna (2020). In the Italian case, the publication of *The Bee and the Architect*, which I discuss in the appendix to this volume, was particularly relevant (Cicotti et al. 2011/2024).

[50] In the critical reconstruction of historical epistemology, understood in a broad sense, I have shown how debates on the non-neutrality of scientific knowledge rightly fall within this canon (Badino et al. 2022). On this point, see also the appendix to this volume.

forms of knowledge. These aspects will be the subject of the next section, dedicated to the trajectories through which FHE became politicized.

3.2 The Politicization of Historical Epistemology

As I have mentioned on several occasions, FHE's international success and intellectual influence arose through the canonization of several key authors within the French tradition who continued the lines of research outlined in the previous pages. Here, I will limit myself to discussing the figures of Althusser, Foucault, and Bourdieu,[51] although the legacy of Bachelardo-Canguilhemism is of course much broader and multifaceted.[52] The reasons for this selection lie in the institutional weight that these three authors had in their respective academic institutions [the École Normale Supérieure (ENS), the Collège de France, the University of Vincennes, and the École des Hautes Études en Sciences Sociales (EHESS)] as well as the rapid reception of their thinking both in France and internationally. Thus, I will trace some lines of continuity that these authors had with the historical-epistemological tradition outlined so far.

I have indicated how a process of the socialization and politicization of historical epistemology was already present in various contributions that appeared in the first half of the twentieth century. Bachelardo-Canguilhemism was first spread by Canguilhem through his teaching at the Sorbonne, and before long, these three authors extended this epistemological perspective, proposing unique re-elaborations and creating intersections with other fields of study (in particular, with structuralism and poststructuralism). The cultural and political climate within which their works emerged was that of 1968 and post-1968 Paris. From the ranks of the Althusserian group, Lecourt described how the figure of Canguilhem facilitated the creation of a theoretical "meeting of the minds" between historical materialism and the history of science:

> It is time to close by returning to my initial question: how are we to explain our theoretical "meeting" with Georges Canguilhem's history of science. I have already provided the elements of an answer in passing: it is the unity which he institutes between the history of science and epistemology that brings him close to historical materialism and dialectical materialism—to their specific unity—that is why I have been able to maintain that this unity was "revolutionary." We have seen that it was Gaston Bachelard's work that first disengaged this unity. [...]

[51] Although the pathway traced in this text might seem to suggest otherwise, the specific differences among Bachelard, Canguilhem, Althusser, Foucault, Bourdieu, etc., are far from irrelevant. Despite this, linking them through some common traits under the label FHE can allow us to better understand their methodological contribution to this phase of French thought.

[52] For example, in the last section of Chap. 2, I discussed the largely politicized trajectories of authors such as Desanti and Gilles Châtelet in the context of mathematical thought. Indeed, the constellation of FHE could be further extended to authors such as Piaget, Simondon, Dagognet, and others.

> [...] Canguilhem's attempt is even more interesting: he thinks he can found this unity between epistemology and history of the sciences on another unity which he finds at the end of his labours, that of concept and life. I confess that I cannot follow him on this road which, as Pierre Macherey emphasized some time ago, often leads him to a "biologistic" conception of history itself. But I add the following essential specification: the necessarily idealist effects of such a conception are always-already erased, countered, in his work by the polemical import of his vitalism. That is why I can say without reservation that our route is a common one. (Lecourt 1972/1975, 186)

The success of FHE as a label that, as seen in Chap. 1, was revived and made famous by Lecourt, derived from the fact that it had allowed a set of representative aspects of an entire philosophical period, characterized by a common Bachelardo-Canguilhemism, to be gathered under a single concept.

In "Life: Experience and Science"—written on the occasion of the publication of the English translation of *The Normal and the Pathological*—Foucault identified two main lines of French philosophy that were in opposition to each other. On the one hand was a philosophy "of the subject," represented by authors such as Sartre and Merleau-Ponty, which aimed to radicalize the Husserlian phenomenological gesture, "which before long was to encounter the questions of [Heidegger's] *Being and Time*." On the other hand was the tradition that Foucault called the "philosophy of the concept," made up of authors such as Bachelard, Cavaillès, Canguilhem, and Koyré (Foucault 1985/2003, 466). Foucault located himself within this latter tradition of thought and, speaking of Canguilhem, added more.

> Whence a paradox: this man, whose work is austere, deliberately delimited and carefully tailored to a particular domain in a history of science that in any case is not regarded as a spectacular discipline, was in some way present in the debates in which he took care never to appear. But, take away Canguilhem, and you will no longer understand very much about a whole series of discussions that took place among French Marxists; nor will you grasp what is specific about sociologists such as Pierre Bourdieu, Robert Castel, Jean-Claude Passeron, what makes them so distinctive in the field of sociology; you will miss a whole aspect of the theoretical work done by psychoanalysts and, in particular, by the Lacanians. Furthermore, in the whole debate of ideas that preceded or followed the movement of 1968, it is easy to find the place of those who were shaped in one way or another by Canguilhem. (Foucault 1985/2003, 465–466)

Bourdieu, echoing Lecourt and Foucault, also recognized the strategic role played by the tradition of FHE in offering a whole generation of young philosophers an alternative to the existentialism that was completely dominant in the French academic field at that time. Also in this case, the intellectual and social role played by Canguilhem was recognized as "a totemic emblem" of unconventional thought:

> Then [came] an epistemology and a history of science represented by authors such as Gaston Bachelard, Georges Canguilhem and Alexandre Koyré. Often of lower-class or provincial origin, or brought up outside France and its academic traditions, and attached to peripheral university institutions, like the École des Hautes Études or the Collège de France, these marginal and temporally dominated authors, hidden from common perception by the celebrity of the dominant figures, offered a recourse to those who, for various reasons, sought to react against the fascinating but rejected image of the total intellectual, present on all the front lines of thought. [...]

> Georges Canguilhem, a contemporary at the École Normale of Sartre and Raymond Aron, from whom he was separated by his lower-class provincial origin, could be invoked by the occupants of opposite positions in the university field. As an exemplary *homo academicus*, he was to serve as an emblem for professors who occupied positions in the institutions of reproduction of the teaching corps entirely homologous to his own, such as Francois Dagognet; but as the advocate of a tradition of the history of science and epistemology which, at the height of the triumph of existentialism, represented the heretical refuge of seriousness and rigour, he would be consecrated, with Gaston Bachelard, as the *maître à penser* of philosophers more remote from the heart of the academic tradition, such as Althusser, Foucault and some others. It was as if his at once central and minor position in the university field and the entirely rare, even exotic dispositions that had predisposed him to occupy it had designated him to play the role of a totemic emblem for all those who sought to break with the dominant model and who constituted themselves as an "invisible college" by rallying around his name. (Bourdieu 2004/2007, 10–11)

The developments of FHE, particularly through the contributions of Foucault, Althusser, and Bourdieu, appear to be inseparable from a truly political stance, and in the following pages, I describe how these three authors received and employed the concepts drawn from this tradition.

3.2.1 Structuralist Marxism and Epistemology: Althusser

As alluded to in Chap. 1, Althusser was responsible for some central conceptual tools used in historical epistemology in the context of the debates on the renewal of Marxism following the Twentieth Congress of the Communist Party of the Soviet Union (CPSU), in 1956. In particular, Althusser and his collaborators argued for a discontinuism that linked Bachelard's theses with those of Koyré, creating a precise idea of FHE (Fichant and Pêcheux 1969).[53] Between 1960 and 1964, Althusser published a series of articles, collected in *For Marx*, in which he adopted the Bachelardian concept of epistemological rupture, adapting it to create an interpretative key for demonstrating the scientificity of the historical materialism formulated by Marx in *The German Ideology*. The concept of epistemological break was the foundation for the structuralist interpretation of Althusserian Marxism. Althusser aimed to dismantle the humanistic legacy, which in his view was characterized by undue subjectivism and individualism, by valorizing the works of the young Marx. In his perspective, the epistemological break took on an even more radical role than in Bachelard's because it represented the line of demarcation, or rather a qualitative distinction, between science and ideology.

In Althusser's interpretation, this moment marked the clear transition between the "Young Marx" as an ideologist and the later Marx as the founder of a scientific discipline, historical materialism, which—in its most complete form—would lead to

[53] Consider that in *Reading Capital*, he acknowledged "the obvious or concealed debts which bind us to our masters in reading learned works, once Gaston Bachelard and Jean Cavaillès and now Georges Canguilhem and Michel Foucault" (Althusser and Balibar 1965/1970, 16).

the theses set out in *Capital* (Althusser 1965/2005, 31–39). Althusser was driven by the need for a new interpretation of Marx's work that would allow him to take a stand against humanistic positions like that of Roger Garaudy and be of use in the development of an authentic "left-wing" criticism of Stalinist drifts. In *For Marx*, he declared: "I write these lines for my own part and as a Communist, inquiring into our past solely for some light on our present which will then illuminate our future" (Althusser 1965/2005, 22).[54] Starting from this premise, Althusser constructed a solid epistemological framework, radicalizing Bachelard's "*rupture épistémologique*" [epistemological rupture] into a "*coupure épistémologique*" [epistemological break]:

> Every recognized science not only has emerged from its own prehistory, but continues endlessly to do so (its prehistory remains always contemporary: something like its *Alter Ego*) by rejecting what it considers to be error, according to the process which Bachelard called "the epistemological *break* [*rupture*]."
>
> I owe this idea to him, but to give it (to use a metaphor) the sharpest possible cutting-edge, I called it the "epistemological break [*coupure*]." And I made it the central category of my first essays. (Althusser 1974/1976, 114)

But what are the characteristics of this terminological and conceptual radicalization? In his work through the articles that make up *For Marx*, Althusser had supported a new periodization of Marx's work that emphasized a radical discontinuity between the "Young Marx" the ideologist and the "mature Marx." Marxist humanism had, in fact, based its political and epistemological positions on readings of Marx's early works, which Althusser considered "ideological." The intellectual renewal that had spread in both the French Communist Party and various communist organizations following the Twentieth Congress of the CPSU had given the possibility of moving away from the dogmatic readings of the classics of Marxist thought.[55] Asking what the specifics of Marx's thought were implied confronting the problem of Marx's relationship with the philosophies he had adopted, such as those of Hegel and Feuerbach. Althusser had, in fact, identified the works published between 1840 and 1844 as "youthful works." These included the articles published in *Die Rheinische Zeitung* expressing a "liberal-rationalist moment" (until 1842), along with the texts published between 1842 and 1844 (*1844 Manuscripts*), the expression of a "communalist-rationalist" moment (Althusser 1965/2005, 34). In the first of these moments, Althusser claimed, Marx faced a "Kantian-Fichtean" problematic, whereas in the second, his theorizations fell within the horizon of a Feuerbachian anthropological problematic. Althusser's innovations, compared to

[54] His approach to Bachelard's *histoire périmée et sanctionnée* also emerges from this quote.

[55] Compared to the years preceding this renewal, Althusser declared: "we spent the best part of our time in agitation when we would have been better employed in the defence of our right and duty to know, and in study for production as such. For we did not even take this time. We knew nothing of Bogdanov and the Proletkult, or of Lenin's historic struggle against political and theoretical Leftism; we were not even intimately familiar with Marx's mature works, as we were only too eager and happy to rediscover our own burning passions in the ideological flame of his early works" (Althusser 1965/2005, 23).

his contemporaries, were not only his recognition of the disunity of Marx's thought but also his insistence that the Young Marx "*was never strictly speaking a Hegelian*": "the thesis that the Young Marx was a Hegelian, though widely believed today, is in general a myth" (Althusser 1965/2005, 34).

This phase was followed by the "works of the break," relating to the year 1845, which included the *Theses on Feuerbach*—in which "short intuitive flashes" were condensed—and *The German Ideology*, which marked the beginning of Marx's scientific period. For Althusser, *The German Ideology*, in which Marx formulated historical materialism understood as the science of history, marked the critical moment of Marx's break with his ideological past. But as is clear, the completion of this break, as well as its theoretical consequences, could not all be accomplished in a single instance in *The German Ideology*. Rather, these effects were to be found in what Althusser called Marx's "mature writings," published between 1845 and 1857, such as *The Communist Manifesto*, *The Poverty of Philosophy*, and *Value, Price and Profit*. Finally, in Althusser's view, Marx achieved full "maturity" in the period between 1857–1883, during the process of writing *Capital*. Thus arose the central problem relating to Althusser's take on FHE:

> The question of the specific difference of Marxist philosophy thus assumed the form of the question as to whether or no [*sic*] there was an *epistemological break* in Marx's intellectual development indicating the emergence of a new conception of philosophy—and the related question of the precise *location* of this break. Within the field defined by this question the study of Marx's Early Works acquired a decisive theoretical importance (does this break exist?) as well as a historical importance (where is it located?). (Althusser 1965/2005, 32)

According to this quote, the "break" identified by Althusser was clearly not only epistemological but also historical, in that it was capable of marking a fundamental chronological passage in Marx's work. This historical dimension was also highlighted in relation to the theory of science and knowledge in general. The term *epistemological break* was used not only to identify the boundary between "science" and "ideology" but also to describe the essential historical passage that marked the emergence of knowledge from its "prehistory" (Althusser 1974/1976, 114) as well as "to designate the mutation in the theoretical problematic contemporary with the foundation of a scientific discipline" (Althusser 1965/2005, 32). This epistemological position revealed an intimate political meaning that cannot be overlooked: "between those who recognize the fact of the 'break' and those who want to reduce it to nothing, there exists an opposition which, it must be acknowledged, is ultimately *political*" (Althusser 1974/1976, 107). Only through the adoption of a discontinuous philosophy of history was the possibility of a revolution, whether scientific or political, conceivable.

But the structure of Althusser's approach can also be attributed to the critical work carried out by his collaborators of that time who contributed to constructing the model. As mentioned previously, these collaborators supplemented Althusser's main classes with additional seminars and lectures (often typed and distributed in other universities) and in some cases published in the "Théorie" series directed by Althusser at the Parisian publisher Maspero.

A group of young scholars, including Balibar, Macherey, Establet, and Rancière, had formed around Althusser's courses at the ENS, and they collaborated to create the famous *Reading Capital* seminar held between January and April 1965, published in two volumes the same year. In this text, the authors continued the interpretive line opened in *For Marx* by arguing, through a careful reading of *Capital*, in favor of the "scientificity" of Marx's historical-materialist method.

With some of his students and closest collaborators (Macherey, Pêcheux, Fichant, Regnault, Balibar, and Badiou), Althusser also developed the "Cours de philosophie pour scientifiques" [Philosophy Course for Scientists] held in the 1967/1968 academic year. In this course, the reflections developed in previous years were more directly applied to the case of scientific practices in a course dedicated not to a philosophical audience but rather to professional scientists. As heirs of Bachelardo-Canguilhemism, Althusser and his collaborators had a double objective: on the one hand, to show how scientific practices themselves were permeated with a spontaneous philosophy and, on the other, to introduce nonphilosophers to some of the key problems of historical epistemology. Such key problems included the relationship between philosophy and the sciences (Althusser), the construction of the scientific object (Macherey), the social practices and the history of science (Pêcheux), the relationship between epistemology and history (Fichant), the analysis of the concept of precursor (Regnault), the structure of the experimental method (Balibar), and the concept of model (Badiou) (Althusser et al. 1967, 3). This large circle of Althusserians attended Canguilhem's courses at the Sorbonne at the same time, encouraging a process of intellectual hybridization between structuralist Marxism and historical epistemology. Furthermore, Foucault and Bourdieu were among the young scholars who followed some of Althusser's courses and seminars at the ENS.

The idea of an introductory course in philosophy for nonphilosophers was based on a specific political objective. The ideological mission of this operation was to deconstruct the idea that philosophy was an activity for bourgeois, privileged, and cultured elites. On the contrary, Althusser and his colleagues wanted to show how philosophical activity could have—following Marx's 11th thesis on Feuerbach—a transformative impact on reality and how, through philosophy, one could "make things happen" [*faire bouger le choses*] (Macherey 2008, 2). This course began with an observation about the relationship between scientific disciplines and philosophy:

> We are all aware, in one way or another, that the division of scientific research into distinct disciplines is an absolute necessity for research, though it can also be an obstacle to research. We all know that the relationship between distinct disciplines can have beneficial effects for both science and philosophy. This relationship between scientific disciplines is partly carried out "spontaneously" [...] and it can be partly carried out in a non-spontaneous form through the intervention of philosophy. (Althusser et al. 1967, 1)

The first point of contact between Althusser and his students with FHE could be found in the way they addressed the relationship between philosophy and science with a markedly Bachelardo-Canguilhemian imprint. The very coining of the expression "spontaneous philosophy of scientists" was aimed at analyzing the forms of "ideological struggle and the forms it can take among the intellectuals of scientific practice" (Althusser 1967/1990, 71).

The section of the course run by Althusser was later published with the title *Philosophy and the Spontaneous Philosophy of the Scientists*. In this text, he defined the sciences as "theoretical practice"—that is, what philosophy deals with in its role as "the theory of theoretical practice." Althusser, while recognizing that philosophy is composed of various branches and research objects, maintained in a Bachelardian way that philosophy is conducted as—and can be defined only as a function of—a reflection starting from the sciences. Also, in this case, against neo-positivist epistemologies, Althusser argued for a "philosophy which can serve the sciences" "rather than enslave them" (Althusser 1967/1990, 73). In Canguilhemian terms, however, this did not mean eliminating epistemological reflection within scientific reflection: "philosophical *questions* are not scientific *problems*. Traditional philosophy [...] does not provide answers to scientific, or other, *problems*" (Althusser 1967/1990, 81). In that sense, "The philosophy that we propose does not seek the 'salvation' of science. It invites scientists to distrust any philosophy that seeks the 'salvation' of science" (Althusser 1967/1990, 112).

With his concept of epistemological break, Althusser did not therefore argue that scientists or scientific practices were immune to ideological forms,[56] but rather, he argued that these forms are of a philosophical and nonscientific nature. In this sense, for Althusser, philosophy was what had to perform the key task of identifying the ideology dwelling in scientific practices.

Starting from the writings contained in *For Marx*, Althusser had clearly distinguished between "*Theory* (with a capital T)," which would correspond to Marxist "philosophy" (i.e., dialectical materialism), and a "theory" (lowercase), which instead would describe the body of concepts that constitute a science. The simple term *philosophy* was used to indicate "*ideological* philosophies" (Althusser 1965/2005, 162, 168). Similarly, and with Bachelardian echoes, in *On the Reproduction of Capitalism*, Althusser also made a clear distinction between a "common-sense philosophy" and "*Philosophy* in the strong sense." Authentic "Philosophy" was distinguished from that of common sense because of its relation to "knowledge of the rational necessity of the course of the world or evolution of history" (Althusser 1995/2014, 12).

On the basis of a Marxist conception, Althusser drew attention to the practical dimension of both philosophical activity and scientific activity to bring out a real theory of philosophical praxis understood as "class struggle in theory." In a completely coherent manner, Althusser admitted that "*Philosophy has not always existed*" (Althusser 1995/2014, 13). At least two contributing factors were necessary for it to emerge:

Philosophy is observed to exist in societies in which

[56] In this context, Althusser proposed the following definition of *ideology*: "An ideological proposition is a proposition that, while it is the symptom of a reality other than that of which it speaks, is a false proposition to the extent that it concerns the object of which it speaks" (Althusser 1967/1990, 79).

1) social classes (and therefore the state) exist;
2) science (or one science) exists.

> Let us be more precise. By science, we mean, not a list of empirical findings (*connaissances*), which can be quite long (thus the Chaldeans and Egyptians were familiar with a considerable number of technical procedures and mathematical results), but an abstract, ideal (or, rather, idea-dependent [*idéel*]) discipline that proceeds by way of *abstractions* and *demonstrations*: for example, the Greek mathematics founded by Thales (or those designated by this no doubt mythical proper name). (Althusser 1995/2014, 13)

Science and the existence of social classes are, therefore, the conditions of possibility of "Philosophy."

In their research activity, scientists spontaneously produce philosophical reflections even though they are not fully aware of them. The expression "spontaneous philosophy of scientists" was elaborated in this context, aimed at emphasizing the ideological tensions within scientific practices. These practices are determined, on the one hand, by "practical ideologies"[57]—both scientific and coming from other practices—and, on the other, by their immersion in a "dominant ideological system"[58] (Althusser 1967/1990, 94–95).

This ideological dimension becomes more visible particularly in moments defined as "crises of science," or those moments in which "science confronts scientific problems which cannot be resolved with existing theoretical means, or (and) that call into question the coherence of the earlier theory" (Althusser 1967/1990, 109–110). In these situations, theoretical clashes within science emerge with greater force, giving rise to possible controversies for which "we might speak either

[57] With Thesis 19, Althusser provided the following definition of *practical ideologies*: "*Practical ideologies* are complex formations which shape notions-representations-images into behaviour-conduct-attitude-gestures. The ensemble functions as practical norms that govern the attitude and the concrete positions men adopt towards the real objects and real problems of their social and individual existence, and towards their history" (Althusser 1967/1990, 83). Practical ideologies can be of various types and correspond to each type of specific practice: "religious, moral, juridical, political, etc. ideology" (Althusser 1967/1990, 84). Therefore, practical ideologies of a scientific type were defined by Althusser as the spontaneous ideology of the practice of scientists: "All science teaching, whether it wants to or not, conveys an ideology of science and of its findings—that is, a certain knowing-how-to-act-in-relation-to-science and its findings, based on a certain idea of the place of science in society, and on a certain idea of the role of intellectuals who specialize in scientific knowledge and therefore of the division between manual and intellectual labour" (Althusser 1967/1990, 94–95).

[58] The dominant ideological system corresponds to the culture of the dominant classes, which tends to operate according to a hegemonic logic: "Culture [meaning the culture of humanities in its different branches: literature, logic, history, ethics, philosophy, etc.] is the elite and/or mass ideology of a given society. *Not the real ideology of the masses* (for, as a result of class oppositions, there are several tendencies within culture) but the ideology that the ruling class seeks to inculcate, directly or indirectly, through education or other means, and on a discriminatory basis (one culture for the elite, one for the popular masses), into the masses they dominate. We are speaking here of an enterprise of *hegemonic* character (Gramsci): obtaining the *consent* of the masses through the diffusion of ideology (through the presentation and inculcation of culture). The dominant ideology is always imposed on the masses against certain tendencies in their own culture which are neither recognized nor sanctioned, but do resist" (Althusser 1967/1990, 93).

of a contradiction between a new problem and the existing theoretical means, or (and) of a disturbance of the entire theoretical edifice" (Althusser 1967/1990, 110).

According to Althusser, there are three main types of reactions to this kind of crisis. The first is that of scientists who, without leaving the scientific field, continue their work, convinced that the current crisis is nothing but a temporary phenomenon on the longer scientific journey. Although this reaction does not seem to rely on any philosophical form, in reality, it is based on a relatively naive, materialistic philosophy originating in scientific practice itself. The second reaction is of the opposite nature. Some scientists caught in such a critical juncture may question not just specific parts of disciplinary theories and practices but the entire rational edifice that supports science. In this situation, scientists look outside the scientific field for possible explanations for this crisis by developing forms of philosophical reflection that, however, are often marked by an "ideological exploitation" of a "spiritualist" type (Althusser 1967/1990, 111–112). The last reaction takes a "critical" form; that is, it generates an approach that calls attention to the naively dogmatic approach implicit in unreflective scientific practice, reactivating interest in investigating the foundations of science. What emerges are figures of philosopher-scientists who "turned against the philosophy" that was dormant within them, "condemning it as 'dogmatic,' 'mechanistic,' 'naive,' and, in a word, as 'materialist'; in short, they condemn it as a bad philosophy of science" (Althusser 1967/1990, 113). Consequently, they set out to provide science with a good philosophy of science, invoking their experience in scientific practice as their philosophical credentials:

> it is from *within science* that they claim to speak of science, that they set out to manufacture, *with scientific arguments* borrowed from the sciences—physics, psycho-physiology, biology—the good philosophy of science that science is claimed to need. [...] *A scientific philosophy of science* made *by scientists*. (Althusser 1967/1990, 113).

Although the spontaneous philosophy of scientists tends to take on more visible or "spectacular" characteristics during these "crisis" phases, in reality, it is profoundly active during the "'normal' course of scientific practice," in which it takes on "silent and invisible" forms (Althusser 1967/1990, 116). According to Althusser, philosophy intervenes to take on a clarifying and organizing role for this type of spontaneous philosophical reflection of science so as to create "lines of demarcation" that would allow scientists to acquire complete mastery of their theoretical practice (Althusser 1967/1990, 99). In this sense, Thesis 20 of the text stated: "The primary function of philosophy is to draw a line of demarcation between the ideological of ideologies on the one hand, and the scientific of the sciences on the other" (Althusser 1967/1990, 83). The boundary that philosophy constructs is aimed at bringing out "the 'space' where the ideological and the scientific merge but where they can and must be separated, to recognize each in its functioning and to free scientific practice from the ideological domination that blocks it" (Althusser 1967/1990, 99). In Bachelardian terms, Althusser defined this demarcation of the scientific from the ideological as a process of eliminating epistemological "obstacles" (Althusser 1967/1990, 100). It was therefore a reflection aimed at creating a boundary for "common sense" and a consequent emancipation of the philosophical reflections implicit in scientific practices.[59]

For Althusser, the spontaneous philosophy of scientists was composed of two main elements. The first, called "Element 1," is of "intra-scientific" origin and contains the sphere of beliefs—formalized in philosophical theses—produced by scientists about their own scientific practices. It is therefore a self-representation that scientists make of their own working activity, according to a "materialist and objectivist" philosophical approach that takes forms of this type:

> 1) belief in the real, external and material existence of the *object* of the scientific knowledge; 2) belief in the existence and objectivity of the *scientific knowledges* that permit knowledge of this object; 3) belief in the correctness and efficacy of the procedures of scientific experimentation, or *scientific method*, capable of producing scientific knowledge. (Althusser 1967/1990, 132–133)

Spontaneity, in this case, is immanent to scientific practice, which determines the emergence of certain forms of its conceptualization.

"Element 2," however, is of an "extra-scientific" nature and is characterized by philosophical theses that do not have an endogenous origin (as in Element 1) but are rather the result of a "reflection *on* scientific practice by means of philosophical Theses elaborated *outside* this practice by the religious, spiritualist or idealist-critical 'philosophies of science' manufactured by philosophers or scientists" (Althusser 1967/1990, 133). As in the previous one, these conceptions found beliefs that guide scientific practices.

According to Althusser, the main reason for the presence of ideological elements within scientific thought is the fact that Element 1 is constantly controlled by Element 2 in the contemporary situation. The latter conceptualizes the experience of scientific practices through "values" formulated in terms of philosophical theses that are external to the scientific field and that, "by exploiting the sciences, uncritically serve a certain number of objectives pertaining to practical ideologies" (Althusser 1967/1990, 133).

Althusser's proposal was to provide a materialistic, dialectical, and historical philosophy of scientific knowledge that could support the practical activity of scientists in order to remove the ideological residues contained in their spontaneous philosophy. However, the dominance of one element over another is not a transhistorical fact. In various phases of the history of science, the relationships between these elements have taken different forms. In the eighteenth century, for example, materialism provided the sciences with notable impulses for scientific activity.

---

[59] In *For Marx*, in fact, Althusser declared: "Left to itself, a spontaneous (technical) practice produces only the 'theory' it needs as a means to produce the ends assigned to it: this 'theory' is never more than the reflection of this end, uncriticized, unknown, in its means of realization, that is, it is a *by-product* of the reflection of the technical practice's end on its means. A 'theory' which does not question the end whose *by-product* it is remains a prisoner of this end and of the 'realities' which have imposed it as an end. [...] This point is crucial if we are to identify the most dangerous ideological menace: the creation and success of so-called theories which have nothing to do with real theory but are mere *by-products* of technical activity" (Althusser 1965/2005, 171 n.7).

However, Althusser's objective was not a return to the materialist ideals of the Enlightenment, because they were characterized by a latent naivety due to their lack of awareness of the historical determination of their contents. However, according to Althusser's analysis, these contents depended on the "juridical, moral and political 'consciousness' of the *intellectuals* of a rising class which is confident that it can take power thanks to the obviousness of Truth and Reason, and which has already put Truth in power" (Althusser 1967/1990, 138). For this reason, Althusser considered Enlightenment materialism to be still characterized by a form of "historical (juridical, moral, political) idealism" that ultimately derived from "their idealist conception of the omnipotence of scientific truth" (Althusser 1967/1990, 138). On the contrary, for Althusser, the new philosophy that was needed to support scientific practice was that of Marxist historical and dialectical materialism. Indeed, only the analysis of social, economic, and class relations allows for understanding the processes through which scientists' spontaneous philosophies were formed and therefore make the appropriate epistemological breaks permitting the qualitative, theoretical, and historical distinction between "science" and "ideology" to be made.[60] This epistemological position was, therefore, intrinsically linked to a historicization of traditional epistemology:

> to understand the content of an SPS [spontaneous philosophy of the scientist], it is indispensable to return to the history of the sciences and to the history of the spontaneous philosophies, which simultaneously depend upon the history of the sciences and on the history of philosophy. (Althusser 1967/1990, 136)

As part of the course on the "Spontaneous Philosophy of Scientists," Fichant, Pêcheux, Balibar, and Badiou spurred the initial exploration of the role of FHE. In their introduction to *Sur l'histoire des sciences*, Pêcheux and Balibar identified a "fundamental point of no return" in the dispute between continuists and discontinuists, in which Bachelard and Koyré, in particular, had played a decisive role (Fichant and Pêcheux 1969, 8–9). The emphasis on the discontinuist dimension was rooted, on the one hand, in the underlying political approach and, on the other, in the eclectic integration of Koyré and Bachelard's positions. Years later, Balibar would write about the positions of Fichant and Pêcheux: "In interpreting their articulation, they were greatly inspired by Koyré, whose conception of the *Scientific Revolution* was opposed, from the beginning, to both factual and theoretical positivism" (Balibar 1994, 100–101). Balibar also highlighted how Fichant and Pêcheux implemented a Bachelardo-Koyréian model based on the "rectification of Koyré's

[60] Also note that "Theoretical practice falls within the general definition of practice. It works on a raw material (representations, concepts, facts) which it is given by other practices, whether 'empirical,' 'technical' or 'ideological.' In its most general form theoretical practice does not only include scientific theoretical practice, but also pre-scientific theoretical practice, that is, 'ideological' theoretical practice (the forms of 'knowledge' that make up the prehistory of a science, and their 'philosophies'). The theoretical practice of a science is always completely distinct from the ideological theoretical practice of its prehistory: this distinction takes the form of a 'qualitative' theoretical and historical discontinuity which I shall follow Bachelard in calling an 'epistemological break'" (Althusser 1965/2005, 167–168).

ideas," in light of the Bachelardian concepts of "discontinuity and epistemological recurrence" and the Althusserian idea of "ideology" (Balibar 1994, 102). As I have already repeated several times, thanks to Lecourt, we have the first introductory monograph on Bachelard's thought and in which the expression *épistémologie historique* was resemantized in light of these debates:

> But historical epistemology already teaches us that science progresses through jolts, through sudden mutations, through reorganizations of its principles: in short, through dialectics. For this reason, the history of science must itself be dialectical: it will deal particularly with those "critical" moments in which the foundations of a science are reorganized. (Lecourt 1969, 80)

Lecourt then stressed, in a completely Althusserian manner, "how Bachelard was led to distinguish between two critical moments" (Lecourt 1969, 80) constituting the diachronic dimension of scientific knowledge:

> – the moment in which, at least on one point, in a specific field, the fabric of the pre-existing ideology is torn apart and scientificity is established. This is what he calls the moment of *rupture*;
>
> – the moment in which, once scientificity has already been reached, the specific science reorganizes its bases; this moment is called recasting, or reorganization. (Lecourt 1969, 80)

At the same time, Badiou, in *The Concept of Model: An Introduction to the Materialist Epistemology of Mathematics*, focused instead on the analysis of mathematics—in line with the tradition of mathematical thought that I have defined as the "minor canon" of FHE—with the aim of founding a materialist epistemology (Badiou 1969). To do this, the first move was to oppose the distinction between "empirical science" and "formal science" prescribed "by bourgeois epistemology" that Carnap had proposed in *The Logical Foundations of the Unity of Science* (Badiou 1969/2007, 5–6). The objective of Badiou's text was to distinguish three levels on which the word *model* had been used: ideological (bourgeois economics), philosophical (logical positivism), and scientific (mathematics).

As we have seen, to account for the discontinuities present in Marx's works, Althusser had radicalized the Bachelardian "rupture" into a "break." Similarly, in his 1969 text *Lenin and Philosophy*,[61] Althusser reworked another Bachelardian concept, the geographical metaphor of "epistemological regions," transforming it into a more radical opposition between "continents": "If in fact we consider the great scientific discoveries of human history, it seems that we might relate what we call the *sciences*, as a number of *regional* formations, to what I shall call the great theoretical *continents*" (Althusser 1969/1971, 38). As noted above, for the Althusserians of the École Normale in Paris, the "*epistemological break*" was also used to represent "the mutation in the theoretical problematic contemporary with the foundation of a scientific discipline" (Althusser 1965/2005, 32).

In light of this, for Althusser, the discontinuity suggested by Bachelard's regional metaphor was not strong enough to represent the diversity of forms of scientific reason that arise from the founding act of a discipline. The metaphor of the continent,

[61] This was a text presented at a lecture in 1968 at the Société Francaise de Philosophie and published in the "Thèorie" series directed by Althusser.

instead, allowed for the larger disciplinary conformations and their epistemological distance to be distinguished in terms of scientific criteria. At the same time, however, Althusser maintained the expression "epistemological regions" to describe the units that make up these continents. If, as Althusser claimed, before Marx there were only two continents, "the *continent of Mathematics* with the Greeks (by Thales or those designated by that mythical name) and the *continent of Physics* (by Galileo and his successors)," then Marx "opened up to scientific knowledge a new, third scientific continent, the continent of History, by an epistemological break whose first still uncertain strokes are inscribed in *The German Ideology*, after having been announced in the *Theses on Feuerbach*" (Althusser 1969/1971, 39).

Within these disciplinary divisions, the break made by Lavoisier in founding chemistry established a "regional science within the continent of physics" (Althusser 1969/1971, 39). In the same way, the various epistemological breaks within biology made by Darwin and Mendel, integrated with molecular chemistry, redefined the regional divisions within the continent of physics. Logic, for its part, would belong to the mathematical continent (Althusser 1969/1971, 39).

3.2.2 Episteme, Discursivity, and Power in Foucault

Foucault also interpreted the stimuli of Bachelardo-Canguilhemism in an original way (Gutting 1989, 52–54). After attending the ENS, he studied under Canguilhem, who directed Foucault's main doctoral thesis, *Folie et déraison. Histoire de la folie à l'âge classique*. His complementary thesis, on Kant's anthropology, was supervised by Jean Hyppolite. Foucault quickly became one of the key intellectuals of French philosophical culture thanks to the publication of works such as *The Birth of the Clinic* (1963), *The Order of Things* (1966), and *The Archaeology of Knowledge* (1969) and was elected to the prestigious Collège de France in 1969.[62] In this same period, he was responsible for setting up the philosophy department of the Université de Vincennes, an experimental university launched in the wake of the French May 1968 movement in response to student demands. Intellectuals of the highest regard were part of this institution, some of whom were very close to the FHE approach, such as Michel Serres, Badiou, Rancière, Regnault, Balibar, François Châtelet, and, among the sociologists, Jean-Claude Passeron (a close collaborator and coauthor of Bourdieu) (Soulié 1998).

As an heir to the Canguilhemian approach, in *History of Madness* (1961), Foucault developed the idea of historical epistemology in an original way, as a reconstruction of conceptual filiations. Already in this text, the first indications of

[62] For a sociological and historical reconstruction of Foucault's intellectual development, see Eribon (1989/1991) and Pestaña (2006).

some of the key themes of his perspective, both during his 1960s investigations of knowledge systems and his later analyses of power relations, are recognizable.[63]

In *History of Madness*, Foucault analyzed in detail the evolution and transformation of concepts such as "mental illness," "madness," and "unreason" between the Middle Ages and the modern era. His objective was to highlight the relationships between the formation process of these concepts at an epistemic level and the historical, social, and cultural conditions from which they emerged.

Compared to Canguilhem's historical epistemology—which was dedicated to the reconstruction of the conceptual transformations within positive knowledge—Foucault shifted the focus of his investigation to a search for the conditions that permitted these forms of knowledge to emerge.[64] In this perspective, the internal transformations of medical and psychopathological knowledge were read as part of a broader cultural-cognitive system. In this sense, these sciences' intrinsic rationality or capacity to produce veridical discourses was no longer the object of analysis; instead, Foucault sought to trace how Western reason was founded upon the exclusion of its Other. *History of Madness* described the emergence of modern rationality through a careful analysis of what it negated, or through the analysis of what it necessarily excluded in order to be founded. This methodology, focused on investigating the essence of cultural phenomena through what they exclude, or what traces remain of what lies invisible in the background of social phenomena, played a fundamental role in all of Foucault's production as he typically focused on limit-experiences (such as madness or prison) in which this dialectic could emerge with greater clarity (see Paltrinieri 2012, 94–96, 110). Foucault's objective was to go back to the original gesture that set in motion the dialectic of inclusion/exclusion by establishing the boundary between reason and unreason [*déraison*]:

> Madness becomes a form related to reason, or more precisely madness and reason enter into a perpetually reversible relationship which implies that all madness has its own reason by which it is judged and mastered, and all reason has its madness in which it finds its own derisory truth. Each is a measure of the other, and in this movement of reciprocal reference, each rejects the other but is logically dependent upon it. (Foucault 1961/2006, 28–29)

The emergence of positive knowledge—or effective scientific forms—appears, so to speak, *après coup*; nothing but an effect of this gesture allows reason to elect madness as its object of investigation precisely because of the aforementioned boundary. Foucault's theoretical objective was to shift the axis of inquiry from the sphere of discursive production to a reconstruction of the historical-cultural experience of madness—or, rather, to a reconstruction of how knowledge, practices, discourses, and the constitution of epistemic objects are interwoven. *History of Madness* was a research project aimed at reconstructing the meaning of a limit-experience—that is, a simultaneous investigation into both the foundations and limits of Western culture. In this text—though not yet completely formulated—

[63] For an in-depth analysis of Foucault's early research on madness, see Basso (2022).

[64] For an analysis of the relationship between Canguilhem and Foucault, especially regarding the problem of "norms," see Macherey (2009).

Foucault's strictly epistemological analysis of the variation of concepts was already intersecting with that of disciplinary power and the analysis of the role played by discursive formations such as the ethical, aesthetic, philosophical, literary, scientific, and others.

With the publication of *The Birth of the Clinic*, *The Order of Things*, and *The Archaeology of Knowledge*, Foucault developed his so-called archeological method. Following Bachelard and Canguilhem, Foucault argued that the history of science is characterized by dispersions and ruptures rather than homogeneity and continuity (Gutting 1989, chapter 1). Archeology's specific task is to identify these moments of discontinuity in the composition of knowledge (both historical and regional) (Braunstein 2008). However, while Bachelardo-Canguilhemian epistemology worked at the level of the analysis of intradiscursive formations within scientific practices—that is, how the production of internal discourses in scientific knowledge occurs—Foucault's archeology rather aimed to bring out the structure of interdiscursive formations to show the dynamics through which veridical discourses[65] are constituted (see Davidson 2001).

In *The Birth of the Clinic*, Foucault continued and extended Canguilhem's reflections on the philosophy of medicine developed in *The Normal and the Pathological* (see Macherey 2009, 102–109). Here, Foucault continued investigating limit-experiences as they can more easily reveal the historicity of the transcendental forms upon which all forms of knowledge are based. These forms, in fact, represent the dialectically negative space that gives meaning to the positive configurations of our culture. As in *History of Madness*, Foucault's objective was to investigate the negative space constituted through the emergence of positive knowledge.

In *The Birth of the Clinic*, Foucault applied his archeological methodology to the analysis of the "medical gaze." The fundamental subject of this work was death and the transformation of its status from a mere antithesis or denial of life to an apparatus that doctors used to understand "the truth of life" and the essence of illness: "Life, disease, and death now form a technical and conceptual trinity" (Foucault 1963/1976, 144). Foucault identified this historical-epistemological transition in the epistemological rupture that took place between the end of the eighteenth and the beginning of the nineteenth centuries with the birth of pathological anatomy, understood as a form of knowledge that took the sick body and the dead body as its objects of study. In this phase, therefore, a reversal of perspective in the "medical gaze" came about through the progressive use of pathological anatomy—compared to the normal anatomy of the healthy body—which was at the basis of the birth of the clinic as a space—that is, as the place where medical knowledge was transmitted from a practical point of view. Death, the sick body, and what was against nature became part of the practical knowledge of treatment, acquiring a positive status with respect to clinical knowledge.

---

[65] For an in-depth discussion of Foucault's relationship with the concept of truth and its ethical implications, see Lorenzini (2023).

Thus, death and illness became an object of strategic and functional knowledge for the protection of life. According to Foucault, before this radical revolution, a disease was analyzed only starting from its external symptomatology, conceptualized as a superficial phenomenon through "ordered and defined kinships" (Foucault 1963/1976, 136). With the emergence of anatomo-clinical experience,

> the medical eye must see the illness spread before it, horizontally and vertically in graded depth, as it penetrates into the body, as it advances into its bulk, as it circumvents or lifts its masses, as it descends into its depths. (Foucault 1963/1976, 136)

Compared to the past, the medical gaze followed a new trajectory, from the symptom to the tissues in which the primary cause of the disease was to be found, or from "the manifest to the hidden" (Foucault 1963/1976, 135). After the revolution of pathological anatomy, the doctor's gaze was "no longer that of a living eye, but the gaze of an eye that has seen death" (Foucault 1963/1976, 144).

The Order of Things was the work that brought Foucault international fame. In this work, he constructed "an archeology of the human sciences," no longer by looking at the margins of knowledge but, instead, by focusing on the internal methods that organize knowledge. Here, Foucault made extensive use of the concept of *episteme* for the first time, understood as a broad space of conditions of possibility within which the positivities of science emerge, or rather, the way criteria for truth and falsity are formed for a given historical period. Foucault carried out this investigation by following the processes through which general grammar was transformed into linguistics, natural history into biology, and the science of wealth into economics.[66]

Here, Foucault applied his archeological method to the analysis of the internal positivities of scientific knowledge, revealing the consolidation process of the modern *episteme* within which the human sciences were constituted. In post-Kantian terms, the objective was to show how the categories through which the world is experienced are intrinsically historical-cultural as well as connected to the ways the contents of these ordering systems exist:

> The fundamental codes of a culture—those governing its language, its schemas of perception, its exchanges, its techniques, its values, the hierarchy of its practices—establish for every man, from the very first, the empirical orders with which he will be dealing and within which he will be at home. (Foucault 1966/1994, xxii)

Foucault's archeology was situated within a "middle region" of an epistemological field, "between the already 'encoded' eye and reflexive knowledge," from which a "pure experience of order and of its modes of being" could be had (Foucault 1966/1994, xxii–xxiii). On the basis of its development of a method to bring out "the ordering codes" from within a culture and in relation to a specific historical time, *The Order of Things* is considered one of the classic texts of French structuralism.

The Archaeology of Knowledge was the final act of this phase of Foucault's thought. It was a purely methodological work in which Foucault, in light of the

[66] For a critical reconstruction of these transformation processes, see Gutting (1989, 139–226).

analyses developed in his previous works, sought to clarify what exactly the archeological method was. Analyzing interdiscursive formations meant looking at how scientific discourse is embedded in the broader network of forms of knowledge. "Knowledge" [*savoir*], in Foucault's conceptualization, is distinct from "knowledge" [*connaissance*]. The term "knowledge" [*connaissance*] refers to discursive formations relating to classes of objects that, through a process of rationalization, are judged knowable or that presuppose a clear distinction between the known object and the knowing subject. The concept of "knowledge" [*savoir*] refers to a broader level of analysis and designates the process through which a knowing subject undergoes modifications precisely in his act of knowing (Hacking 2002, 77; Revel 2009, 90). This level of analysis was thus aimed at investigating the effects of the organization of knowledge as it produced subjectivities, objects, and discourses.

It was therefore a question of developing a method capable of identifying the specific *epistemes* at each stage of knowledge transformation. Taking up a variation on the anthropological approach developed by authors such as Meyerson, Metzger, Koyré, and Canguilhem, Foucault's archeology aimed at identifying what he called the "historical *a prioris*," which define the space of what is visible and what can be stated in a given historical moment (see Taylan 2018, chapter 3). These historical (and concrete) a prioris are situated halfway between the analysis of the pure and universal conditions of knowledge (in the Kantian sense) and empirical experience. The Foucauldian historical a priori therefore has a dual nature. On the one hand, it performs a structural function in that it determines the space of possibility for the creation of specific forms of knowledge. In this sense, a set of "historical a prioris" determines the *episteme* of an era. On the other hand, it fulfills a critical function with respect to classical structuralism (proposed by authors such as Ferdinand de Saussure, Roman Jakobson, Claude Lévi-Strauss, Roland Barthes, and Althusser). While structuralism tended to conceive structural determinants as unchangeable, the Foucauldian a priori was subjected to constant transformation (e.g., by bringing out new criteria for the truth or falsehood of propositions). Thus, the historical a priori is, at the same time, an effect of the dynamics of changing knowledge as well as its condition of existence. In *The Order of Things*, Foucault claimed the following:

> Quite obviously, such an analysis does not belong to the history of ideas or of science: it is rather an inquiry whose aim is to rediscover on what basis knowledge and theory became possible; within what space of order knowledge was constituted; on the basis of what historical *a priori*, and in the element of what positivity, ideas could appear, sciences be established, experience be reflected in philosophies, rationalities be formed, only, perhaps, to dissolve and vanish soon afterwards. (Foucault 1966/1994, xxiii)

Davidson effectively summarized Foucault's theoretical position by arguing that "*science* stands to *epistemology* what *knowledge* [*savoir*], which I think Foucault often uses interchangeably with *discursive formations*, stands to *archeology*" (Davidson 2001, 193). In this sense, epistemology's object of investigation is science, and archeology targets "discursive formations." If the task of epistemology is to reconstruct the analysis of veridical discourses, archeology presents itself as a meta-epistemology that foregrounds the processes through which the distinction between true and false emerges within a specific context of knowledge. In other words, it targets how the discursive formations that make up the "*epistemes*" of a specific historical period are constituted. By the term *episteme*, Foucault meant "the total set of relations that unite, at a given period, the discursive practices that give rise to epistemological figures, sciences, and possibly formalized systems" (Foucault 1969/1972, 191). Defined this way, the term does not represent a type of "knowledge" or "rationality," because these already presuppose the subject–object division. On the contrary, *episteme* means the set of discursive regularities that describe the complex of relationships between the sciences in a given era:

> The description of the episteme presents several essential characteristics therefore: it opens up an inexhaustible field and can never be closed; its aim is not to reconstitute the system of postulates that governs all the branches of knowledge (*connaissances*) of a given period, but to cover an indefinite field of relations. Moreover, the episteme is not a motionless figure that appeared one day with the mission of effacing all that preceded it: it is a constantly moving set of articulations, shifts, and coincidences that are established, only to give rise to others. As a set of relations between sciences, epistemological figures, positivities, and discursive practices, the episteme makes it possible to grasp the set of constraints and limitations which, at a given moment, are imposed on discourse [...] it is what, in the positivity of discursive practices, makes possible the existence of epistemological figures and sciences. Lastly, we see that the analysis of the episteme is not a way of returning to the critical question ("given the existence of something like a science, what is its legitimacy?"); it is a questioning that accepts the fact of science only in order to ask the question what it is for that science to be a science. In the enigma of scientific discourse, what the analysis of the episteme questions is not its right to be a science, but the fact that it exists. And the point at which it separates itself off from all the philosophies of knowledge (*connaissance*) is that it relates this fact not to the authority of an original act of giving, which establishes in a transcendental subject the fact and the right, but to the processes of a historical practice. (Foucault 1969/1972, 191–192)

In this sense, *The Archeology of Knowledge* first recognized the existence of a domain dealing with intradiscursive formations called "epistemology." This is the space in which, for example, the analyses relating to Canguilhem's veridical discourses would be placed (see Chap. 2). Foucault, however, through his archeological project, aimed to introduce a further level of analysis—that is, the interdiscursive relationships constituting "knowledge" [*savoir*]. Archeology's objects of investigation are discursive formations—that is, heterogeneous sets of utterances that at the same time describe, regulate, and bring into being objects, practices, and cognitive aspects. This sphere includes the conceptualizations and observation procedures of the sciences, as well as its legal regulations, administrative rules, social norms, etc. The complex of discursive formations describes what is sayable and unsayable within a given historical period—or how something becomes the object of a form of knowledge or, to use Hacking's expression, what can be a candidate for "truth-or-falsehood" (Hacking 2002, 77).

All of Foucault's works mentioned so far were united in their discontinuist approach (Braunstein 2002). Starting from *History of Madness*, he had analyzed the fundamental moments of rupture in the history of the conceptualization of "madness" from the Middle Ages to the modern age. *The Birth of the Clinic* was also a moment of discontinuity in which the "medical gaze" and the "clinical object" emerged at the same time. Passing through the research of *The Order of Things*, this path was completed with *The Archaeology of Knowledge*, which, as outlined above, was an attempt to systematize this first phase of Foucault's thought.[67]

Pointing out the necessity of detachment from those historical perspectives of *longue durée*, Foucault argued that

> in place of the continuous chronology of reason [...] there have appeared scales that are sometimes very brief, distinct from one another, irreducible to a single law [...] and which cannot be reduced to the general model of a consciousness that acquires, progresses, and remembers. (Foucault 1969/1972, 8)

Therefore, "the notion of discontinuity assumes a major role in the historical disciplines" (Foucault 1969/1972, 8). In the past, the historian's task was precisely to delimit and reduce these moments of fracture that had to be filled in through his analysis. On the contrary, now the historian shifts their attention from "vast unities like 'periods' or 'centuries' to the phenomena of rupture, of discontinuity" (Foucault 1969/1972, 4). This approach was, once again, a direct consequence of the Bachelardo-Canguilhemism of France in those years:

> There are the *epistemological acts* and *thresholds* described by Bachelard: they suspend the continuous accumulation of knowledge [...]; they direct historical analysis away from the search for silent beginnings, and the never-ending tracing-back to the original precursors, towards the search for a new type of rationality and its various effects. There are the *displacements* and *transformations* of concepts: the analyses of G. Canguilhem may serve as models; they show that the history of a concept is not wholly and entirely that of its progressive refinement, its continuously increasing rationality, its abstraction gradient, but that of its various fields of constitution and validity, that of its successive rules of use, that of the many theoretical contexts in which it developed and matured. (Foucault 1969/1972, 4)

For Foucault, these moments of discontinuity have a triple function for the historian. First, such moments of rupture are no longer gaps in the documentary material but rather a deliberate operation of the historian. Second, these interruptions are the result of the philosopher's description and not something that they must eliminate from it. The historian "is trying to discover the limits of a process, the point of inflexion of a curve, the inversion of a regulatory movement, the boundaries of an oscillation, the threshold of a function, the instant at which a circular causality breaks down" (Foucault 1969/1972, 9). Finally, discontinuity becomes the concept that is continually investigated by the historian through their own practice:

> The notion of discontinuity is a paradoxical one: because it is both an instrument and an object of research; because it divides up the field of which it is the effect; because it enables the historian to individualize different domains but can be established only by comparing those domains. (Foucault 1969/1972, 9)

This is the defining motif of Foucault's *épistémologie historique*—of Bachelardo-Canguilhemian origin—which clearly distinguishes it from historiographical models

[67] Some have also argued that Foucault, particularly in *The Archaeology of Knowledge*, proposed a model capable of overcoming the debate between historical continuism and discontinuism. On this point, see Kusch (1991), who also discussed Foucault's relationship with other authors affiliated with FHE and other historiographic trends, such as the *Annales* and the history of ideas.

of the *longue durée* or of the history of ideas with respect to which "archaeology is much more willing than the history of ideas to speak of discontinuities, ruptures, gaps, entirely new forms of positivity, and of sudden redistributions" (Foucault 1969/1972, 169).

In the wake of epistemological regionalism, Foucault also used spatial metaphors to represent the internal divisions of "knowledge" [*savoir*]. In the tenth chapter of *The Order of Things*, for example, he maintained that "The domain of the modern *episteme* should be represented rather as a volume of space open in three dimensions" (Foucault 1966/1994, 378). The first of these dimensions describes the space of the mathematical and physical sciences, whose principle is the deductive and linear order of evident or verified propositions. The second dimension is occupied by the sciences of language, life, and the production and distribution of wealth, whose object is the study of the causal correlations between discontinuous but analogous elements. Lastly, the third philosophical dimension is dedicated to the structuring of "regional ontologies"[68] through which the foundations of the empirical sciences can be sought. For this reason, Foucault described *The Order of Things* as "a strictly 'regional' study" (Foucault 1970/1994, x). Likewise, *The Archaeology of Knowledge* also explicitly referred to epistemological regionalism: "In archaeological analysis comparison is always limited and regional. Far from wishing to reveal general forms, archaeology tries to outline particular configurations" (Foucault 1969/1972, 157). This attention to the spatial dimension focuses the analysis "not so much [on] the genesis of our sciences as an epistemological space specific to a particular period" (Foucault 1970/1994, xi). In the study of "discursive regularities," the objective is to account for the peculiarities of the different forms of rationality established by the "*epistemes*": "It is understandable in these conditions that we should distinguish carefully between scientific domains and archaeological territories: their articulation and their principles of organization are quite different" (Foucault 1969/1972, 183).

Foucault's regionalist approach, which investigated reason at a spatial level, also played a significant role in leading him beyond the project of archeology: "Once knowledge can be analysed in terms of region, domain, implantation, displacement, transposition, one is able to capture the process by which knowledge functions as a form of power and disseminates the effects of power" (Foucault 1976/1980, 69). Indeed, Foucault's investigation did not stop at analyzing how discursive forms bring objects of knowledge into being; he also traced the material changes deriving from them in terms of social and cultural practices. In subsequent works such as *Discipline and Punish* (1975) and the vast list of courses held at the Collège de France starting in 1971, Foucault progressively abandoned the term *episteme* to use the broader expression "order of discourse," marking a transition in his intellectual development from the "archeological" to the "genealogical" method. This transformation in Foucauldian thought was rooted in the progressive extension of his field of

[68] I have previously highlighted how Bachelard had already established the idea of "regional epistemologies," starting from the Husserlian idea of "regional ontologies."

investigation from a domain strictly focused on the production of discourses to an analysis of how discourses interact with the dynamics through which practices, strategies, and institutions are formed and function (see Davidson 2001).

The emergence of objects within certain discursive formations leads to the creation of institutional apparatuses that are combined with the establishment of epistemic spaces such as "the asylum," "the clinic," and "the prison." Discursive formations operate within these spaces; that is, they translate what has been established as sayable, visible, and enunciable into concrete practices. An object of knowledge also becomes an object of specific practices in these peculiar spaces. For Foucault, the act of knowing is inseparable from the act of disciplining and regulating, which, in the horizon of praxis, is transposed into concrete actions that can be of an educational, coercive, or therapeutic nature, among others.

In Foucault's archeological texts, the transformation of discursive formations into praxis referred to a political dimension of the effects of knowledge. Dealing with this problematic while extending the archeological project to a genealogical one was the specific aim of his subsequent works. Whereas epistemology investigated "intradiscursive" relationships and archeology analyzed "interdiscursive" relations, with "genealogy" he shifted to the study of "extradiscursive" relationships (Davidson 2001, 205). And while science was the object of epistemology and "knowledge" [*savoir*] that of "archeology," "genealogy" opened Foucault's investigation to the relationships between knowledge and power (see Braunstein 2002, 952). Starting from this position, various lines of investigation were developed that worked to expose the dynamics of how disciplinary power functions and its impact on the level of biopolitics, leading Foucault to produce analyses of power that focused on governmentality and the microphysics of power as interpretative keys for understanding the processes of subjectivation.

3.2.3 Bourdieu: Between Sociology and Historical Epistemology

Bourdieu also trained in the cultural and intellectual context of the ENS in Paris that was marked by Bachelardo-Canguilhemism. After earning a degree in philosophy, he maintained a close relationship with Canguilhem, with whom he developed a doctoral project (never completed) dedicated to analyzing the temporal structures of emotional life, starting from Husserl's thought. While in Algeria for his military service, Bourdieu was drawn to the social sciences, abandoning the conventional philosophical path. The Algerian years provided him with a series of ethnographic case studies that he used to investigate fundamental aspects of models of cultural production and the organization of society.

Like Althusser and Foucault, Bourdieu also dealt with the legacy of classical structuralism and the epistemological reflections of Bachelardo-Canguilhemism. In particular, Bourdieu elaborated a structural-constructivist position for the social sciences that was both epistemological and methodological. He was unsatisfied with the rigid structural determinism advanced by Lévi-Strauss' studies of kinship systems, so Bourdieu focused on developing an approach that could provide an integrated explanation of the degree of structural conditioning as well as the ability of social agents to interact with and modify the structures within which they move. This thematic core was developed through intense social research applied to a wide spectrum of fields of cultural production, including art, journalism, religion, economics, education, and science. Here, I limit myself to outlining the aspects of the most immediate relevance to historical epistemology.

After returning from Algeria, Bourdieu established himself as a key intellectual for the 1968 generation, publishing a series of works with Passeron,[69] including *Les étudiants et leur études* (1964), *The Inheritors: French Students and Their Relations to Culture* (1964), and *Reproduction in Education, Society and Culture* (1970), explicitly dedicated to analyzing scholastic inequalities and the relationships between educational systems and social reproduction (which were extremely significant issues for the French May 1968 movement) (Bourdieu and Passeron 1964a, 1964b, 1970). In the 1980s, he returned to reflecting on these themes, publishing *Homo Academicus* (1984), in which he addressed the impact of May 1968 on the structures of the academic field, and *State Nobility: Elite Schools in the Field of Power* (1989), dedicated to analyzing the system of social reproduction underlying the French *Grandes écoles* (Bourdieu 1984, 1989). In these studies, Bourdieu reflected on the relationships between ways of organizing knowledge and the social, economic, and political structures of contemporary societies.

Despite his changing his intellectual trajectory from philosophy to the social sciences, Bourdieu's thought remained strongly linked to the discussion of epistemological issues. The problem of the roots, social function, and validity of the sciences (both social and natural sciences) was one of the nerve centers of his analyses. In Bourdieu's works, the social structure of science was not only an object of sociological investigation but also a matter of epistemological concern—as a means for justifying the practice of sociological research (the problem of reflexivity).

In 1968, Bourdieu, in collaboration with Passeron and Jean-Claude Chamboredon, published *The Craft of Sociology*. This text—a manual written for the epistemology of sociology classes that Bourdieu and Passeron taught at the EHESS—trained at least two generations of specialists, opening the way to historical investigations of the formation of the social sciences and also encouraging new forms of critical attention nourished by the sociology of sociological training and the sociology of science (Brian 2012). This fact is even more significant given that this text was published in the years in which American sociology, marked by a heavily empiricist approach, was largely hegemonic throughout the world. Indeed, as can clearly be seen from Beate Krais's interview with Bourdieu in 1988, *The Craft of*

[69] Bourdieu and Passeron were close collaborators until 1972, when they went their separate ways. Elsewhere, I have traced the evolution of their intellectual relationship, paying particular attention to their respective positions within social science epistemology (Ienna 2018).

Sociology was openly seen as providing an alternative epistemology of the social sciences to contrast the theoretical tendencies of the American sociological approach.[70]

The researchers gathered around Bourdieu proposed an integration of theoretical tools coming simultaneously from historical epistemology (French and otherwise) and the social sciences (such as those of Marx, Durkheim, Mauss, and Weber). On this point, in fact, Bourdieu declared: "I tried to transpose into the field of social sciences a whole epistemological tradition represented by Bachelard, Canguilhem, Koyré" (Bourdieu and Krais 2005/1991, 248). Also in this case, the resulting model was an eclectic integration of Bachelardian theory combined with the reflections of Canguilhem and Koyré, which represented a typical posture of FHE in this period.[71]

In this text, Bourdieu and his coauthors adapted the classical theories of Bachelardo-Canguilhemism to sociology. Through the concept of *métier* ["craft" or "profession"], and in alignment with the principles of Bachelard's "applied rationalism," they argued for the inseparability of the analytical level of "scientific practices" and "epistemology" as a meta-level of reflection upon them.[72] Failing to adopt this methodological stance would risk "forc[ing] on researchers a split image of scientific work" (Bourdieu et al. 1968/1991, 1). This split would, on the one hand, produce an image of an "epistemic reason" completely detached from the analysis of scientific practices and, on the other, subject the exercise of the profession to the social division of research work. Because, as Bachelard argued, "science does not have the philosophy it deserves" (Bachelard 1953, 20), Bourdieu and his colleagues proposed to analyze the social sciences with a series of tools taken from FHE in order to facilitate an internal renewal of these disciplines.

Following the thesis of epistemological regionalism, Bourdieu and his colleagues argued that a specificity of sociological rationalism existed as much as one existed for electrical rationalism (Bachelard 1949) or the rationalism of the life sciences (Canguilhem 1994). The book's objective was precisely to foreground the scientific criteria regulating this epistemological region so that sociology could prove itself to

[70] As a positioning strategy in the international sociological field, Bourdieu disparagingly called Parsons, Merton, and Lazarsfeld the "Capitoline triad" of sociology. He also stated that during this period, he refused to attend a series of lectures by Lazarsfeld at the Sorbonne (see Bourdieu and Krais 2005, v–xix).

[71] For example, the discontinuist perspective was a fixed point of this volume: "In other words, if it is true that every scientific theory clings to the 'given' as a historically constituted, provisional code which, for the space of an epoch, constitutes the sovereign principle of an unequivocal distinction between the true and the false, the history of a science is always discontinuous because the refinement of the interpretive grid never goes on forever but always culminates with the substitution of one grid for another" (Bourdieu et al. 1968/1991, 29).

[72] From the text's first lines referring to Comte, the impossibility of maintaining a separation between the theoretical analysis and that of the concrete functioning of the mechanisms of scientific practice was made explicit, a thesis further justified by Bachelard's idea of open or applied rationalism, which presupposes a constant dialogue between science and technology.

be a science like the others (in its relationship with/opposition to the natural sciences) (Bourdieu et al. 1968/1991, 8).[73]

The unity of this region was determined by the ways sociology constructed its object and by the methods and tools used to study it. The specificity that distinguished this form of rationalism was also derived from the type of "epistemological obstacles" it had to overcome to demonstrate itself to be a scientific practice. It was an epistemological model that went against the grain of what Bachelard called the "fixism" of the neo-positivist approach:

> This epistemology rejects the formalism and fixism of a single, indivisible Reason in favour of a pluralism of rationalisms linked to the scientific domains that they rationalize. [...] It is thus predisposed to provide a language and theoretical assistance to the social sciences, which, in order to constitute their regional rationalism, have to overcome some particularly daunting epistemological obstacles. (Bourdieu et al. 1968/1991, 81)

For sociology to prove itself as a science like any other, it thus had to overcome a series of epistemological obstacles. The first obstacle was the attribution of validity to knowledge derived from immediate knowledge. The second consisted of breaking with the so-called illusion of transparency—that is, the tendency to claim understanding of a social fact starting only from an analysis of the intentions that produced it (Bourdieu et al. 1968/1991, 15). Third, gaining distance from "artificialism," or "the illusory representation of the genesis of social facts according to which the social scientist can understand and explain these facts merely through 'his own private reflection,'" was necessary because the latter rests "on the presupposition of innate wisdom which, being rooted in the sense of familiarity, is also the basis of the spontaneous philosophy of knowledge of the social world" (Bourdieu et al. 1968/1991, 15). Sociology was, instead, an experimental science that built explanatory theories capable of defining social facts in terms of relationships and variations, excluding the use of essences, substances, invariants, and residues—that is, human or social "natures"—as explanatory principles. Furthermore, this discipline had to radically break with the "temptation to prophesy." The sociologist is often assigned the ability to respond to ultimate questions relating to civilization by establishing a relationship with its audience similar to that which a prophet has with their audience. If the chemist must fight the alchemist that they has within themself, then the sociologist must not be dragged down by the temptation to engage in prophetic sociology.

For this reason, the sociologist must constantly maintain an "epistemological break" with common sense and its related naive language. The inheritance of ideas is linked to the inheritance of words. The naive language of common sense brings with it a "spontaneous sociology" that structures the way social agents view and divide up

[73] Translating Bachelard's applied rationalism to the case of sociology, the authors argued as follows: "Transposed to the social sciences, this philosophy of scientific work [...] can yield the principles of a reflection capable of inspiring and controlling the concrete acts of a truly scientific practice, by defining the specificity of the principles of the 'regional rationalism' characteristic of sociological science" (Bourdieu et al. 1968/1991, 8).

the world, which is opposed by a scientific vision of the social world. The divisions made by common language clearly act as unconscious preconstructions of objects, bringing with them a form of implicit philosophy of society. In sociology, the break with common sense is even more important than in other areas of research because of its unique relationship to its object of study:

> In observation and experimentation, the sociologist enters a relationship with his object which, being a social relationship, is never one of pure knowledge. The data therefore appear to him as living, singular, "all-too-human" configurations that tend to impose themselves as object structures. (Bourdieu et al. 1968/1991, 14)

Following from these observations, sociology needed to posit an "objectivation" of the subject of objectivation—that is, the conditions of the possibility of sociological practice, at the center of its methodology. A question of adopting an "epistemological vigilance" allowed sociologists to distance themselves from the preconstructed objects of the naive vision of common sense.[74] Epistemological vigilance was particularly needed in the case of the social sciences, "where the separation between everyday opinion and scientific discourse is more blurred than elsewhere" (Bourdieu et al. 1968/1991, 13). This "reflexive" dynamic allows the observer's implicit assumptions in the relationship with a research object to emerge. Following Bachelard's precepts, Bourdieu and colleagues argued that reflexivity must shape the *modus operandi* and not be applied in the analysis of the *opus operatum* (Bourdieu et al. 1968/1991, 8); that is, it must be incorporated as a "system of intellectual habits" (Bourdieu et al. 1968/1991, 1–2).

The second part of the text was dedicated to delineating a theory, once again explicitly Bachelardian, of the construction of the research object. From this point of view, and in continuity with the first part, a distinction needed to be made between "common objects" (which are the products of a language derived from common sense) and "scientific objects" (which, on the contrary, represent an achievement against common sense). In this way, FHE was integrated with Durkheim's theses in *The Rules of Sociological Method* and vice versa (Bourdieu et al. 1968/1991, 34). Adopting the theses of Koyré's *Galileo Studies*, the authors suggested that "direct experience played no part, other than as an obstacle, in the birth of classical science" (Bourdieu et al. 1968/1991, 37). All the sciences thus struggled against experience (in the empirical-positivist sense of the term); for this reason, the epistemological vector proceeds from the rational to the real and, therefore, is essentially a "realizing" vector. The authors' ambition was to deconstruct the "dogma of immaculate perception," arguing rather that a "social fact" is always a conquered fact, constructed and verified against the naive experience of the social world.

The construction of the object then had to pass through a problematization of the relationship between hypothesis and experience and the role of scientific instruments and their intrinsic theoretical nature. Through a continuous comparison with the classic texts of epistemology, sociology, and anthropology, the authors examined traditional techniques of inquiry such as statistics and interviews, acknowledging

[74] Bourdieu subsequently transformed this "epistemological vigilance" into the concept of reflexivity, understood as "objectivating the subject of objectivation" (Bourdieu 2001a/2004, 88).

how their preparation forms a fundamental part of the construction of the scientific object.

For example, statistics, one of the most traditional sociological research tools, must be interrogated at times by the social scientists using it in order to clarify their methodological assumptions. In other words, the process through which the object of study is created must always be reflexive (e.g., asking whether there is a real break with the representation of objects of common sense, or, again, questioning how the object of study is defined). Referring to Max Planck, the authors stated that

> "To set up apparatus to make a measurement is in itself to ask nature a question," said Max Planck. Measurement and measuring instruments and, more generally, all the operations of sociological practice, from drawing up questionnaires and coding to statistical analysis, are so many theories in action, inasmuch as they are conscious or unconscious procedures for constructing facts and relations between facts. (Bourdieu et al. 1968/1991, 39)

Starting from this, the sociologist clearly must neither give in to the temptation, on the one hand, to pull the information that they need to confirm their theory directly from the subject's mouth nor, on the other, act in a totally passive manner in front of the data because, in this way, it would do nothing but reproduce the philosophy implicit in spontaneous sociology. When preparing a test, a statistic, or an interview, the object of investigation must be constructed in a way that breaks with common sense by relating factors that remain separate in naive perception. For this reason, the authors mention as a point of reference the Galilean revolution brought about by Durkheim in sociology—that is, the theoretical gesture that he made in studying social facts as things, as systems of quantifiable relations. The observation and experimentation techniques used in sociology, like the tools of other sciences, are only apparently neutral.

Research activity is divided into a more or less structured set of procedures that act as a paradigm that is unconsciously incorporated by most researchers. These operational protocols are the projection of bureaucratic organization into an epistemological space, and its constraints are similarly incorporated into technical tools (Bourdieu et al. 1968/1991, 71). In this sense, sciences are organized around objects that no longer have anything in common with those coming from common-sense perception. Sociology represents the division of scientific labor as "a real division of reality" (Bourdieu et al. 1968/1991, 33). The classic representation of scientific procedure as a cycle of successive phases (observation, hypothesis, experimentation, theoretical elaboration, observation, etc.) is nothing more than a positivist pedagogical illusion that replaces reality with an enchanted image of epistemological acts (especially regarding the "bureaucratized" division of scientific work). This misrepresentation, in fact, leaves out the logical order of epistemological acts that cannot be reduced to chronological order.

This epistemological stance remained unchanged in Bourdieu's subsequent works, in which he developed a style of sociological investigation that was keenly focused on the cognitive dimension of power (and from which he derived the notion of symbolic power). But before proceeding to a detailed analysis of Bourdieu's specific contributions to the study of the "scientific field," I must note how some theoretical elements drawn from the FHE tradition were implemented in Bourdieu's sociohistorical theory and methodology.

Although Bourdieu is largely known—especially in Anglophone contexts—as a theorist of social reproduction, in reality, his contributions also aimed at forging a general theory of change, transformation, and symbolic revolutions.[75] In Bourdieu's social theory, history occupies a decidedly central place, and many cultural historians have adopted his perspective in their work.[76] For example, Eric Hobsbawm stated that by reading Bourdieu, he was able to find a common approach: "For him as for me, it was about knowing how men and women live in a period of historical change" (Hobsbawm 2004, 291). One of the first occurrences of the term *habitus* appeared in Bourdieu's (1967) famous postscript to his French translation of Erwin Panofsky's *Gothic Architecture and Scholasticism*,[77] and through this concept, Bourdieu addressed the question of how innovation in cultural structures is possible through the contributions of individuals endowed with specific dispositions (in the context of Panofsky's book, Abbot Suger contributed to the creation of the Gothic style). In later works—such as *Rules of Art: Genesis and Structure of the Literary Field* (1992), *Homo Academicus* (1984), the lectures at the Collège de France dedicated to Manet collected in *Manet: A Symbolic Revolution* (2013), *On the State* (2012), and his last work on the scientific field, *Science of Science and Reflexivity* (2001)—Bourdieu moved toward the historicization of the social fields[78]

[75] This aspect was already visible in Bourdieu's first ethnographic investigations in Algeria and his native region of Bearn, in which he addressed the problem of cultural transformation. As noted above, Bourdieu's Algerian period corresponded to his transition from philosophy to the social sciences. The major works of this period were *The Algerians* (1958), a series of writings later collected in *Algerian Sketches* (2008), *Travail et travailleurs en Algérie* (1963), and *The Uprooting* (1964).

[76] For a general survey of the relationship of Bourdieu's theory with history, as well as its reception among historians, see Steinmetz (2011) and Gorski (2013). See also Bourdieu's dialogues with French historians: Bourdieu et al. (1985); Bourdieu and Chartier (1988); and Bourdieu (1999).

[77] Together with the concepts of field and capital, the concept of habitus was the third pillar of Bourdieu's social theory. This concept has a long history, starting in Aristotle, passing through Leibniz, Husserl, and Merleau-Ponty up to Panofsky and Norbert Elias. Bourdieu defined habitus as "systems of durable, transposable dispositions, structured structures predisposed to function as structuring structures, that is, as principles which generate and organize practices and representations that can be objectively adapted to their outcomes without presupposing a conscious aiming at ends or an express mastery of the operations necessary in order to attain them. Objectively 'regulated' and 'regular' without being in any way the product of obedience to rules, they can be collectively orchestrated without being the product of the organizing action of a conductor" (Bourdieu 1980/1990, 53).

[78] According to Bourdieu, modern societies are highly differentiated, and social space is made up of relatively autonomous social microcosms called social "fields." In analogy to the language of physics, these social microcosms are fields of forces, structured at any given moment by the balance of forces within them. In this sense, fields are spaces of objective relationships structured according to specific issues at stake in, and irreducible to, those of other fields. The autonomy of a social field is not given once and for all but rather is the result of a historical achievement that must be constantly renewed with respect to "external pressures on the field" (Ienna 2015). For an analysis of general field theory, see the recently published collection Bourdieu (2022a).

(with a specific focus on cultural production) by adopting a discontinuous approach translated into sociological terms.

As an heir of historical epistemology, he developed an approach on the basis of a form of dynamic nominalism that implied a form of historicization of ontology.[79] On this point, for example, Robert Chartier argued:

> This notion of field, I believe, makes it possible to conceive discontinuity. You find again here the problem of nominalism, that is, that we need to have words, whether scientific or not, to denote institutions, objects, practices. These words may remain the same but, behind this stability, they designate specific configurations. (Bourdieu and Chartier 1988/2015, 62)

Bourdieu was, indeed, very careful not to fall into the anachronisms that historians often fall into by not historicizing their own analytical categories. In his view, historians tend to use nonreflexive language in research practice. Responding to Chartier, Bourdieu argued that "the paradox is that historians, for example […] often show an extraordinary naivety in their use of categories" (Bourdieu and Chartier 1988/2015, 11). As an example of this naivety, Bourdieu described the paradoxical nature of the attempt to create a history of medicine, whose idea was historically determined, without taking into consideration that the notion of the field had never stopped changing (i.e., not all healing practices can be called medical). We must therefore methodologically acknowledge that "It is the very categories with which the historic object is constructed that must be the object of historical analysis" (Bourdieu and Chartier 1988/2015, 11). In a round table organized at the Société d'histoire moderne et contemporaine, Bourdieu reiterated his theoretical position by arguing that

> This is an important point, and I think that historians, whom I read with more pleasure than sociologists, do not make enough use of historical reflexivity and are content to say that anachronism must be avoided, whereas they could use their historical culture to interrogate their historical concepts. Durkheim said, roughly, that "the unconscious is history," by which he meant that our unconscious is the forgotten or reformulated product of collective (and not only individual, as Freud said) history. One unconscious—we can probably have several—with which we must struggle in scientific work is what history has deposited in our minds, as sediments, concepts, problems, linguistic and mental reflexes. Only the reflexive use of history can give us freedom from history, give us the possibility of not conveying concepts weighed down by their historicity. (Bourdieu 1999, 18).

As is easy to see, this is a complete integration of Canguilhem's methodological principles for conceptual history into the sociological field. Bourdieu aimed to understand those moments of strong historical discontinuity during the processes of cultural transformation; it was thus an extension of the method of FHE to a broader field of phenomena (such as those relating to the fields of art, religion, power, etc.).

This approach, from a methodological point of view, allowed Bourdieu to develop an innovative historiographical model. He sought to overcome the

---

[79] Here, one can grasp how close Bourdieu was to Ian Hacking, who developed FHE in the Anglophone context (see Hacking 2002, 2004, 2008, 51–52).

dichotomy between "subject" and the "structure" of classical Marxism by paying particular attention to the concept of practice (Bourdieu 1972, 1980). Against any form of epistemic apriorism, he argued that every theoretical expression is the result of socially and historically situated epistemic practices. Historical time is a scientific-epistemic abstraction; that is, it is the result of the practice of historians who have an idealized vision of historical time as an "*a priori* frame" in which practices are situated.[80] On the contrary, for Bourdieu, practice is what determines temporal rhythms:

> Like the body-as-thing of the Cartesian idealist vision, time-as-thing, time of clock-makers and scientists, is the product of a scholastic point of view which has found its expression in a metaphysics of time and history which considers history either as a pregiven reality, a thing in-itself, previous and external to practice, or as the (empty), *a priori* framework for every historical process. We can break with this point of view by reconstructing the point of view of the acting agent, of practice as "temporalization," thereby revealing that practice is not *in* time, but *makes* time (human time, as opposed to biological or astronomical time). (Bourdieu 1997b/2000, 206)

Sociological analysis allows for emphasizing the role of practices in the articulation of social rhythms and vice versa. For Bourdieu, in fact, temporality is always twofold: On the one hand, it is the temporality of social agents and, on the other, it is the temporality of structures. The notion of habitus allowed him to describe how history is embodied by social agents in the form of trajectories, positions, and social dispositions. Instead, social temporality describes history objectified and inscribed in structures. This perspective laid the foundations for overcoming the apparently forced alternative between a structural history and a history based on events [*histoire évenémentielle*] (Bourdieu 1984, 227). The relationship between these two poles influences each other (Bourdieu 1997b/2000, 151–156) and, at the same time, allows us to describe both the change and the reproduction of social and cultural structures. Bourdieu considered social temporality to be an organizing and unifying principle of social communities, as a shared abstraction.[81]

Indeed, on these epistemological bases and historiographical stances, Bourdieu—from the mid-1970s through his last course at the Collège de France in 2000–2001—developed a sociology of science and knowledge in direct continuity with the principles of FHE. In this series of texts, Bourdieu provided a description of scientific activity as a field animated by conflicts and strongly characterized by phases of historical discontinuity.[82]

---

[80] This conception of time is implicitly indebted to Bachelard's theses set out in *Intuition of the Instant* and *The Dialectic of Duration* (see Chap. 2, Sect. 2.2).

[81] On the presence of multiple social temporalities in Bourdieu's theory and the possible application of this concept in the history of science, see Ienna (2025).

[82] Bourdieu's first contribution to the reconstruction of the analysis of the scientific field was in 1975 in the journal *Sociologie et sociétés* (Bourdieu 1975). The following year, a more detailed theoretical explanation of this topic was published in the journal he edited, *Actes de la recherche en sciences sociales* (Bourdieu 1976). Bourdieu returned to work on the scientific field only in the 1990s, with two essays. The first, "The Peculiar History of Scientific Reason," was published in

What Bourdieu was trying to address were the tensions produced by the so-called Sokal affair within the fields of Science and Technology Studies (STS) and cultural studies. He was especially concerned about the relativistic positions coming out of Sociology of Scientific Knowledge (SSK). Among the primary targets of his criticisms was Latour's perspective[83]—which, among other things, emphasized the incompatibility of historical epistemology with the British tradition in SSK (Latour and Bowker 1987)—given the progressive recognition it had been receiving in the French academic field.

1991 in *Sociological Forum* (Bourdieu 1991). The second was from a 1997 conference, titled *Les usages sociaux de la science: pour une sociologie clinique du champ scientifique* (Bourdieu 1997a). But the most significant text, and certainly the most detailed, was the collection derived from his last course at the Collège de France in 2000–2001, *Science of Science and Reflexivity* (Bourdieu 2001a). However, I believe I must point out that in his progressive structuring of the concept of the scientific field, one can detect certain epistemologically oriented nuances. In his contributions from the 1970s, Bourdieu described the logic of its functioning on the basis of a conflictual sociological model in opposition to Mertonian-style sociology because, in Bourdieu's opinion, it was too inclined to provide pacific descriptions of the internal social dynamics of the scientific field. Bourdieu instead targeted Merton's positions starting from the 1990s to counter the growing success enjoyed by authors such as Latour and SSK in general. In these texts, Bourdieu emphasized the self-regulatory capacity of the scientific field—while not denying its conflictual aspects. For a more detailed reconstruction of the Bourdieusian perspective in the sociology of science, see Ienna (2023).

[83] Bourdieu directly attacked at least three aspects of Latour's work. The first was of a sociological and methodological nature related to the tendency developed in *Laboratory Life* (Latour and Woolgar 1979) to favor microsociological research. In Bourdieu's view, this rejected the ambition of contextualizing the "laboratory" within a broader social structure and its power relations—that is, Latour described the physical space of the lab as entirely separate from the world, thus presenting a distorted and falsified sociological image of the scientific enterprise. The second aspect was of a purely epistemological nature. According to Bourdieu, in *Laboratory Life*, Latour and Woolgar adopted a deliberately ambiguous position: "By saying that facts are artificial, in the sense of manufactured, Latour and Woolgar intimate that they are fictitious, not objective, not authentic" (Bourdieu 2001a/2004, 26). Finally, Bourdieu offered a methodological critique regarding the use of a semiotic paradigm as an interpretative key for understanding scientific practices. Here, in addition to *Laboratory Life*, Bourdieu referred to other works by Latour, such as *The Pasteurization of France* (1988): "The semiological vision of the world that induces them to emphasize the traces and signs leads them to that paradigmatic form of scholastic bias, textism, which constitutes social reality as text […]. Science is then, just one discourse or fiction among others, but one capable of exerting a 'truth effect' produced, like all other literary effects, through textual characteristics such as the tense of verbs, the structure of utterances, modalities, etc. (The absence of any attempt at prosopography condemns them to seek the power of the texts in the texts themselves). The universe of science is a world which succeeds in imposing universally the belief in its fictions" (Bourdieu 2001a/2004, 28). Notably, Bourdieu and Latour's relationship was not always so strained. Indeed, Bourdieu published one of Latour's first articles in his journal *Actes de la recherche en sciences sociales* in 1977—coauthored with the semiologist Paolo Fabbri—titled "La rhétorique de la science." At the same time, in the first edition of *Laboratory Life*, some of Bourdieu's theoretical tools were mentioned on various occasions in connection with the concept of the "circle of credit" (Latour and Woolgar 1979, 205–206). The analysis of the relationship between Bourdieu and Latour deserves a more in-depth treatment that, however, would go beyond the scope of this volume.

According to Bourdieu, the social sciences had a duty to resolve a problem that they themselves had raised—that is, whether one could say that science is a product of history while claiming that it produces transhistorical truths (Bourdieu 2001a/2004, 1). The tendency to historicize reason without relativizing it was one of the main theoretical characteristics of historical epistemology, and it was endorsed by Bourdieu precisely as an antidote to the increasingly relativist outcomes of SSK. Indeed, the central issue of Bourdieu's final Collège de France course was precisely how for science could be both a product of history and yet capable of producing transhistorical truths.[84] Bourdieu's authentic Bachelardism emerged particularly on this antirelativist point and on the project of constructing a "historicist rationalism"[85] (Bourdieu 1997b/2000, 106–108) that largely echoed Bachelard's "applied rationalism." Bourdieu's epistemological stance was based on overcoming a series of false binary oppositions: logicism vs. realism; dogmatism vs. skepticism; idealist-constructivism vs. realist-positivism; logicist-absolutism vs. historicist-relativism; and idealism vs. realism (Bourdieu 1997b, 2001a):

> We have to acknowledge that reason did not fall from heaven as a mysterious and forever inexplicable gift, and that it is therefore historical through and through; but we are not forced to conclude, as is often supposed, that it is reducible to history. It is in history, and in history alone, that we must seek the principle of the relative independence of reason from the history of which it is the product. (Bourdieu 1997b/2000, 109)

In Bourdieu's historicist rationalism, the principles supporting the relative independence of reason are to be found in the historical and social structure of knowledge. The degree of objectivity of scientific products is determined by how conflicts within science are regulated. Even the epistemological principles and ethical-communicative norms shared by science are not the fruit of "an immaculate conception" but rather the product of the "peculiar history of the scientific field" (Bourdieu 1991). These are the rules and regularities inscribed in social structures and habitus—that is, a complex interweaving of dispositions embodied by social agents (Bourdieu 2001a/2004, 71).

At the same time, however, according to the methodological principle of the "illusion of transparency," the validity of a fact or a theory cannot be judged from the analysis of the intentions that produced it (Bourdieu et al. 1968/1991, 15). Consequently, the validity of a product of the scientific field remains irreducible to its conditions of production. Scientific knowledge is nothing other than that which survives socially and culturally determined objections from the past, present, and future. In this perspective, the scientific field historically acquired precisely the structure from its internal self-regulation—according to the specific nomos of "interest in disinterest" and the "search for truth"—which guarantees the truthfulness of the results produced in it, by virtue of their ability to pass the scrutiny of history and of the scientific community. Clearly, the resulting idea of objectivity is not based on the actions of a single scientist but rather based on that of a coordinated community:

[84] Here, we can see another point of contact with Hacking's work (Hacking 2008, 51–52).

[85] For more on "historicist rationalism" in the context of historical epistemology, see Ienna (2022b).

> Objectivity depends on the "the agreement of a class of observers about what is recorded by the measuring devices in a very precise experimental situation." So one can say that there is no objective reality independent of the conditions of its observation, without calling into question the fact that what manifests itself, once these conditions are determined, retains a character of objectivity. (Bourdieu 2001a/2004, 74)

Only after such a process can a true idea be imposed on competitors in the scientific field by virtue of its intrinsic strength. Every scientific act is, in fact, a social construction: Even Newton and Einstein, Bourdieu claimed, cannot be classified as "individual geniuses," because they worked in collectives with tools that embodied objectified history in the scientific field. A single mathematical formula, for example, allows the practical use of centuries of history, even if in an unreflective way. Consider a university physics student capable of mastering a historically given amount of knowledge thanks to the efficient form that it takes by being crystallized into formulas. Scientific tools themselves are historical products, the result of theories that were accepted, ratified, and objectified in the *field*. In this idea is a strong affinity with Kuhnian thought. For an appropriate analysis of a practical action in the scientific field, the dialectical interplay between these two ways of preserving and validating history must be considered.

Individual constructions thus turn out to be historical and collective constructions, regulated by the principles with which the scientific field has been equipped through its internal norms, so that common objectives can be pursued with the possibility of excluding those who transgress such norms. Universalization and departicularization, which are possible thanks to this process through which theories undergo social circulation, are of an inseparably social and intellectual nature. For this reason, one can conceptualize this as a socio-logical process of verification. In this perspective, a scientific fact is recognized as such only when the totality of the structure of the scientific field has cooperated in its formation: In this sense, scientific reason is both historical and social.

From a historiographical point of view, Bourdieu was very concerned about the risk of anachronism. What we call science today is the result of a long historical process of achieving relative autonomy. Claiming that the "scientific field" emerged in a specific historical period does not mean that everything preceding that moment had no scientific value (Bourdieu 2001a/2004, 47). Rather, supporting this position means recognizing how, prior to this event, there was substantial overlap, or a lack of distinction, in the field of cultural production between scientific, religious, magical, political, and artisanal practices, among others. The emergence of modern science was situated within a more general process of reorganizing social space, leading to the emergence of areas in relative autonomy from one another. In Bourdieu's view, the Scientific Revolution was precisely the event that created the distinction between scientific knowledge and other spheres of cultural production (art, religion, politics, and so on). At the basis of the conquest of the relative autonomy of the scientific field was the mathematization of knowledge, which—especially with post-Galilean and Newtonian physics—resulted in a series of convergent phenomena of both a social and intellectual nature.

1. Social inclusion/exclusion criterion: The use of mathematics determines the social distinction between amateurs and professionals.
2. The transformation of the concept of explanation: The physicist produces explanations through mathematics and then moves on to testing through experimentation.
3. The replacement of Aristotelian substances with functional relationships and structures: The logic of symbolic manipulation guides the physicist's research activity.

Despite the specificity of each field of cultural production, their "relative" autonomy indicates that they are not dependent on the rest of the social space.[86] The boundaries of the fields are, in fact, relatively blurred: "We may think of a field as a space within which an effect of field is exercised" (Bourdieu and Wacquant 1992/1992, 100). Bourdieu's sociology uses the concept of field as a functional instrument of inquiry to reveal the intrinsic conflicts and power relations underlying the structure of society. The field, in Bourdieu's formulation, is a microcosm within the social space that is relatively autonomous compared to others. This relative autonomy is a product of a process of "differentiation in the ways of knowing the world" such that "each field corresponds to a fundamental point of view on the world" (Bourdieu 1997a, 105). Bourdieu defined the specificity of the scientific field in the following way:

> The structure of the scientific field at any moment is defined by the state of the power distribution between the protagonists of the struggle (agents, or institutions), i.e., by the structure of the distribution of the specific capital, the result of previous struggles which is objectified in institutions and dispositions and commands the strategies and objective chances of the different agents or institutions in the present struggles. (Bourdieu 1975, 100)

In this sense, the distribution of scientific capital (understood as the possession of knowledge, the material and symbolic means to produce research, and a position within the resource distribution structure) is the result of the previous phases of negotiation and struggle in the field that have been directly objectified in the institutionalized forms and complex structure of regulations that determine the objective strategies and possibilities within the field in the present struggles of the social agents involved. The struggle between scientists is thus anything but peaceful:

> The scientific field is an armed struggle among adversaries who possess weapons whose power and effectiveness rises with the scientific capital collectively accumulated in and by the field (and therefore in the incorporated state, in each of the agents) and who agree at least to appeal to the verdict of experience, the "real," as a kind of ultimate referee. This "objective

[86]For Bourdieu, the historical emergence of a field stems from its conquest of an autonomy on the basis of at least three fundamental criteria. First, a *specific stake* must be recognized by the social agents participating in that field. *Specific axiomatics*, interests, a *form of capital*, and a specific form of authority are then connected to it. The second criterion is a common interest on the part of the participants in that field (despite being in constant struggle among themselves) in the reproduction and perpetuation of that field. Lastly is the criterion of "field effects," which demonstrates the impossibility of understanding a specific work in a field without knowing the sociohistorical conditions of its production relative to that field.

> reality" to which everyone explicitly or tacitly refers is ultimately no more than what the researchers engaged in the field at a given moment agree to consider as such, and it only ever manifests itself in the field through the representations given of it by those who invoke its arbitration. (Bourdieu 1997b/2000, 112–113)

Internal changes within the field are generated by changes relating to its boundaries, which in turn call into question the definition of the field and its stakes. History is therefore determined by a progressive alternation of "symbolic revolutions" linked as cause or effect to "the sudden arrival of new entrants endowed with new resources" (Bourdieu 2001a/2004, 36) that make the evolution of the field possible. From the perspective of the practices of individual social agents, the scientists who revolutionize the "paradigmatic" state of normal science are always heretics situated in a liminal area between scientific disciplines. This type of heterodox positioning is often what allows scientific innovation to go beyond the simple reproduction of disciplinary orthodoxy (Bourdieu 2001a/2004, 43).[87] The border is therefore the place where that "essential tension" between tradition and iconoclasm that Kuhn spoke about is most intense (Kuhn 1977).

Following a Canguilhemian approach, Bourdieu maintained that "One of the particularities of scientific revolutions is that they introduce a radical transformation while conserving previous achievements. So, they are revolutions that conserve past gains—without being conservative revolutions aimed at overthrowing the present to restore the past" (Bourdieu 2001a/2004, 64). This aspect, however, has an essential correlate in the structure of the scientific community that creates a constant challenge between conservatives—those in a dominant position in the field—and innovators:

> Scientific revolutions have the effect of transforming the hierarchy of importances: things considered unimportant may be reactivated by a new way of doing science, and, conversely, whole sectors of science may be overtaken, rendered obsolete. The struggles within the field are struggles to be or remain contemporary. Someone who introduces a new legitimate way of doing things shakes the power relations and introduces time. If nothing happened, there would be no time; the conservative agents would like to abolish time, to eternize the present state of the field, the state of the structure that is favourable to their interests because they occupy the dominant positions within it, whereas the innovators, without even seeking to "compete" with anyone, introduce change by their mere intervention and bring about the specific temporality of the field. (Bourdieu 2001a/2004, 64)

In this passage, using an approach analogous to Canguilhem's, Bourdieu puts the history of conceptual transformations in relation to the social and power structures from within which it emerges and from which it is inseparable. Starting from this, for Bourdieu, "scientific revolutions overthrow the hierarchy of the social values

[87] Note also that scientific disciplines produce a "disciplinary habitus" that operates as a "historical transcendental," as a scheme of perception and evaluation (Bourdieu 2001a/2004, 86). The embodiment of this disciplinary habitus acts as an incorporated system of censure for the sayable and unsayable within the sphere of scientific production. In this case, Bourdieu played with the duplicity of the term *discipline*, in one sense understood as a scientific-disciplinary sector and in another understood as the imposition of rules on action, of patterns of perception that operate as censorship. For Bourdieu, the field was always a field of struggles where social agents, endowed with different habitus, clash to preserve or transform the structures of power relations.

attached to the various forms of scientific practice" (Bourdieu 2001a/2004, 64), and because of this, they take on a "symbolic" character. The originality of Bourdieu's thought lay in the way he underlined that the cognitive dimension of power is characterized precisely by its "symbolic" aspect (Bourdieu 2001b/1991).

In conclusion, I point out that one of the main innovations brought about by Bourdieusian sociology was opening the history of the social sciences and humanities to sociological investigation. However, this aspect remains largely neglected in the research lines of fields such as STS and SSK, even as it continues to be a fertile area of research pursued by various scholars of Bourdieusian orientation.[88] In this context, the sociology of academic and intellectual fields proposed in works such as *Homo Academicus* (Bourdieu 1984) and *The Political Ontology of Martin Heidegger* (Bourdieu 1988), and research on the dynamics of the international circulation of ideas (Bourdieu 2002, 2022b; Ienna and Santoro 2016), have been invaluable sources of inspiration.

References

Abadía, Oscar Moro. 2008. Beyond the Whig History Interpretation of History: Lessons on 'Presentism' from Hélène Metzger. *Studies in History and Philosophy of Science Part A* 39 (2): 194–201. https://doi.org/10.1016/j.shpsa.2008.03.005.

———. 2009. Thinking about 'Presentism' from a Historian's Perspective: Herbert Butterfield and Hélène Metzger. *History of Science* 47 (1): 55–77. https://doi.org/10.1177/007327530904700104.

Althusser, Louis. 1965. *Pour Marx*. Paris: Maspero. *For Marx*. Translated by Ben Brewster. London: Verso, 2005.

———. 1967. *Philosophie et philosophie spontanée des savants*. Paris: Maspero. *Philosophy and the Spontaneous Philosophy of the Scientists, and Other Essays*, ed. Gregory Elliott and Translated by Warren Montag. London: Verso, 1990.

———. 1969. *Lénine et la philosophie*. Paris: Maspero. *Lenin and Philosophy and Other Essays*. Translated by Ben Brewster. London: New Left Books, 1971.

———. 1974. *Éleménts d'autocritique*. Paris: Librairie Hachette. Elements of Self-Criticism. In *Essays in Self-Criticism*, Translated by Grahame Lock, 101–161. London: New Left Books, 1976.

———. 1995. *Sur la reproduction*. Paris: PUF. *On the Reproduction of Capitalism: Ideology and Ideological State Apparatuses*. Translated by G. M. Goshgarian. London: Verso, 2014.

Althusser, Louis, and Étienne Balibar. 1965. *Lire le capital, v. 1*. Paris: Maspero. *Reading Capital*. Translated by Ben Brewster. London: Verso, 1970.

Althusser, Louis, Pierre Macherey, Michel Pêcheux, Michel Fichant, François Regnault, Étienne Balibar, and Alain Badiou. 1967. *Cours de philosophie pour scientifiques*. Paris: École Normale Supérieure. MS 169.

Baby, Jean, et al. 1935. *À la lumiere du marxisme: essais*. Vol. 1. Paris: Editions sociales internationales.

Bachelard, Gaston. 1927. *Essai sur la connaissance approchée*. Paris: Vrin.

[88] Some recent examples of this trajectory include C. Fleck (2011); Heilbron (2015); C. Fleck et al. (2018); Heilbron et al. (2018); and Sapiro et al. (2020). On this point, see also Ienna et al. (2024).

———. 1940. *La Philosophie du non: essai d'une philosophie du nouvel esprit scientifique.* Paris: PUF. *The Philosophy of No: A Philosophy of the New Scientific Mind.* Translated by G. C. Waterston. New York: Orion Press, 1968.
———. 1949. *Le rationalisme appliqué.* Paris: PUF.
———. 1953. *Le Matérialisme rationnel.* Paris: PUF.
Badino, Massimiliano, Gerardo Ienna, and Pietro D. Omodeo. 2022. *Epistemologia Storica: Correnti, temi e problemi.* Rome: Carocci.
Badiou, Alain. 1969. *Le concept de modèle: introduction à une épistémologie matérialiste des mathématiques*. Paris: Maspero. *The Concept of Model: An Introduction to the Materialist Epistemology of Mathematics*, ed. and Translated by Zachary Luke Fraser and Tzuchien Tho. Melbourne: re.press, 2007.
Balibar, Étienne. 1994. *Lieux et noms de la vérité.* Le Château: Éditions de l'Aube.
Baracca, Angelo, Silvio Bergia, and Flavio Del Santo. 2017. The Origins of the Research on the Foundations of Quantum Mechanics (and Other Critical Activities) in Italy during the 1970s. *Studies in History and Philosophy of Science Part B: Studies in History and Philosophy of Modern Physics* 57:66–79. https://doi.org/10.1016/j.shpsb.2016.11.003.
Barnes, Barry. 1974. *Scientific Knowledge and Sociological Theory*. London: Routledge.
Basso, Elisabetta. 2022. *Young Foucault: The Lille Manuscripts on Psychopathology, Phenomenology, and Anthropology, 1952–1955*. Translated by Marie Satya McDonough. New York: Columbia University Press.
Bensaude-Vincent, Bernadette. 2008. Meyerson critique ou héritier de Comte? *Dialogue* 47 (1): 3–23. https://doi.org/10.1017/S0012217300002353.
Bharucha, Zac. 2018. *The Radical Science Movement in the U.K. 1968-1978. Struggles Against the Impact of Capitalist Ideology on Science, Technology and Social Relations of Science*. Poland: Amazon Fulfillment.
Bird, Alexander. 2008. The Historical Turn in the Philosophy of Science. In *The Routledge Companion to Philosophy of Science*, ed. Stathis Psillos and Martin Curd, 67–77. London: Routledge.
Bloor, David. 1976. *Knowledge and Social Imagery*. Chicago: University of Chicago Press.
Bontems, Vincent. 2013. Le "non-cartésianisme." La méthode non-cartésienne selon Gaston Bachelard et Ferdinand Gonseth. In *Qu'est-ce qu'être cartésien?* ed. Delphine Kolesnik-Antoine, 567–580. Lyon: ENS Éditions. https://doi.org/10.4000/books.enseditions.8883.
Bourdieu, Pierre. 1967. Postface. *Architecture gothique et pensée scolastique*, by Edwin Panofsky. Translated by Pierre Bourdieu, 133–167. Paris: Editions de Minuit.
———. 1972. *Esquisse d'une théorie de la pratique précédé de Trois études d'ethnologie kabyle.*. Geneva: Droz. *Outline of a Theory of Practice*. Translated by Richard Nice. Cambridge: Cambridge University Press, 1977.
———. 1975. La spécificité du champ scientifique et les conditions sociales du progrès de la raison. *Sociologie et sociétés* 7 (1): 91–118. https://doi.org/10.7202/001089ar.
———. 1976. Le champ scientifique. *Actes de la recherche en sciences sociales* 2 (2): 88–104. https://doi.org/10.3406/arss.1976.3454.
———. 1984. *Homo academicus*. Paris: Éditions de Minuit. *Homo Academicus*. Translated by Peter Collier. Stanford: Stanford University Press, 1988.
———. 1980. *Le sens pratique*. Paris: Minuit. *The Logic of Practice*. Translated by Richard Nice. London: Polity Press, 1990.
———. 1988. *L'Ontologie politique de Martin Heidegger*. Paris: Éditions de Minuit. *The Political Ontology of Martin Heidegger*. Translated by Peter Collier. Stanford: Stanford University Press, 1991.
———. 1989. *La Noblesse d'état. Grandes écoles et esprit de corps*. Paris: Éditions de Minuit. *State Nobility: Elite Schools in the Field of Power*. Translated by Lauretta C. Clough. Stanford: Stanford University Press, 1996.
———. 1991. The Peculiar History of Scientific Reason. *Sociological Forum* 6 (1): 3–26.

———. 1997a. *Les usages sociaux de la science: pour une sociologie clinique du champ scientifique*. Paris: Quae.

———. 1997b. *Méditations pascaliennes*. Paris: Éditions du Seuil. *Pascalian Meditations*. Translated by Richard Nice. Stanford: Stanford University Press, 2000.

———. 1999. Les Historiens et la sociologie de Pierre Bourdieu. Tableronde de la S.H.M.C., le 6 mars 1999. *Bulletin de la S.H.M.C. Supplément à la revue d'histoire moderne et contemporaine* 3 (3–4): 4–27.

———. 2001a. *Science de la science et Réflexivité*. Paris: Raisons d'agir. *Science of Science and Reflexivity*. Translated by Richard Nice. Cambridge: Polity Press, 2004.

———. 2001b. *Langage et pouvoir symbolique*. Paris: Éditions Fayard. *Language and Symbolic Power*, ed. John B. Thompson and Translated by Gino Raymond and Matthew Adamson. Cambridge: Polity Press, 1991.

———. 2002. Les conditions sociales de la circulation internationale des idées. *Actes de la recherche en sciences sociales* 145 (5): 3–8.

———. 2004. *Esquisse pour une auto-analyse*. Paris: Raisons d'agir. *Sketch for a Self-Analysis*. Translated by Richard Nice. Cambridge: Polity Press, 2007.

———. 2022a. *Microcosmes. Théorie des champs*. Paris: Raison D'agir.

———. 2022b. In *Retour sur la réflexivité*, ed. Jérôme Bourdieu and Johan Heilbron. Paris: EHESS.

Bourdieu, Pierre, Jean-Claude Chamboredon, and Jean-Claude Passeron. 1968. *Le métier de sociologue*. Paris: Mouton Bordas. *The Craft of Sociology: Epistemological Preliminaries*, ed. Beate Krais and Translated by Richard Nice. Berlin: de Gruyter, 1991.

Bourdieu, Pierre, and Roger Chartier. 1988. *Le Sociologue et l'historien*. Paris: Agone. *The Sociologist and the Historian*. Translated by David Fernbach. Cambridge: Polity, 2015.

Bourdieu, Pierre, Roger Chartier, and Robert Darnton. 1985. Dialogue à propos de l'histoire culturelle. *Actes de la recherche en sciences sociales* 59:86–93.

Bourdieu, Pierre, and Beate Krais. 2005. Entretien avec Pierre Bourdieu – recueilli par Beate Krais en décembre 1988. In *Le métier de sociologue*, by Pierre Bourdieu, Jean-Claude Chamboredon, and Jean-Claude Passeron, 5th ed., v-xix. Berlin: Mouton de Gruyter. "Meanwhile I have come to know all the diseases of sociological understanding," in *The Craft of Sociology: Epistemological Preliminaries*, ed. Beate Krais and Translated by Richard Nice, 247–259. Berlin: de Gruyter, 1991.

Bourdieu, Pierre, and Jean-Claude Passeron. 1964a. *Les étudiants et leur études*. Paris: La Haye.

———. 1964b. *Les héritiers: les étudiants et la culture*. Paris: Minuit. *The Inheritors: French Students and Their Relation to Culture*. Translated by Richard Nice. Chicago: University of Chicago Press, 1979.

———. 1970. *La reproduction: Éléments d'une théorie du système d'enseignement*. Paris: Minuit. *Reproduction in Education, Society and Culture*. Translated by Richard Nice. London: Sage, 1990.

Bourdieu, Pierre, and Loïc J. D. Wacquant. 1992. *Réponses: pour une anthropologie réflexive*. Paris: Éditions du Seuil. *An Invitation to Reflexive Sociology*. Cambridge: Polity Press, 1992.

Braunstein, Jean-François. 2002. Bachelard, Canguilhem, Foucault. Le "style français" en épistémologie. In *Les philosophes et la science*, ed. Pierre Wagner, 920–963. Paris: Gallimard.

———. 2008. *L'histoire des sciences. Méthodes, styles et controverses*. Paris: Vrin.

———. 2016. Foucault, Canguilhem et l'histoire des sciences humaines. *Archives de Philosophie* 79 (1): 13. https://doi.org/10.3917/aphi.791.0013.

Brian, Éric. 2012. Où en est la sociologie générale? (Seconde partie). *Revue de synthèse* 133 (3): 401–444.

Bukharin, Nikolai, ed. 1971. *Science at the Cross Roads. Papers Presented to the International Congress of the History of Science and Technology*. London: Frank Cass & Co.

Cammelli, Michele. 2007. Canguilhem e il problema delle norme in sociologia. Riflessioni di epistemologia biopolitica. *Storicamente* 3:1–22. https://doi.org/10.1473/stor397.

———. 2011. Présentation. Le fascisme et les paysans (1935). In *Oeuvres Complètes, 1, 1926-1939, by Georges Canguilhem*, 515–534. Paris: Vrin.
Canesso, Annagiulia. 2024. *Il governo del normale. Canguilhem e il pensiero sociologico francese*. Milan: Meltemi.
Canguilhem, Georges. 1935. *Le fascisme et les paysans*. Paris: Comité de Vigilance des Intellectuels Antifascistes.
———. 1952. *La connaissance de la vie*. Paris: Vrin. *Knowledge of Life*, ed. Paola Marrati and Todd Meyers and Translated by Stefanos Geroulanos and Daniela Ginsburg. New York: Fordham University Press, 2008.
———. 1955. Le problème des régulations dans l'organisme et dans la société. In *Écrits sur la medicine*, Paris: Seuil. The Problem of Regulation in the Organism and in Society. In *Writings on Medicine*, ed. and Translated by Stefanos Geroulanos and Todd Meyers, 67–78. New York, Fordham University Press, 2012.
———. 1958. Qu'est-ce que la psychologie? *Revue de Métaphysique et de Morale* 63 (1): 12–25.
———. 1966. *Le normal et le pathologique*. Paris: PUF. *The Normal and the Pathological*. Translated by Carolyn R. Fawcett. New York: Zone Books, 1991.
———, ed. 1970. *Introduction à l'histoire des sciences. Textes choisis*. Paris: Hachette.
———. 1977. *Idéologie et rationalité dans l'histoire des sciences de la vie*. Paris: Vrin. *Ideology and Rationality in the History of the Life Sciences*. Translated by Arthur Goldhammer. Cambridge, MA: MIT Press, 1988.
———. 1994. *Études d'histoire et de philosophie des sciences concernant les vivants et la vie*. Paris: Vrin.
———. 2011. *Oeuvres complètes. Vol. I, Écrits philosophiques et politiques (1926-1939)*. Paris: Vrin.
Canguilhem, Georges, and Gabriel Gohau. 1978. Le concept d'idéologie scientifique. *Raison présente* 46 (1): 55–60. https://doi.org/10.3406/raipr.1978.1928.
Carrier, Martin. 1988. Some Aspects of H. Metzger's Philosophy of Science. *Corpus: Revue de philosophie* 8:135–150.
Castelão-Lawless, Teresa. 2004a. Kuhn's Missed Opportunity and the Multifaceted Lives of Bachelard: Mythical, Institutional, Historical, Philosophical, Literary, Scientific. *Studies in History and Philosophy of Science Part A* 35 (4): 873–881.
———. 2004b. Les épistémologies de Bachelard, de Popper, de Polanyi et de Kuhn dans le contexte des études critiques de la science. In *Ri-cominciare. Percorsi e attualità dell'opera di Gaston Bachelard*, ed. Francesca Bonicalzi and Carlo Vinti, 205–217. Milan: Jaca Book.
Castellana, Mario. 1974. *Il surrazionalismo di Gaston Bachelard*. Naples: Glaux.
———. ed. 2009. Introduzione. Storia della Scienza, Epistemologia ed Ermeneutica in Hélène Metzger. In *Il metodo filosofico nella storia delle scienze: testi 1914-1939*, by Hélène Metzger, 23–55. Manduria: Barbieri Selvaggi Editori.
———. 2021. In *Il surrazionalismo di Gaston Bachelard con due saggi inediti*, ed. Pietro Console. Lecce: Milella.
Castelli Gattinara, Enrico. 1996. *Epistemologia e storia. Un pensiero all'apertura nella Francia fra le due guerre mondiali*. Milan: FrancoAngeli.
Cavazzini, Andrea. 2015. Althusser/Bachelard: Une Coupure Et Ses Enjeux. *Revue de synthèse* 136 (1–2): 117–134. https://doi.org/10.1007/s11873-015-0272-4.
Chimisso, Cristina. 2001. Hélène Metzger: The History of Science between the Study of Mentalities and Total History. *Studies in History and Philosophy of Science Part A* 32 (2): 203–241. https://doi.org/10.1016/S0039-3681(01)00006-1.
———. 2008. *Writing the History of the Mind: Philosophy and Science in France, 1900 to 1960s*. Aldershot: Ashgate.
———. 2015. Narrative and Epistemology: Georges Canguilhem's Concept of Scientific Ideology. *Studies in History and Philosophy of Science Part A* 54 (December): 64–73. https://doi.org/10.1016/j.shpsa.2015.08.016.
———. 2019. *Hélène Metzger, Historian and Historiographer of the Sciences*. London: Routledge.

Chimisso, Cristina, and Gad Freudenthal. 2003. A Mind of Her Own: Hélène Metzger to Émile Meyerson, 1933. *Isis* 94 (3): 477–491. https://doi.org/10.1086/380655.

Ciccotti, Giovanni, Marcello Cini, Michelangelo de Maria, and Giovanni Jona-Lasinio. 2011. *L'Ape e l'Architetto. Paradigmi scientifici e materialismo storico*. Milan: FrancoAngeli. *The Bee and the Architect: Scientific Paradigms and Historical Materialism*, ed. Gerardo Ienna and Pietro Daniel Omodeo and Translated by Giuliana Giobbi. Venice: Verum Factum, 2024.

Comte, Auguste. 1870. *Lettres d'Auguste Comte à M. Valat. Professeur de mathématiques, ancien Recteur de l'Académie de Rhodez, 1815-1844*. Paris: Dunod.

Cornu, Auguste, et al. 1937. *À la lumiere du marxisme: essais*. Vol. 2. Paris: Editions sociales internationales.

Dahms, Hans-Joachim. 2016. Thomas Kuhn's *Structure*: An "Exemplary Document of the Cold War Era"? In *Science Studies during the Cold War and Beyond*, ed. Elena Aronova and Simone Turchetti, 103–125. New York: Palgrave MacMillan.

Davidson, Arnold. 2001. *The Emergence of Sexuality: Historical Epistemology and the Formation of Concepts*. Cambridge: Harvard University Press.

Debailly, Renaud. 2015. *La critique de la science depuis 1968. Critique des sciences et études des sciences en france après Mai 68*. Paris: Hermann.

Desanti, Jean-Toussaint. 1984. Gaston Bachelard ou "la surveillance intellectuelle de soi". *Revue Internationale de Philosophie* 38 (3): 272–286.

Duroux, Françoise. 1993. L'immaginaire biologique du politique. In *Georges Canguilhem, philosophe, historien des sciences: actes du colloque, 6-7-8 décembre 1990*, ed. Georges Canguilhem, 49–57. Paris: A. Michel.

Durkheim, Émile. 1895. *Les Règles de la méthode sociologique*. Paris: Alcan. *The Rules of Sociological Method*, ed. Steven Lukes and Translated by W. D. Halls. New York: Free Press.

Eribon, Didier. 1989. *Michel Foucault*. Paris: Flammarion. *Michel Foucault*. Translated by Betsy Wing. Cambridge MA: Harvard University Press, 1991.

Fichant, Michel, and Michel Pêcheux. 1969. *Sur l'histoire des sciences*. Paris: Maspero.

Fleck, Christian. 2011. *A Transatlantic History of the Social Sciences: Robber Barons, the Third Reich and the Invention of Empirical Social Research*. London: Bloomsbury Publishing.

Fleck, Christian, Matthias Duller, and Victor Karády. 2018. *Shaping Human Science Disciplines: Institutional Developments in Europe and Beyond*. Cham: Springer.

Foucault, Michel. 1961. *Folie et déraison. Histoire de la folie à l'âge classique*. Paris: Gallimard. 1972. *History of Madness*, ed. Jean Khalfa and Translated by Jonathan Murphy and Jean Khalfa. Abingdon: Routledge, 2006.

———. 1963. *La naissance de la clinique*. Paris: PUF. *The Birth of the Clinic: An Archaeology of Medical Perception*. Translated by A. M. Sheridan. New York: Routledge, 1976.

———. 1966. *Les mots et les choses*. Paris: Gallimard. *The Order of Things: An Archaeology of the Human Sciences*. Translated by Alan Sheridan. New York: Vintage, 1994.

———. 1969. *L'archéologie du savoir*. Paris: Gallimard. *The Archaeology of Knowledge and Discourse on Language*. Translated by A. M. Sheridan Smith. New York: Pantheon, 1972.

———. 1970. Preface à l'édition anglaise. In *Dits et ecrits*, vol. 2. Paris: Gallimard. Foreword to the English Edition. In *The Order of Things: An Archaeology of the Human Sciences*, Translated by Alan Sheridan, ix-xiv. New York: Vintage, 1994.

———. 1976. Question à Michel Foucault sur la géographie. In *Dits et ecrits*, vol. 3. Paris: Gallimard. Questions on Geography. In *Power/Knowledge: Selected Interviews and Other Writings, 1972-1977*, ed. and Translated by Colin Gordon, 63–77. New York: Pantheon, 1980.

———. 1985. La vie: L'experience et la science. *Revue de Méthaphysique et de Morale* 90 (January–March): 3–14. Life: Experience and Science. In *Aesthetics, Method, and Epistemology: Essential Works of Foucault 1954-1984*, vol. 2, ed. and Translated by James D. Faubion, 465–478. New York: New Press, 2003.

Freudenthal, Gad. 2009. Note biografiche di Gad Freudenthal. In *Il metodo filosofico nella storia delle scienze: testi 1914-1939, by Hélène Metzger*, ed. Mario Castellana, 11–21. Manduria: Barbieri Selvaggi Editori.

Freudenthal, Gideon, and Peter McLaughlin, eds. 2009. *The Social and Economic Roots of the Scientific Revolution: Texts by Boris Hessen and Henryk Grossman*. Dordrecht: Springer.

Fruteau de Laclos, Frédéric. 2009. *L'épistémologie d'Emile Meyerson. Une anthropologie de la connaissance*. Paris: Vrin.

Gandolfi, Giulia, Gerardo Ienna, and Charles Wolfe, eds. 2024. *Canguilhem face à la biopolitique*. Paris: Éditions matériologiques.

Giannetto, Enrico. 2020. Préface. Hélène Metzger et la théo-physique de Newton. In *Attraction universelle et religion naturelle, by Hélène Metzger*, ed. Mario Castellana and Enrico Giannetto, 5–20. Paris: Hermann.

Gil, Didier. 2004. Bachelard ou Kuhn. In *Ri-cominciare. Percorsi e attualità dell'opera di Gaston Bachelard*, ed. Francesca Bonicalzi and Carlo Vinti, 219–231. Milan: Jaca Book.

Golinski, Jan. 1987. Hélène Metzger and the Interpretation of Seventeenth Century Chemistry. *History of Science* 25 (67): 85–97.

Gorski, Philip S., ed. 2013. *Bourdieu and Historical Analysis*. Durham: Duke University Press.

Guerraggio, Angelo. 2010. *Il '68 e la scienza in italia*. Milan: Centro Pristem, Università Bocconi.

Gutting, Gary. 1989. *Michel Foucault's Archaeology of Scientific Reason: Science and the History of Reason*. Cambridge: Cambridge University Press.

Hacking, Ian. 2002. *Historical Ontology*. Cambridge MA: Harvard University Press.

———. 2004. Science de la science chez Pierre Bourdieu. In *La liberté par la connaissance: Pierre Bourdieu (1930-2002)*, ed. Daniel de La Roche and Jacques Bouveresse, 147–162. Paris: Odile Jacob.

———. 2008. *The Scientific Reason*. Taipei: National Taiwan University Press.

Heilbron, Johan. 2015. *French Sociology*. Ithaca, NY: Cornell University Press.

Heilbron, Johan, Gustavo Sorá, and Thibaud Boncourt, eds. 2018. *The Social and Human Sciences in Global Power Relations*. Cham: Springer International. https://doi.org/10.1007/978-3-319-73299-2.

Hessen, Boris. 1931. The Social and Economic Roots of Newton's "Principia." In *Science at the Cross Roads*, ed. N. I. Bukharin, 151–212. London: Kniga. *Les racines sociales et économiques des Principia de Newton*, ed. and Translated by Serge Guérout. Paris: Vuibert, 2006.

Hobsbawm, Eric. 2004. Sociologie critique et histoire sociale. Translated by Patrick Hersant. In *La liberté par la connaissance: Pierre Bourdieu (1930-2002)*, ed. Jacques Bouveresse and Daniel Roche, 281–292. Paris: Odile Jacob.

Ienna, Gerardo. 2015. Teoria generale dei campi. La struttura logico-epistemologica. *Paradigmi* 33 (May–August): 161–171.

———. 2018. Sociologia ed epistemologia delle frontiere disciplinari: un percorso fra Bourdieu e Passeron. *Sociologia e ricerca sociale* 115:15–36.

———. 2020. Fisici italiani negli anni '70. Fra scienza e ideologia. *Physis* 55 (1–2): 415–442.

———. 2022a. Corrationalisme: La dimension sociale de l'épistémologie bachelardienne. Translated by Charles Alunni. *Bachelard Studies - Études Bachelardiennes* 1 (1): 105–136.

———. 2022b. Razionalismo storicistico: per un'epistemologia politica del terzo escluso. *Paradigmi* 15 (2): 189–212.

———. 2022c. The International and Interdisciplinary Circulation of Boris Hessen's Theses. In *Manuscripts and Documents on the History of Physics: A Historical Materialist Textbook*, ed. Pietro D. Omodeo and Sean Winkler, 75–130. Venice: Verum Factum.

———. 2023. Bourdieu and the Social History of Scientific Reason. In *Handbook for the Historiography of Science*, ed. Mauro L. Condé and Marlon Salomon, 145–171. Cham: Springer International.

———. 2025. Multiple Social Temporalities of the Scientific Revolution. A Bourdieusian Perspective in Political Epistemology. In *Early Modern Political Epistemology: Key Studies (Intersection series)*, ed. Pietro D. Omodeo, Senthil Babu, and Claus Zittel. Leiden: Brill. (*forthcoming*).

Ienna, Gerardo, Carmelo Lombardo, Lorenzo Sabetta, and Marco Santoro. 2024. Il "collettivo riflessivo" come stile di pensiero. In *Sulla riflessività*, by Pierre Bourdieu, ed. and Translated by Gerardo Ienna, Carmelo Lombardo, Lorenzo Sabetta, and Marco Santoro. Milan: Meltmi.
———. 2019. Boris Hessen at the Crossroads of Science and Ideology. From International Circulation to the Soviet Context. *Societate Si Politica* 13 (1): 37–63.
———. 2021. The 1931 London Conference: The Rise of British Marxism and the Interdependencies of Society, Nature and Technology. *HoST - Journal of History of Science and Technology* 15 (1): 107–130.
Ienna, Gerardo, and Marco Santoro. 2016. Per una scienza delle relazioni internazionali in campo culturale. *Studi culturali* 13 (1): 61–69.
Jaubert, Alain, and Jean-Marc Lévy-Leblond. 1973. *[Auto] critique de la science*. Paris: Seuil.
Jorland, Gerard. 1981. *La science dans la philosophie. Les recherches épistémologiques d'Alexandre Koyré*. Paris: Gallimard.
Koyré, Alexandre. 1965. *Newtonian Studies*. London: Chapman & Hall.
Kuhn, Thomas S. 1962. *The Structure of Scientific Revolutions*. Chicago: University of Chicago Press.
———. 1977. *The Essential Tension. Selected Studies in Scientific Tradition and Change*. Chicago: University of Chicago Press.
———. 1994. Paradigms of Scientific Evolution: Thomas S. Kuhn. In *The American Philosopher: Conversations with Quine, Davidson, Putnam, Nozick, Danto, Rorty, Cavell, MacIntyre, and Kuhn*, ed. Giovanna Borradori, 153–167. Chicago: University of Chicago Press.
———. 2000. *The Road Since Structure. Philosophical Essays, 1970-1993, with an Autobiographical Interview*. Chicago: University of Chicago Press.
Kusch, Martin. 1991. *Foucault's Strata and Fields. An Investigation into Archaeological and Genealogical Science Studies*. Dordrecht: Springer.
Latour, Bruno, and Geof Bowker. 1987. A Booming Discipline Short of Discipline: (Social) Studies of Science in France. *Social Studies of Science* 17 (4): 715–748.
Latour, Bruno, and Steve Woolgar. 1979. *Laboratory Life*. 2nd ed. Princeton: Princeton University Press.
Laugier, Sandra. 2015. Duhem, Meyerson et l'épistémologie américaine postpositiviste. In *L'épistémologie française, 1830-1970*, ed. Michel Bitbol and Jean Gayon, 67–90. Paris: Éditions Matériologiques.
Lecourt, Dominique. 1969. *L'Épistémologie historique de Gaston Bachelard*. 2002 ed. Paris: Vrin.
———. 1972. *Pour une critique de l'épistémologie: Bachelard, Canguilhem, Foucault*. Paris: Maspero. *For a Critique of Epistemology*. In *Marxism and Epistemology: Bachelard, Canguilhem, Foucault*. Translated by Ben Brewster. London: New Left Books, 1975.
Lévy-Bruhl, Lucien. 1910. *Les fonctions mentales dans les sociétés inférieures*. Paris: Alcan.
———. 1922. *La mentalité primitive*. Paris: Alcan.
Lorenzini, Daniele. 2023. *The Force of Truth: Critique, Genealogy, and Truth-Telling in Michel Foucault*. Chicago: University of Chicago Press.
Macherey, Pierre. 2008. Althusser et le concept de philosophie spontanée des savants. Lecture to the study group "La philosophie au sens large." Savoirs Textes et Langage UMR, 8163. http://philolarge.hypotheses.org/files/2017/09/21-05-2008.pdf.
———. 2009. *De Canguilhem à Foucault, la force des normes*. Paris: La Fabrique.
Mazauric, Simone. 2019. Canguilhem et le marxisme. *La Pensée* 399 (3): 93–108.
Melhado, Evan. 1988. Metzger, Kuhn, and Eighteenth-Century Disciplinary History. *Corpus: Revue de philosophie* 8:111–134.
Merton, Robert K. 1973. The Normative Structure of Science. In *The Sociology of Science: Theoretical and Empirical Investigations*, ed. Norman W. Storer, 267–278. Chicago: University of Chicago Press.
Metzger, Hélène. 1918. *La genèse de la science des cristaux*. Paris: Albert Blachard.
———. 1923. *Les doctrines chimiques en France du début du XVIIéme à la fin du XVIIIéme siècle*. Paris: PUF.

———. 1926. *Les concepts scientifiques*. Paris: Alcan.

———. 1938. In *Attraction universelle et religion naturelle chez quelques commentateurs anglais de Newton*, ed. Mario Castellana and Enrico Giannetto, 2021. Paris: Hermann.

———. 1987. In *La Méthode philosophique en histoire des sciences. Textes 1914-1939*, ed. Gad Freudenthal. Paris: Fayard.

Meyerson, Émile. 1908. *Identité et realité*. Paris: Alcan. *Identity and Reality*. Translated by Kate Lowenberg. London: George Allen & Unwin. 1930.

———. 1921. *De l'explication dans les sciences*. Paris: Payot. *Explanation in the Sciences*. Translated by Mary-Alice Sipfle and David A. Sipfle. Dordrecht: Kluwer Academic Publishers, 1991.

———. 1931. *Du cheminement de la pensée*. Paris: Alcan.

Mulkay, Michael. 1979. *Science and the Sociology of Knowledge*. London: G. Allen & Unwin.

Omodeo, Pietro Daniel. 2020. Ideologies of the Scientific Revolution: The Rise and Fall of a Historiographical Concept. In *Encyclopedia of Early Modern Philosophy and the Sciences*, ed. Dana Jalobeanu and Charles T. Wolfe, 1–10. Dordrecht: Springer.

Paltrinieri, Luca. 2012. *L'expérience du concept. Michel Foucault entre épistémologie et histoire*. Paris: Éditions de la Sorbonne.

Pestaña, José Luis Moreno. 2006. *Convirtiéndose en Foucault: Sociogenésis de un Filósofo*. Barcelona: Montesinos. *En devenant Foucault sociogenèse d'un grand philosophe*. Translated by Philippe Hunt. Broissieux: Éditions du Croquant, 2006.

Quet, Mathieu. 2013. *Politiques du savoir. Sciences, technologies et participation dans les années 1968*. Paris: Édition des archives contemporaines.

Revel, Judith. 2009. *Dictionnaire Foucault*. Paris: Ellipses.

Rheinberger, Hans-Jörg. 2010. *On Historicizing Epistemology*. Stanford: Stanford University Press.

Sapiro, Gisèle, Marco Santoro, and Patrick Baert, eds. 2020. *Ideas on the Move in the Social Sciences and Humanities: The International Circulation of Paradigms and Theorists*. Cham: Palgrave Macmillan.

Schmalzer, Sigrid, Daniel S. Chard, and Alyssa Botelho, eds. 2018. *Science for the People. Documents from America's Movement of Radical Scientists*. Amherst: University of Massachusetts Press.

Sfara, Emiliano. 2018. *Georges Canguilhem inédit Essai sur une philosophie de l'action*. Paris: L'Harmattan.

Shapin, Steven. 1992. Discipline and Bounding: The History and Sociology of Science as Seen Through the Externalism-Internalism Debate. *History of Science* 30 (4): 333–369. https://doi.org/10.1177/007327539203000401.

Shapin, Steven, and Simon Schaffer. 1985. *Leviathan and the Air-Pump: Hobbes, Boyle, and the Experimental Life*. 2011 ed. Princeton: Princeton University Press.

Soulié, Charles. 1998. Histoire du département de philosophie de Paris VIII. Le destin d'une institution d'avant-garde. *Histoire de l'éducation* 77 (1): 47–69.

Stadler, Friedrich. 2014. History and Philosophy of Science: Between Description and Construction. In *New Directions in the Philosophy of Science*, ed. Maria Carla Galavotti, Dennis Dieks, Wenceslao J. Gonzalez, Stephan Hartmann, Thomas Uebel, and Marcel Weber, 747–767. Cham: Springer International. https://doi.org/10.1007/978-3-319-04382-1_52.

Steinmetz, George. 2011. Bourdieu, Historicity, and Historical Sociology. *Cultural Sociology* 5 (1): 45–66. https://doi.org/10.1177/1749975510389912.

Talcott, Samuel. 2019. *Georges Canguilhem and the Problem of Error*. Dordrecht: Springer.

Taylan, Ferhat. 2018. *Concepts et rationalités. Héritages de l'épistémologie historique, de Meyerson à Foucault*. Paris: Éditions Matériologiques.

Vinti, Carlo. 1977. *L'epistemologia francese contemporanea. Per un razionalismo aperto*. Rome: Città Nuova Editrice.

Werskey, Gary. 1988. *The Visible College: A Collective Biography of British Scientists and Socialists of the 1930s*. 2nd ed, 1988. London: Free Association Books.

———. 2007. The Marxist Critique of Capitalist Science: A History in Three Movements? *Science as Culture* 16 (4): 397–461. https://doi.org/10.1080/09505430701706749.

Zambelli, Paola. 2016. *Alexandre Koyré in incognito*. Florence: Olschki.

Chapter 4
Is there a Canon of Italian Historical Epistemology?

Abstract The Italian intellectual and academic scene has been a lively site for the development of meta-scientific reflections. Many works have explored the intersections of epistemology and the history of science, developing integrated reflections on them. Moreover, in recent years, various attempts have been made to legitimize certain individual thinkers or Italian historical-philosophical currents within debates on historical epistemology.

These appendices extend this research by proposing the first formalization of a hypothetical, specifically Italian canon in historical epistemology. My objectives are as follows: (1) to call attention to some of the main contributions in this context; (2) to show how, in various cases, they have been interconnected with international debates; and (3) to foreground how Italian contributions to historical epistemology have been characterized by political connotations. Within Italian historical-philosophical production, the study of the sciences from both historical and epistemological perspectives has been hindered by the enormous legacy of Benedetto Croce's thought. Indeed, precisely in reactions to the Italian tradition of historicism and neo-idealism have the roots of Italian historical epistemology been found. I propose a path that, starting from Federigo Enriques and passing through Antonio Gramsci, Ludovico Geymonat, Giulio Preti, Galvano della Volpe, Paolo Rossi, and Aldo Giorgio Gargani, arrives at the hotly contested debates of the so-called Italian Science Wars of the 1970s regarding the neutrality or non-neutrality of scientific knowledge.

As numerous publications can attest, the Italian intellectual and academic scene has been a lively site for the development of meta-scientific reflections (Pancaldi 1980; Dalla Chiara 1981; Agazzi 1986; Castellana 1988; Parrini 2004; Pogliano 2007; Barotta and Montuschi 2019). Many of these works have explored the intersections of epistemology and the history of science, developing integrated reflections on them. Moreover, in recent years, various attempts have been made to legitimize certain individual thinkers or Italian historical-philosophical currents within debates on historical epistemology (Cavazzini 2011; Vagelli 2015; Castellana 2019; Minazzi 2021/2022).

G. Ienna, *Genesis and Development of French Historical Epistemology*, Studies in History and Philosophy of Science 64, https://doi.org/10.1007/978-3-031-88699-7_4

Without making any claim of completeness, in these appendices, I extend this research by proposing the first formalization of a hypothetical, specifically an Italian canon in historical epistemology. My objectives are, in particular, as follows: (1) to call attention to some of the main contributions in this context; (2) to show how, in various cases, they have been interconnected with international debates; and, finally, (3) to foreground how Italian contributions to historical epistemology have been characterized by various political connotations.

Within the field of Italian historical-philosophical production, the study of science from both historical and epistemological perspectives has been hindered for some time by the enormous legacy of Benedetto Croce's thought. Indeed, precisely in reactions to the Italian tradition of historicism and neo-idealism can the roots of Italian historical epistemology be found. In the following pages, I propose a path that, starting from Federigo Enriques and passing through Antonio Gramsci, Ludovico Geymonat, Giulio Preti, Galvano della Volpe, Paolo Rossi, and Aldo Giorgio Gargani, arrives at the hotly contested debates of the so-called Italian Science Wars of the 1970s regarding the neutrality or non-neutrality of scientific knowledge.

4.1 Federigo Enriques and his Conflict with Neo-Idealism and the Vienna Circle

As with Mach, Poincaré, Duhem, and others discussed previously, certain authors active in Italy between the end of the nineteenth century and the first decades of the twentieth viewed the search for the foundations of particular sciences as absolutely inseparable from historical-epistemological analysis. In this period, many professional scientists dedicated themselves to producing epistemological reflections on, and historical reconstructions of, their disciplines.[1] Among these, the most significant author for historical epistemology was Federigo Enriques, a mathematician, historian, and philosopher of science. During his lifetime, Enriques simultaneously conducted research on mathematics, contributing to the development of algebraic geometry, and in the field of historical epistemology. His most significant work in this latter area was *Problems of Science* (1906/1914),[2] which presented the results of an intense period of historical-epistemological reflection starting from the early 1890s. Other influential works included *Razionalismo e storicismo* (1909), *Scienza*

[1] Important figures of this period include physicist Vito Volterra and mathematicians Giuseppe Peano, Giovanni Vailati, and Bruno de Finetti. Aldo Mieli, in particular, spurred the development of the history of science in Italy through his connection to the Centre de Synthèse, directed by Henri Berr, and as founder of the journal *Archeion*, which, following the rise of Nazism, was moved to France and took the name *Archives internationales d'histoire des sciences*.

[2] A significantly revised and expanded edition was published in 1908.

e razionalismo (1912), *Per la storia della logica* (1922), and *Il significato della storia del pensiero scientifico* (1934).

The critical positivism proposed by Enriques—a sort of renewed form of neo-rationalism formulated in light of the progress of contemporary science—addressed both Kantian and positivist perspectives. Like Poincaré, Enriques's reflection was driven by the epistemological effects triggered by non-Euclidean geometries; in addition, like Mach, he engaged with internal questions regarding the history of mechanics and, later, the emergence of the new theories of relativity and quantum mechanics in physics.

A constant point in his discussions was bringing together—following the controversy over the so-called bankruptcy of science—philosophy and science through historical knowledge (anticipating the famous debate on the "two cultures"). This project was carried forward not only through his published works but also by the foundation of the Società Filosofica Italiana in 1906 (of which he was president from 1907 to 1913), the creation of the famous *Rivista di Scienza*: *Organo internazionale di sintesi scientifica* in 1907 (later renamed *Scientia*), and his active participation in various editions of the International Congress of Philosophy.

Enriques's epistemological—and cultural-political—concepts started from the idea of the historical unity of knowledge contained in the concept of synthesis, which had already been occupying Henri Berr since 1900 with the project of the *Revue de Synthèse Historique* in France. For Enriques, the idea of synthesis was mobilized in polemical opposition to the tendency toward particularism in scientific activity as he thought it risked losing sight of the relationship between the unity of knowledge and action (both individual and social) (*Rivista di scienza.* 1907, 1). Starting from this general idea, the task of epistemology for Enriques was to reconstitute the synthetic unity of science through a reconstruction of rationality from a historical-genetic point of view—a stance that aroused the interest of the members of the Vienna Circle.

Nemo profeta in patria, Enriques's epistemology faced substantial resistance in Italian intellectual circles, which only many years later began to acknowledge his importance. The reasons for this were rooted mainly in a bitter controversy in which Enriques strongly opposed the neo-idealist historicism of Croce and Giovanni Gentile, which held a hegemonic position in Italy in the first decades of the twentieth century.[3]

Indeed, Croce's caustic judgments on the philosophical status of scientific knowledge, which he considered inconsistent, are well known. For example, in an article published in *La critica*, he wrote: "the men of science are [...] the embodiment of a mental barbarism produced by the substitution of schemes for concepts, scraps of news for the philosophical-historical system" (Croce 1908, 171). In Croce's perspective, only philosophy and the humanities could draw on the truth, while the individual sciences operated with groupings of conventional and convenient "pseudo-concepts." Croce's philosophical stance was resolutely opposed to the

[3]For a sociohistorical reconstruction of this point, see Boschetti (2024).

reception of positivism in Italy. As a consequence, the development of research in the philosophy of science and sociology, whose origins were rooted in the positivist tradition, was disincentivized.

The controversy between Enriques and Italian neo-idealist historicism extended across several issues. Some of these were purely intellectual (and will be discussed below), while others were of a political-cultural nature relating to the reform of the Italian educational system (both school and university) as well as the creation of an Italian philosophical community.[4] Croce started the conflict when, in 1908, he commissioned the young Gentile to write a long critical review of Enriques's *Problems of Science* for his journal, *La critica*. Gentile's article, in addition to commentary on Enriques's book, also contained a broader attack on the intellectual project of the *Rivista di Scienza* and on the project of developing a historically informed epistemology:

> Historical questions are foreign to epistemology, which defines the concepts of the elements of knowledge and therefore defines knowledge, regardless of its historical process. [...] Epistemology, which, like philosophy, is a vision of the eternal and not of the temporary, considers every objective under the form of the eternal [*sub specie aeterni*]. (Gentile 1908, 439)[5]

Enriques's epistemological project was founded on the proposal of a "scientific philosophy" structured around several nuclei central to historical epistemology. The first of these is a dynamic, or historical-evolutionary, conception of scientific thought, understood as a progressive convergence toward the truth. Indeed, for Enriques, rationalism is not to be defined abstractly; rather, it is necessary to grasp its characteristics in concrete experimental activity and in the dynamics of its psychological-subjective and historical establishment (Enriques 1906/1914). For this reason, his project was completely opposed to attempts to define a theory of knowledge through recourse to metaphysics (Enriques 1906/1914, 30) as well as to those—as Croce had proposed—that reduced science to a pure game of conventions.[6]

In particular, the second cornerstone of Enriques's theory in the early stages of his intellectual trajectory (until around 1920) was his definition of *experimental rationalism*, which emerged from an analysis of the epistemological peculiarities of "the specific methodologies by which different scientific theories operate" (Castellana

[4]The dispute with the Italian Hegelians also continued within the Italian Philosophical Society and during the planning of the fourth International Congress of Philosophy, organized by Enriques in Bologna in 1911. This controversy was so central to Italian culture that it was discussed not only in the usual institutional environments for philosophy but also in the pages of newspapers. For further details, see Ciliberto (1981), Parrini (2004, 57–76), Pompeo Faracovi (2007, 2015), and Gambetti (2014).

[5]For further insights into Gentile's ahistorical conception of science, see Minazzi (2021/2022, 253–257).

[6]On this point, see Parrini's critical reconstruction of a comparative analysis between Croce and Enriques's interpretations of the epistemological theses of authors such as Mach, Avenarius, Le Roy, and Poincaré (Parrini 2004, 58–60).

2005, 308; Enriques 1975). Contrary to Croce's positions, Enriques claimed that an "implicit philosophy" was in the activity of the different sciences and that epistemology is what must account for it (Enriques 1912, 144). With this, Enriques aimed to valorize the epistemological status of "error" in the theory of knowledge (Castelli Gattinara 2017), which led to the analysis of the evolution of scientific thought through concepts such as "rectification" and "approximation" without, however, giving in to the temptations of conventionalism (with strong similarities to Bachelardian epistemology). For Enriques, scientific reason is unitary and universal, a necessary presupposition to make sense of a historical investigation aimed at identifying invariants and structural homologies. In history, the processes of this rational development's taking place can be seen; errors, in this sense, are nothing more than stages in historical movements that lead toward the truth (Paolo Rossi 1986, 215).

Despite the strong opposition that Enriques faced in Italy, he was considered a key author in international contexts. His works were translated into various languages, including German, French, English, Russian, and Spanish.[7] The significance of his work was shown by the fact that it was considered "a truly unique case" in the landscape of twentieth-century epistemology because it was recognized as a reference point for opposing intellectual traditions (Castellana 2019, 67).

On the one hand, Enriques's theoretical project was identified as a precursor to the tendencies developed by the Vienna Circle. For example, in *The Scientific Conception of the World*, Hahn, Neurath, and Carnap included him among the few predecessors of their project (Hahn et al. 1929/1973, 304). Enriques's works were reviewed by the members of the Vienna Circle, and the *Rivista di Scienza* hosted their contributions several times. In particular, Neurath was most influenced by Enriques's perspectives (Neurath 1937, 310), and they also exchanged letters and an invitation to participate in the *International Encyclopedia of Unified Science* (Stöltzner 1998; Parrini 2004, 80). Similarly, at the first scientific philosophy conference in 1935, Enriques was considered "at a level similar to that of Bertrand Russell. Both were claimed as the fathers of the movement of logical empiricism and modern scientific philosophy" (Stöltzner 1998, 463; Polizzi 1982).

On the other hand, Enriques was also recognized as a precursor to, and inspiration for, the founding fathers of French historical epistemology such as Bachelard, who saw Enriques's perspective as a true epistemological "reversal" (Bachelard 1934/1984, 50). Bachelard repeatedly referred to Enriques in various works, including *Essai sur la connaissance approchée*, *The New Scientific Spirit*, and *The Formation of the Scientific Mind*, and the parallels between their epistemological perspectives have been noted on more than one occasion (Castellana 2005, 2021; Castelli

[7] For a complete bibliography, see Enriques (2001, 141–197).

Gattinara 2017). During the same period, other authors from this tradition, such as Rey, Winter, Brunschvicg, Gonseth, and Piaget, drew inspiration from his work.[8]

A turning point took place during the ninth International Congress of Philosophy, held in Paris in 1937, known as the Congrès Descartes. As I noted in the first chapter, during this meeting, the first major clash took place between the neo-positivist or logical empiricist positions developed in Vienna, Berlin, and Poland and the tradition of historical epistemology (Alunni and Brian 2002, 131; Castellana 2019, 63–92). In his speech, titled "Il problema della ragione," Enriques directly attacked both the a priori perspective of Kantianism—which presupposed the existence of eternal truths, contrary to Enriques's perspective founded on the idea of a constant convergence toward the truth—and the positions of logical empiricism developed by the Viennese—which, by denying the possibility of synthetic a priori judgments, reduced scientific propositions to sensible experience.

Enriques, in fact, understood knowledge to be a "process of mutual adaptation of sensible reality to reason, and of reason to reality," thus opening up a constructivist perspective (Enriques 1992, 105). Already in *Problems of Science*, he had described the progress of knowledge as "an alternation of inductive and of deductive phases," which, viewed more generally, would be the "law of scientific development" (Enriques 1906/1914, 99). The inductive phase "obtains new hypotheses from new associations, and from the verification of these it goes on to more extended and more exact associations and hypotheses," while the deductive phase engages in "*logical processes* of transformation of concepts and hypotheses" (Enriques 1906/1914, 99).

In the text of his speech at the Congrès Descartes, Enriques openly stated that the unity and synthesis of rationality must be found in history. In a key passage, he stated: "The new epistemology will find its foundation and its method precisely in the historical critique of scientific concepts" (Enriques 1992, 106). This position is justified by the idea that

> reason does not exist *sub specie aeternitatis*, but is something that develops and is revealed through the historical evolution of thought. We must, above all, look to the evolution of scientific thought for the intimate laws that fall within the field of reason. (Enriques 1992, 105)[9]

In this sense, the complexity of the existence of rationality cannot be reduced to the a priori search for its logical language, because only in historical processes can its foundations be located. For this reason, during the later part of his career, Enriques dedicated himself more actively to the history of scientific thought to make the general epistemological structures constituting scientific reason evident. This turning point in his work and thought must be read in conjunction with his encounter, and conflict, with the tradition of the Vienna Circle—at least with some of its

[8] There is a substantial critical literature on the interconnections between Enriques and Francophone epistemology. See, in particular, Castellana (2005, 2016, 2019, 2020) and Alunni and André (2015).

[9] Note the implicit—but obvious—polemical reference to the passage from Gentile's review cited above.

participants—which instead had imagined an ahistorical epistemological model. In 1932, in *Storia del pensiero scientifico*, Enriques once again renewed the project of a historical epistemology by arguing that

> current science, when we try to fix it in a dogmatic exposition and almost isolate it in its static present, seems to be only one abstract moment in the evolution of scientific thought. To concretely understand ideas and problems, justify hypotheses or theories, account for their many contradictions and the insufficiencies that we must temporarily accept, we need to leave the present and ask ourselves how difficulties can be resolved in the progress of scientific research and why they have been passed down to us from the past: the present then remains only a passing point in the evolution of science. In this way, the push for progress leads from science to the philosophy of science, and from this to history. (Enriques and De Santillana 1932, 6)

On this basis, a Franco-Italian alliance in historical epistemology was established that focused largely on mathematics and physics.[10] If, as mentioned previously, Enriques conceived the evolution of science as a progressive convergence toward the truth on the basis of the constant rectification of scientific theories, then one must necessarily ask whether he imagined a continuist or discontinuist model of scientific development.

Obviously, this point is controversial and has been the subject of debate. On the one hand, commentators such as Paolo Rossi have pointed out how, in many passages, Enriques focused his argument on strong continuist claims (Rossi 1986, 211–213). For Enriques, by virtue of the synthetic unity of rationality, the progress of science is always continuous, guaranteeing structural stability to scientific knowledge. This theoretical position translated into a specific methodological stance for historiographical research. In his view, the history of science must focus its attention on tracing the filiations between concepts by searching for their historical antecedents. To bring out the continuities between apparently distinct phases of scientific evolution, the history of science must demonstrate the connections between newly emerging ideas and those that preceded them (Enriques 1934, 64). Although this view of Enriques's perspective is well supported by references, other commentators, such as Castellana, have rectified this interpretation by showing how, in other passages, Enriques envisioned discontinuous turning points within the existing continuity (Castellana 2004, 15–50; 2019, 149–178). Indeed, already in *Problems of Science*, he had clearly stated that "The progress of knowledge, although it is founded on a continuous development, has a qualitative meaning which is discontinuous" (Enriques 1906/1914, 51). This double view was motivated by the intersection of two scales of observation on the evolution of scientific thought. With this expression, Enriques sustained that, on the one hand, continuity is guaranteed by the fact that science has unifying tendencies aimed at producing progressively more-rigorous, more-general, more-extensive, and more-comprehensive results regarding

[10]On this point, a large critical literature has developed focused on certain "hidden harmonies" within the history of epistemology that shed light both on the emergence of the so-called analytic-continental divide and the initial phases of the constitution of historical epistemology. See Castellana (2004, 2005, 2019), Alunni and André (2015), Alunni (2018), and Ienna (2021).

natural phenomena. This tendency guarantees the stability of this form of knowledge over time. On the other hand, innovations continually emerge in local, limited contexts through specific theories that are discontinuous with respect to previous theories. Even though the tendency of scientific knowledge, taken as a whole, is a continuous convergence toward the truth, this convergence is produced through specific, qualitative ruptures that bring radical transformations to the conception of scientific concepts, theories, and objects.

4.2 The Post–World War II Period and Gramsci's Perspectives

As outlined in the previous section, the cumbersome hegemony—and subsequent inheritance—of neo-idealist historicism in the Italian philosophical field long discouraged the emergence of the history and the philosophy of science (and of the social sciences in general).[11] Enriques and other authors of the time dedicated themselves to this kind of research despite being largely relegated to science departments or otherwise marginalized compared to the positions of Croceans. The institutionalization of university chairs explicitly dedicated to the philosophy and the history of science in humanities faculties would, in fact, happen later.[12]

With the fall of fascism, the liberation of Italy, and the end of World War II, various authors laid the foundations for a challenge to the hegemony of Croce's neo-idealism by moving along different theoretical-intellectual trajectories. The first of these was rooted in the reception process of neo-positivist epistemology and scientific philosophy in general. As previously mentioned, starting from the 1920s and 1930s, the prestigious Italian journal *Scientia* had published various articles by members of the Vienna Circle. Ludovico Geymonat was very interested in Comtian positivism, dedicating *Il problema della conoscenza nel positivismo* to this topic in 1931.[13] He was also fascinated by neo-positivism, to which he devoted *La nuova filosofia della natura in Germania*, published in 1934, and *Studi per un nuovo*

[11] Gentile, as minister of public education in the Fascist government from 1922–1924, gave an initial impetus to the institutionalization of the history of science in Italy by including it as a subject in scientific high schools. However, to the great disappointment of Enriques and Mieli, he refused to create university positions dedicated to these topics. During fascism, key institutions for the development of the history of science were founded in Italy, including the Treccani Institute (1925), the Museum of the History of Science in Florence (1925–1930), and, thanks to Gentile, the Domus Galileiana in 1941 (Govoni 2004, 42–46; Cimino and Sava 2007).

[12] The first position in the philosophy of science in Italy, created in 1956, was held by Ludovico Geymonat. For the history of science, however, not until a 1980 public competition were six professorships established, assigned to Enrico Bellone, Vincenzo Capelletti, Paolo Galluzzi, Carlo Maccagni, Giorgio Tabarroni, and Maurizio Torrini (Cimino and Sava 2007, 252).

[13] This was Geymonat's first work, resulting from the reworking of his university thesis, which he had completed under the direction of the epistemologist Annibale Pastore (Minazzi 2021/2022, 246).

razionalismo, published symbolically on Italian Liberation Day, April 25, 1945. Attracted by the latter tradition, he had spent some of the winter term of 1935 in Vienna in close contact with Moritz Schlick and other members of the Vienna Circle[14] (Geymonat 1983, 9–10; Agazzi et al. 1989, 13–15), later returning to Italy to become one of the main promoters of neo-positivism.

Another key figure in this phase was Giulio Preti. Preti had approached meta-scientific themes starting from a phenomenological background[15] before becoming interested in Dewey's American pragmatism, neo-positivism (for which he was one of the major importers to Italy), and the history of logic.[16] These were years in which philosophical eclecticism was mobilized as a strategy in the search for alternatives to Italian neo-idealist historicism. I will return to the developments emerging from the epistemological reflections of these authors in the next section.

While Geymonat and Preti were reacting against the ideas of Croce and Gentile, a second reaction to neo-idealism emerged in the work of Antonio Gramsci. The publication of the *Prison Notebooks* between 1948 and 1951 by Felice Platone, under the guidance and political direction of Communist Party Secretary Palmiro Togliatti,[17] marked a fundamental turning point for postwar Italian culture. Indeed, its publication generated a vast renewal in the field of Italian humanities and social sciences, though only more recently has the impact of Gramsci's theses in the field of historical epistemology been appreciated.[18]

Despite having been a convinced Crocean in his younger days, Gramsci made various criticisms and rectifications of the perspective of historicism in the tenth notebook, titled *The Philosophy of Benedetto Croce*. Gramsci believed that the most mature results of Crocean philosophy represented the maximum expression of bourgeois thought and revisionism. Influenced by Hegel, Croce had presented an antipositivist philosophy of history that denied that facts could be the objective material of historical narratives. For Croce, history was the history of the

[14]On this point, see Minazzi (2010), which also contains some correspondence between Geymonat and various authors, including Schlick.

[15]After writing his thesis in philosophy at the University of Pavia on Husserl, Preti became involved in Antonio Banfi's philosophical school in Milan.

[16]Preti published various essays on these themes, some of which can be found in the collection *Saggi filosofici*, edited by Mario Dal Pra (Preti 1976a, 1976b). For an in-depth examination of Preti's trajectory, see Minazzi (2011a). *Idealismo e positivismo* (1943) was also an important work in which he proposed a model for integrating Husserl's perspective with neo-positivism (Coniglione 2011, 60–61).

[17]The first edition of the *Prison Notebooks* was organized according to the themes covered by the author. Only in 1975 was a second edition published, by Valentino Gerratana, which reorganized the material chronologically (see Gerratana 1997). English translations of the *Prison Notebooks* are scattered across a range of works, though Columbia University Press has published three volumes of chronologically arranged critical editions, covering notebooks one to eight. Most of the citations here are from notebooks 10 and 11, and the translations, when they exist, are often partial, with parts of the same text often spread across different works.

[18]On this point, see Nieto-Galan (2011), Omodeo (2016, 2018, 2020), and Badino and Omodeo (2020).

abstractions of the spirit, and for that reason, Gramsci wrote that "His philosophy remains a 'speculative' philosophy" (Gramsci 1977b/1995, 347). In contrast, Gramsci proposed a re-evaluation of the deeper meaning of historical materialism on the basis of a philosophy of praxis, starting from Marx's *A Contribution to the Critique of Political Economy* and *Theses on Feuerbach* (Omodeo 2020). What interested Gramsci was the unfolding of history in concrete clashes between classes and in the struggle for the conquest of cultural hegemony (Gramsci 1977b/1995, 395–396; see Badaloni 1962, 173).

These interests led Gramsci to address questions of an epistemological-scientific nature in his eleventh notebook, *Introduction to the Study of Philosophy*. Partly following Croce's antipositivist approach, Gramsci attacked forms of Soviet Marxism that relied on a strongly deterministic, economistic, and single-issue approach. In particular, Gramsci's target was the model of "vulgar materialism" presented by Bukharin in the widely read *Historical Materialism: A System of Sociology*. This work, in the wake of Lenin's *Materialism and Empirio-Criticism*, had proposed a materialist radicalization of Marxism through the elimination of the subjective elements of the theory of knowledge, thus siding against conventionalist epistemologies and empiriocriticism (Omodeo 2016, 6–62). In contrast to this theory of knowledge based on matter as its objective referent, Gramsci proposed a historical-epistemological model:

> Matter as such therefore is not our subject but how it is socially and historically organised for production, and natural science should be seen correspondingly as essentially an historical category, a human relation. […] And might not the idealistic conception according to which nature is none other than the economic category be reduced, once cleansed of its speculative superstructures, into the terms of the philosophy of praxis and demonstrated to be historically linked to and a development of that philosophy? (Gramsci 1977a/1971, 465–466)

In accordance with the centrality that he attributed to the philosophy of praxis, Gramsci aimed to show how both the subjective and objective poles of knowledge are produced from historically and socially situated relationships rather than an ahistorical transcendental.[19] In this sense, Gramsci argued that both the "mechanical materialism" (built on an objectivist basis) and the subjectivist conception had fallen prey to forms of ideological-religious belief (Gramsci 1977a, 1486). In his view, both models are extremes that must be opposed by a form of "monism" that is neither materialist nor idealist. Instead, it is a philosophy of praxis that is capable of integrating these two poles on the basis of an "identity of opposites in concrete historical act" and that therefore holds together, on the one hand, "human activity (history-spirit) in the concrete" and, on the other, nature "transformed," "organised,"

[19]This is why for Gramsci nature itself cannot be considered as a static, objective referent but must be analyzed in its dialectical relationship with humans: "Thus Man does not enter into relations with the natural world just by being himself part of the natural world, but actively, by means of work and technique" (Gramsci 1977a/1971, 352). Contrary to Hegelian dialectics, which starts from abstract being such as Idea or Spirit, Marxist dialectics originates in nature and in its transformative and conflictual relationship with humans (Dobb 1969). On this point, see also the section on alienated labor in the *Economic and Philosophic Manuscripts of 1944* (Marx 1932/1977).

and "historicised" by the same human activity (Gramsci 1977a/1971, 372). Returning to the famous passage from the preface to Marx's *A Contribution to the Critique of Political Economy*, Gramsci argues that

> the question of the "objectivity" of knowledge [...] can be treated by starting from the proposition [...] that "men become conscious (of the conflict between the material forces of production) on the ideological level" of juridical, political, religious, artistic and philosophical forms. (Gramsci 1977a/1971, 371)

These positions were in contrast with the Soviet Diamat, which—on the basis of Engels's *Dialectics of Nature* and Lenin's *Materialism and Empirio-Criticism*—had constructed a form of ontological reductionism in which both the natural and social worlds were governed by the same dialectical rules. This position was taken further to the extreme in Bukharin's model, which proposed to adopt the same form of determinism that could be used in the study of physical phenomena for the study of human phenomena[20] (Omodeo 2016, 62).

Gramsci opposed this conception by proposing a culturalist conception of Marxism that completely overturned these interpretations. This project passed through the rectification of Bukharin's deterministic interpretation of the relationships between structure and superstructure in order to overcome its rigid economic reductionism.[21] Gramsci, focusing on modes of ideological production, stressed the decisive weight of superstructural components in maintaining the stability of the relations of production described by the economic "structure" (through which he re-evaluated the cultural and ideal moment of the class struggle). The notion of cultural hegemony was, in fact, coined by Gramsci precisely to explain how the dominant classes manage to impose a vision of the world and "moral-intellectual leadership [*direzione*]" on the subaltern classes. The supremacy of one class over another is obtained not solely through the exercise of domination through force but also through strategies for creating consensus. These are the results of the process in which the subaltern classes incorporate the ideology of the dominant class through "hegemonic apparatuses" such as education, the press, cinema, religion, etc. The Gramscian philosophy of praxis aimed to foreground how structure and superstructure develop together in a way that is "intimately bound together and necessarily interrelated and reciprocal" (Gramsci 1977b/1995, 414).

In this context, science took on a key role in the economy of Gramscian thought because, in his conception, it has a special connection to the superstructure, a relationship "greater in extension and continuity of development" (Gramsci 1977b/

[20] In these sections of the *Prison Notebooks*, Gramsci articulated a strong criticism of positivism, which he considered a reactionary ideology (Paolo Rossi 1977, 230). From his perspective, positivism could be equated to Bukharin's theoretical model and was, consequently, a "sociology of metaphysical materialism" (Gramsci 1977d/2011, 739) (viewed as an inappropriate application of a mechanistic model to the study of society).

[21] To overcome the interpretive rigidity of the terms *structure* and *superstructure*, Gramsci proposed the biological metaphor—also used by Marx in *A Contribution to the Critique of Political Economy*—of the "anatomy" and "skeleton" of a society (Gramsci 1977b/1995, 397; 1977c/2007, 354).

1995, 293). Science, as part of the superstructure, is characterized by an intrinsic ideological dimension:

> notwithstanding all the efforts of scientists, science never appears as a bare objective notion—it always appears in the trappings of an ideology; in concrete terms, science is the union of the objective fact with a hypothesis or system of hypotheses which go beyond the mere objective fact. (Gramsci 1977b/1995, 293)

Despite this, Gramsci's position did not lead to a form of epistemic relativism, because scientific knowledge itself is equipped with the tools to distinguish "the objective notion from the system of hypotheses" through abstraction (Gramsci 1977b/1995, 293). For this reason, even though some theories or scientific discoveries originate within a certain ideological context (and are shaped by it), this does not mean that different social groups cannot make use of their objective agency by rejecting the underlying ideology.

In agreement with the idea of the mutual conditioning of structures and superstructures, Gramsci criticized the idea proposed by mechanistic materialism that only technical-scientific instruments (the means of production) bring about the emergence of scientific forms. On the contrary, Gramsci insisted on looking at "the ensemble of social relations of production" (Gramsci 1977c/2007, 340). Indeed, the notions of science and technique needed to be reformulated because both had been conceived in a reductionist form by Bukharin. By "technique," "one should understand not only the *ensemble* of scientific ideas applied industrially (which is the normal meaning of the word) but also the 'mental' instruments, philosophical knowledge" (Gramsci 1977a/1971, 353). In the same way, Gramsci also denounced how the concept of science (which needed to be "critically destroyed" (Gramsci 1977a/1971, 438)) had been constructed by flattening the epistemological specificities of various disciplines to those of physics. Positivists claimed that the natural sciences were the sciences par excellence and therefore were the basis for their model. In this claim, Gramsci saw the grounds for a form of epistemological pluralism (or, to use a Bachelardian term, *regionalism*):

> It has to be established that every research has its own specific method and constructs its own specific science, and that the method has developed and been elaborated together with the development and elaboration of this specific research and science and forms with them a single whole. (Gramsci 1977a/1971, 438–439)

Thus, all forms of epistemological reductionism are to be avoided:

> To think that one can advance the progress of a work of scientific research by applying to it a standard method, chosen because it has given good results in another field of research to which it was naturally suited, is a strange delusion which has little to do with science. (Gramsci 1977a/1971, 439).

Throughout the *Prison Notebooks*, Gramsci frequently mentioned the concept of scientific progress, which he interpreted in a historicist way. What science claims is never definitively, objectively, true. If it were, Gramsci argued, there would be no research; there would be only the dissemination of already-known results. He thus derived the thesis that science itself is a "historical category" precisely because it is "continuously correcting [*rettifica*]" human relationships with nature through progressive modifications of culture, conceptions of the world, and technological instruments (Gramsci 1977b/1995, 292). All forms of knowledge, including scientific truths, emerge from praxis and can, therefore, be only "provisional," never "extra-historical" or "extra-human" (Gramsci 1977a/1971, 406, 445).

On various occasions in the *Prison Notebooks*, Gramsci argued that science is historically determined, and he outlined some ways to reflect on the model of its evolution over time. The political tone with which Gramsci addressed epistemological questions was based on a particular form of presentism that he initially took from Croce, reworking it in a manner consistent with his philosophy of praxis. For the Croce of *Theory and History of Historiography*, "every true history is contemporary history" (Croce 2007/1921, 12). This contemporaneity is determined by the point of observation, the interests, and the stance of the historian. According to Croce, authentic history is never a collection of facts or news; on the contrary, it is "the synthesis of intuition and concept"—or, in idealist-historicist terms—"thought in its concreteness" (Croce 2007/1921, 76). Gramsci took up this presentist conception, depriving it, however, of its idealistic nature and bringing it back to the Marxist canon. For him, "history is always contemporary history, i.e. politics" (Gramsci 1977b/1995, 383). With this clarification on Croce's position, Gramsci asserted that what allows history to be truly authentic—that is, not merely "bookish and erudite"—is its political nature, "the identification of history and politics" (Gramsci 1977b/1995, 383). Indeed, the historian is always part of a community and, therefore, in their active intellectual reconstruction of the past, they operate within a conflictual context of struggle for political-cultural hegemonies (Omodeo 2018, 222). In this sense, Gramsci's philosophy of praxis was intended as a "historical methodology" (Gramsci 1977d/2011, 739).

Marxism, in its Gramscian historicist form (Badaloni 1962), went on to be widely recognized as the Italian path of Marxism and was the main cultural line of the Italian Communist Party (Rossi 2009, 120–121). And among those who embraced the Gramscian perspective in the process of elaborating a cultural history—with a significant impact on the history of science—was certainly Eugenio (Garin 1997).[22]

4.3 Historical-Epistemological Alternatives to Italian Neo-Idealism

New stimuli emerging in the post–World War II Italian intellectual climate spurred the development of new ways of historicizing epistemology; these new forms, in dialogue with Marxism, were proposed as alternatives to Gramsci's historicist model. In this context, the contributions of authors such as della Volpe and the aforementioned Geymonat and Preti were most influential.

[22] On this point, see Santucci (1996), Maggi (2009), and Ciliberto (2011).

Della Volpe[23] was considered by many to be "the Italian Althusser" for his ability to hold together an anti-Hegelian and scientific interpretation of Marx's work along with epistemological reflection.[24] A fundamental point of reference in this sense was *La logica come scienza positiva*—republished in 1969 with the less polemical title *La logica come scienza storica*. In this text, della Volpe developed an alternative perspective to both Gramscian-Crocean historicism[25] and the Soviet Diamat (based on Engels's *Anti-Düring* and *Dialectics of Nature*). While Gramsci's perspective had been open to a form of epistemological pluralism and antireductionism, della Volpe instead sought to unify the methods of the physical-natural sciences with those of the historical sciences. However, for della Volpe, the central issue of Marxism was the materialist conception of history rather than dialectical materialism. The former had a methodological function for the social sciences and could be reduced neither to historicism nor, much less, to a mere vision of the world. Marxism, rather, was configured as a description of the logic of historical research (Parolini 1978).

Among della Volpe's objectives was his aim to demonstrate the scientific nature of Marxism under the sign of the foundation of a "*moral Galileanism*" (della Volpe 1969/1980, 178; Parolini 1978, 485). Starting from Marx's *Introduction to A Contribution to the Critique of Political Economy* in 1857, the *Grundrisse* in 1857–1858, and *A Contribution to the Critique of Political Economy* in 1859, della Volpe reconstructed the logical-epistemological framework of the Marxist method later implemented in *Capital*. This method was based on the critique of the forms of epistemological-ontological naturalization, or eternalization, of the economic laws and concepts implemented by classical bourgeois economists. Their model was, in fact, based on an approach to definitions of economic categories such as labor, production, value, surplus value, capital, etc. that were antihistoricist and based on natural law (and thus epistemologically recognizable as "indeterminate [...] abstractions") (della Volpe 1969/1980, 194–195).

Della Volpe instead emphasized how Marx's methodology moved with the circular logical-epistemological structure of the "concrete-abstract-concrete" (della Volpe 1969/1980, 196), which appeared to be completely superimposable on and consistent with the Galilean method based on the "induction and deduction" model. The three phases of "concrete-abstract-concrete" correspond to "the data, the historical-rational situation, the experiment as a reciprocal functionality of data-hypothesis" (Parolini 1978, 472). For della Volpe, both the Marxian and Galilean methods represented "the materialist logic of modern science" (della Volpe 1969/

[23]For a general overview of della Volpe's work, see Vacca (1970).

[24]These authors were united by taking Marx's *Capital* as the theoretical point of reference for their project of demonstrating the scientific nature of the historical-materialistic method. Unlike Althusser, however, della Volpe did not identify an "epistemological break" between an early idealist Marx and a scientist Marx, but on the contrary, he elaborated an interpretation aimed at showing how Marx had already made a methodological break with the Hegelian idealistic model in Marx's youthful writings (Romeo 2002, 250).

[25]The search for an alternative to Gramscian-Crocean historicism would also be the cornerstone of della Volpe's aesthetic theory, put forth in *Critica del gusto* (1960).

1980, 198). Della Volpe's most significant contribution was his ability to draw attention to the notion of determinate abstraction as the fulcrum of Marxian epistemology (Vacca 1970, 33–44). In his perspective, Marx's materialist theory of knowledge was based on the "continuous historical evolution of abstractions"—that is, scientific and cognitive categories—through the three-phase circular movement (della Volpe 1969/1980, 188). Thus, a perspective that ran contrary to bourgeois economics instead constituted its categories through the "simplification, or one-sided and forced abstraction, of their content" (della Volpe 1969/1980, 188).

The "concrete-abstract-concrete" structure described an "antidogmatic" analytical model structured in three phases: (1) It starts from problems inscribed in the concrete present situation as conceptualized by common sense; (2) it passes to the investigation of their historical roots to reconstruct their origins, making certain abstractions regarding the historical situation being investigated (i.e., scientific conceptualizations understood as non–a priori but historical abstractions); (3) it returns to the concrete by evaluating the abstractions created in the second phase in relation to the present situation, putting the dialectical movement back into circulation.

Clearly, della Volpe's epistemology was in strong opposition both to idealism—due to its inclination toward the creation of ideal hypostases—and to positivism—due to its "idolatry of 'the facts' and consequent repugnance toward 'hypotheses'" (della Volpe 1969/1980, 198). The scientific nature of Marxism rested precisely on overcoming idealism in the complete inversion of the Hegelian dialectic according to the concrete-abstract-concrete circle, treating the scientific facts themselves as determinate abstractions (not reducible to the facts as they are understood by positivism). The extension of this Marxian-Galilean method to all sciences went in a different direction compared to the logical-formalist solution advanced by neo-positivists, and della Volpe dedicated an entire appendix to critiquing them (della Volpe 1969/1980, 247–265). Neo-positivism was, on the one hand, completely incapable of justifying its own foundations and, on the other, powerless in the face of the problem of "scientific discovery" (della Volpe 1969/1980, 261).

Della Volpe's school of thought was later developed further, in various forms and trajectories, through the contribution of authors such as Lucio Colletti, Umberto Cerroni, Nicolao Merker, Mario Rossi, Giulio Pietranera, and others (Alcaro 1977).

Moving beyond their initial activities and research projects in the 1930s, Geymonat and Preti also contributed to the development of alternatives to Gramsci. These two authors did not simply work in the same period but also addressed some of the same theoretical-methodological problematics that marked the subsequent development of Italian historical epistemology.[26] Despite their proximity to the neo-positivist paradigm (Preti 1946; Geymonat 1991, 246)—though with relative differences—both were dissatisfied with the "general epistemological deafness in Vienna in relation to the history of science" (Minazzi 2021/2022, 246). Starting from the second half of the 1940s, both of these authors pursued a dual interest in

[26] For detailed comparisons, see Minazzi (1984) and Coniglione (2011).

their work: on the one hand, focusing on purely epistemological questions and, on the other, pursuing an interest in the history of science.

Geymonat devoted himself mainly to the study of the history of mathematics, publishing *Storia e filosofia dell'analisi infinitesimale* (1947) and a section devoted to *Storia della matematica* (1962) in the encyclopedic neo-Enlightenment project of Nicola Abbagnano's *Storia della scienza*. Moreover, he wrote *Galileo Galilei* (1957), widely considered Geymonat's masterpiece as a historian. Preti, for his part, was the author of a series of methodological essays on the historiography of science, including *Due orientamenti nell'epistemologia* (1950), *Considerazioni di metodo sulla storia delle scienze* (1958), and *Storia del pensiero scientifico* (1957).

The similar trajectories that these authors pursued can thus be understood as an attempt to find an intersection between epistemology and history, and precisely in the attempt to resolve this problematic during the post–World War II period, Italian historical epistemology created new trajectories. Preti pursued this connection in *Praxis ed empirismo* (1957), and Geymonat explored the same issues in *Filosofia e filosofia della scienza* (1960) and *Scienza e realismo* (1977), both seeking to integrate neo-positivism and Marxist dialectical materialism, two traditions that were widely considered to be not merely opposed to each other but instead irreconcilable.

In *Studi per un nuovo razionalismo*, Geymonat had already acknowledged the need to integrate neo-positivist epistemological analysis with a historical reconstruction of science. In this work, he emphasized the irreconcilability of classical theories of knowledge with the evolutionary structure of scientific thought that had emerged in the preceding decades. Thus, for Geymonat, the process of scientific transformation was to be read under the sign of a "dialectic," understood as a "rational scheme that precisely illuminates the logical structure of scientific ideas, their crises, the expansion and determination of the meaning of their terms" (Geymonat 1945, 157).

Preti introduced the notion of historical epistemology into Italian for the first time in "Due orientamenti nell'epistemologia" (1950), in which he used this concept in opposition to that of technical epistemology. Following his ambition of founding a scientific philosophy, for Preti, the concept of epistemology[27] corresponded to a "reflection of science on itself"—or, rather, a reflection originating from within concrete scientific practices (cutting across philosophical themes) and not from the abstract reflection of philosophy (which presupposes science as an object distinct from itself) (Preti 1950, 200). Starting from this conception, the terminological distinction between a "technical" epistemology and a "historical" epistemology was coined by Preti, with a double polemical intention aimed at epistemologies based exclusively on logical-formal and a priori apparatuses as well as the old

[27] The concept of epistemology had been introduced for the first time in Italian by Annibale Pastore, who taught both Geymonat and Gramsci at the University of Turin, in *Del nuovo spirito della scienza e della filosofia* in 1906 (Pastore 1906; Castellana 2004, 51–78; 2005, 309), thus at the same time as the first uses of the term in French by Rey and Meyerson (see chap. 1, section 1.3).

tradition of history of science reductively built by collecting biographies of scientists.

"Technical" epistemology referred to those special epistemologies that served to resolve problems emerging within specific disciplines. In particular, Preti was referring to the debates that emerged about the foundations of these disciplines in the wake of the crisis of the physical-mathematical sciences triggered by the emergence of mathematical logic, the theory of relativity, quantum mechanics, and so forth (Preti 1950, 211–212). For Preti, these reflections had indisputable epistemological merits—for example, the effort to critically re-examine certain fundamental categories such as time, space, and causality—but, at the same time, were not entirely adequate for constructing a more complete epistemological reflection. Some expressions of "technical" epistemology that became the primary targets of Preti's critique were the logicist program of Russell and the neo-positivism of the Vienna Circle. For Preti, both of these intellectual movements seemed to be plagued by intrinsic "epistemological fallacies," such as the presumption that included epistemological postulates would be recognized as "positive and objective" research results (as in Russell), or the claim that the methodology of physical-mathematical sciences could be extended to other disciplines, such as political economy or biology (as proposed by the Vienna Circle) (Preti 1950, 212). Contrary to these tendencies, Preti defined *historical epistemology*—or "epistemology of historical orientation"—in the following way:

> Historical epistemology could be defined as "the methodology of the history of science": in this case, it would be a branch of historiography, but with the peculiarity (characteristic of all philosophical production, in which a part always subsumes the whole) of being able to reabsorb the totality of historiography itself, where this is conceived as a science. Or it could be defined as "the historical dynamics of science in its formal unity," that is, it would be a theory of the unity of science, but as a historical tendency and not as an acquired or technically realized "fact." (Preti 1950, 213)

Preti had, in fact, distanced himself from the traditional history of science due to its tendency to produce analyses fraught with internalist and presentist approaches (Barsanti 2007, 130). Historical epistemology, on the contrary, facilitated the understanding of the vast conceptual and cultural frameworks within which certain forms of scientific rationality had emerged, as well as the pursuit of the historical-evolutionary dynamics underlying the unity of science. For this reason, Preti argued that the "history of science," understood as a simple report or erudite assembly of information, needed to be replaced with a "history of scientific thought" because only the latter was capable of accounting for the processes through which the operational categories, purposes, methods, and perspectives of a science are transformed (Preti 1957b, 7–8). For Preti, this approach required the historicization of the Kantian a priori as a function of the construction of a theory of knowledge capable of accounting for the different levels of objectivity that progressively emerged throughout history (Parrini 2004, 183–189; Minazzi 2021/2022, 335–339).

Preti, in his book/manifesto *Praxis ed empirismo*, drew together his reflections on the previous years with the ambition of definitively overcoming the culturally hegemonic Italian idealist approach that had also been well received by the Italian

Communist Party.[28] Preti's proposal was to build a historical-epistemological and political theory that was based on the integration of the Marxian issue of praxis with the perspectives of pragmatism and logical empiricism, without falling into the extremes of idealism or naive realism. Starting from the young Marx (particularly from *The German Ideology*), Preti proposed a form of realism that was based on a circular dialectical structure. He pragmatically assumed the existence of a common sense codified in ordinary linguistic exchanges and in the categorical apparatuses of knowing subjects. From this, Preti saw scientific praxis as a tool for organizing reality that was capable of constituting the specific regional ontologies structuring the historically determined a priori that define the space of possibilities for scientific investigation. Each science shapes its reference ontology according to its methods, categories, and problem-solving models in relative autonomy from one another. What links these regional ontologies is common sense, understood as a pragmatic referent of the verification process to which they refer. From Preti's perspective, this would establish a socially organized epistemological pluralism on a democratic basis. Preti's aim was, indeed, to constitute not so much a transcendental subject of knowledge but rather a collectively organized subjectivity (Preti 1957a, 140).

The ontological region of "historical knowledge," understood as science, played a key role in Preti's argument (Preti 1957a, 152–185). By openly siding against Popper's epistemological devaluation of history, he showed how historical materialism and neo-positivism were not in real contradiction but rather were two epistemological approaches that needed to be integrated. For Preti, the concrete practice of the historian was based on principles of verification that were entirely consistent with the epistemic standards of other disciplines. Instead, idealist historicism and ahistorical interpretations of dialectical materialism had suggested an irreconcilability that, on the contrary, Preti strove to overcome.

Compared to the model advanced by Preti, Geymonat followed a different path in search of a way to integrate the materialist-dialectical perspective with neo-positivism. Taking inspiration from Enriques's project,[29] Geymonat located this synthesis in the construction of an epistemology of rationalist commitment capable of integrating historical reflection, philosophical reflection, and scientific research.

[28] The text was published in 1957, the year after the Twentieth Congress of the Communist Party of the Soviet Union (CPSU), which marked the opening—in Italy as in other contexts—of a phase of remodulation for Marxist theory with respect to the orthodoxy of the CPSU. This text generated a large debate in the pages of the major Italian journals of those years, such as *Passato e presente*, *Società*, *Il Mulino*, *Aut-Aut*, *Il Protagora*, and *Almanacco letterario Bompiani*, among others. (Minazzi 2004, 115). It was also violently attacked in a polemical pamphlet titled *Marxismo e neopositivismo* (Cases 1958).

[29] Starting in 1975, Geymonat (together with the physicist Pietro Caldirola and the geneticist Giuseppe Montalenti) took over the direction of the journal *Scientia* (from 1910 the title of the *Rivista di Scienza*) founded by Enriques, spurring a renewal of the Italian rationalist tradition (Caldirola, Geymonat, and Montalenti 1975).

Geymonat had already identified the need to historicize epistemology in his 1951 essay "Convenzionalità e storicità delle teorie scientifiche" (included in *Saggi di filosofia neorazionalistica* of 1953). In that essay, he emphasized how the continual proliferation of sciences, increasingly varied and rich, had forced epistemology to revise its models. Contrary to formalist approaches, it was no longer a question of deducing "the idea of a possible science developed *a priori*" from the great variety of concrete sciences but rather one of "following their real development" to reveal their deeper structures (Geymonat 1953, 66).

Geymonat further developed this perspective through a more systematic adoption of dialectical materialism, starting from *Filosofia e filosofia della scienza* in 1960. In the wake of the disastrous events of the Lysenko affair (Lecourt 1976), Geymonat's primary motivation was to rehabilitate dialectical materialism as an epistemological perspective capable of accounting for both the reality of the objective, external referent of knowledge and the processes of the historical transformation of science (Parrini 2004, 120–135). Geymonat also pursued this historical-epistemological venture through collective projects with a broad cultural orientation centered on the relevance of dialectical materialism. In 1972, Geymonat and Silvano Tagliagambe edited a special issue of *Critica Marxista* called *Sul Marxismo e le scienze* (Geymonat and Tagliagambe 1972), and in 1974, Geymonat coauthored *L'attualità del materialismo dialettico* with Enrico Bellone, Giulio Giorello, and Tagliagambe (Bellone et al. 1974). During this period (1972–1976), Geymonat was working on his monumental 11-volume project of the *Storia del pensiero filosofico e scientifico*, which represented a further aspect of his intellectual interests and range (Prestipino 1973). These works were the centerpiece around which Geymonat's so-called Milan School formed, aiming to definitively dismantle Italian neo-idealist historicism—whose cultural hegemony continued through the heirs of Crocean thought—to establish a form of neo-rationalism.

Geymonat's later works contained essays such as "Metodologia neopositivistica e materialismo dialettico" and "Primi lineamenti di una teoria della conoscenza materialistico dialettica," which were key for the definition of his intellectual pathway (Geymonat 1972, 1974). The results of his reflections during this period then found a more systematic and complete version in *Scienza e realismo* (1977). In all of these works, Geymonat's objective was to develop an epistemological, rationalist, and materialist model capable of accounting for both science's historicity and its objectivity. The cornerstone of his proposal was the redefinition of the concept of truth, overcoming both the "character of absoluteness" traditionally attributed to it and the "agnosticism" emerging from conventionalist epistemologies (Geymonat 1977, 147). In this sense, he asserted, "science is actually capable of arriving at truths, but [...] relative truths, not absolute truths" (Geymonat 1977, 147). Developing Lenin's theses of *Materialism and Empirio-Criticism*, Geymonant maintained that the truths of science are "relative, transitory, approximate, and always replaceable with others that are more complete and precise" (Geymonat 1977, 74; see also 1972, 37–38).

Connected to this epistemological thesis was another, one that was of an ontological nature. According to this position, reality cannot be conceived in static terms—that is, made up of ultimate elements that, once known, would lead definitively to reality—but rather is characterized by "a dialectical structure" that cannot be enclosed within "*a priori* determinations" (Geymonat 1977, 76). The history of science is what demonstrates this constant process of transforming, expanding, and revolutionizing old notions and cognitive categories. The historical evolution of science, however, should not be observed exclusively from a logical-formal analysis of individual scientific theories, which represent its contingent and discontinuous aspect. Rather, it needs to be approached from the global perspective of the processes of the progressive rectification of the "scientific-technical inheritance" (Geymonat 1977, 33–35; Minazzi 2011b). If the idea of truth as corresponding to reality were true, then methods of in-depth analysis that can augment our knowledge would have no reason to exist: The truth would be given directly, once and for all. Geymonat aims to overcome the classical model of truth, understood as "adequation"—characterized by a strongly static nature—by constructing a dynamic model based on the concept of approximation in which "the historical dimension becomes part of the notion of truth itself" (Geymonat 1977, 74–75). For Geymonat, dialectics was the intellectual tool both for "studying reality […] in its historical becoming" (Geymonat 1977, 49) and for systematizing multiplicity within our scientific-technical heritage (Geymonat 1977, 50–53).

Given this epistemological model, Geymonat needed to formulate a criterion that would allow a distinction to be made between "mere arbitrary conventions" and those that could actually claim an effective "cognitive value" (Geymonat 1977, 66; Parrini 2004, 127). Although in different ways compared to Gramsci or Preti's perspectives, Geymonat also sought a solution to this problem in his description of the relationship between theory and praxis in terms of dialectical unity. In this sense, Geymonat declared: "practice is, for the most part, permeated with theory, and theory is permeated with practice, so that it would be extremely contrived to consider them as completely distinct from each other" (Geymonat 1977, 67). Given this assumption, the inseparability of theory and practice is what has to be considered as a criterion of truth for the objectivity of scientific research (despite its fundamental perfectibility). This does not mean, however, that Geymonat surrendered to a pragmatist epistemology from which he clearly distanced himself. If pragmatism centers its epistemology on a model on the basis of subjective praxis and the practical usefulness of scientific theories, Geymonat's materialist rationalism conceives praxis as intrinsically social[30] and inseparable from theory. Therefore, Geymonat's model, rather than being focused on practical success, was aimed at explaining the validity and truth of scientific activity (Geymonat 1977, 68–69).

For both Geymonat and Preti, these efforts at a historical-epistemological analysis of scientific reason constituted the fundamental axis—in the wake of what Enriques

[30] "The practice which we have been talking about is always 'social practice' […]. [This] implies, for example, that the validity of a scientific discovery cannot be assessed solely on the basis of the operations carried out by the individual scientist in the laboratory but also, and above all, on the basis of the use made of it in production (which obviously depends on the type of society in which it is applied, corrected, perfected)" (Geymonat 1977, 68–69).

had proposed—for seeking a line of communication between the "two cultures" (Snow 1959; Preti 1974). For both authors, the antidogmatic nature of this project had intrinsic political value because the creation of a scientific culture turned out to be the prerequisite for democratic practices (Preti 1957a) and for a progressive socialist emancipation (Bellone et al. 1974).

4.4 Paolo Rossi, Between the Spiders and the Ants

Whereas the intellectual trajectories described so far have represented models for historicizing epistemology, Paolo Rossi's work instead is an attempt to epistemologize the history of science through a profound reflection on its methodology. As a student of Garin (and during his training, close to the rationalist school of Antonio Banfi),[31] Rossi subsequently developed a research program that led him to become one of the main promoters of the history of ideas both in Italy and abroad.[32] Situated in the postwar climate of Italian cultural renewal, Rossi mobilized the models of authors such as Koyré and Arthur O. Lovejoy as an antidote to Croce's idealist approach and Gentile's historicism, which he considered "disappointing" (Rossi 1969).

The main thrust of Rossi's historiographical proposal was his claim that the evolution of scientific knowledge must be observed in its cultural context. He paid particular attention to the process through which the "images of science"—consolidated in elite intellectual culture as well as in subcultures and common sense—affect the concrete developmental process of scientific knowledge. By "image," he meant the complex system of beliefs, values, discourses, conceptualizations, etc. relating to the nature and definition of what science is, and what its functions are, for both the community and the individual (Rossi 1977, 7–10). However, scientists are not the only creators of these images of science; philosophical, political, and literary culture more generally do so as well.[33]

In this context, his work on reconstructing the birth of modern science, understood as the emergence of a new form of knowledge that "differed structurally from other cultural forms" (Rossi 1997/2001, 3), was particularly relevant. Although he was a strong supporter of the discontinuist model in the history of science, he also raised a number of criticisms of externalist interpretative theories based on Marxism, thus freeing his analysis of the discursive and cultural production of science from its

[31] During his education and in the first phase of his intellectual trajectory, he was involved in the development of debates close to Marxism, from which he gradually distanced himself.

[32] Some of his works were translated into different languages, arousing great interest abroad. Particularly well known and well received was *Francis Bacon: From Magic to Science*, which revitalized the interpretation of Bacon's work at the international level.

[33] In addition to his 1977 volume with this phrase as the title, he continued to insist on the concept of images of science throughout his historiographical research. In this regard, see his strategic references to this concept in Paolo Rossi (1986, 1997/2001).

socioeconomic roots. On this point, he stressed that "the specific theories that are the *kernel* of any science are not actually the reflection of specific socio-historical conditions" (Rossi 1997/2001, 6). Throughout his historiographical research, Rossi tried to demonstrate through this interpretive approach just how much the Scientific Revolution should be interpreted as a revolution of a purely intellectual nature (Rossi 1997/2001, 5–6).

Toward the end of the 1950s and into the 1960s, Rossi's research focused, in particular, on analyzing the establishment of the boundary between science and pseudoscience (with particular attention to magic and religion). Seeking the foundations of modern science in medieval and Renaissance philosophical and scientific culture, he wrote on the art of memory and the role of technical-practical knowledge for the emergence of modern science. He dedicated a number of internationally recognized works to these topics, including *Francis Bacon: From Magic to Science*, *Logic and the Art of Memory*, and *Philosophy, Technology and the Arts in the Early Modern Era* (Rossi 1957/2009, 1960/2006, 1962/1970).

The reflections contained in the volume *I ragni e le formiche. Un'apologia della storia della scienza* (1986) were particularly relevant for historical epistemology. The book included a series of essays devoted to historiographical methodology and the history of the historiography of science that had already been published in other venues between 1980 and 1985. The title referred to a famous metaphor used by Bacon that Rossi adapted to describe the lack of connection and dialogue between epistemology and history. Spiders, corresponding to the community of epistemologists, are characterized as having little curiosity about the outside world and tending to work in solitude. They draw

> from themselves a very thin web of arguments and minute distinctions, they take pleasure in relentlessly advancing subtle objections, they become submerged by the tangle of threads to which they have given life, and find themselves with shattered knowledge in their hands. (Paolo Rossi 1986, 6)

Whit this metaphor Rossi intended to represent the idea that epistemology had found itself at an impasse due to its inclination to describe, a priori, and through complex theoretical structures, what science should be and what it should do. In the eyes of spiders, history would be reduced, at best, to merely a place in which to "take their prefabricated cognitive models for a walk" (Paolo Rossi 1986, 7).

For their part, the ants, corresponding to the community of historians of science, are characterized by a strong curiosity about nature, meticulously gathering the resources needed for their work. In relation to the material collected, their plan of action leads them to the following attitude: "they accumulate and store it with great constancy, but without bothering to select it; they have no light to guide them, and they grope their way forward [...] they continually oscillate [...] between bewilderment and enthusiasm" (Paolo Rossi 1986, 6). Thus, historians of science have the opposite problem of epistemologists in that they tend not to realize the rational and transhistorical value of the material that they work with and are therefore less capable of detecting the specificities of science than those of other forms of knowledge. The ants confuse the mere collection and ordering of data with the elaboration of a narrative about the evolution of solid science (Paolo Rossi 1986, 7).

Rossi contrasted this division of research work, characterized by problems of disciplinary prejudices, with the need to put epistemology and the history of science back into dialogue. Following the same Baconian metaphor, the community of "bees" would be the figures capable of bringing and keeping together "the sharpness of spiders and the tirelessness of ants" (Paolo Rossi 1986, 6). As a historian, however, Rossi did not dedicate much space to the elaboration of an explicit and programmatic definition of what an epistemologically informed history of science would need to do.[34] Rossi, both in this work and others, preferred to show—through the analysis of exemplary cases—how the phases through which various images of science came together throughout history can be reconstructed. In addition, he focused on how they came to bear decisive weight in determining a series of elements, including the classification principles for the sources of knowledge, the formation of scientific models and the regulatory apparatus of science, the legitimation of fields of investigation, the selection of problematics, and the construction of research objects (Paolo Rossi 1986, 57). Contrary to epistemologists involved in the construction of theoretical models ready to be applied to standard cases taken from the history of science, Rossi suggested moving into the gray areas of history, where the boundaries—which epistemologists believe to be relatively solid—appear "more like semi-permeable membranes than uncrossable moats" (Paolo Rossi 1986, 53). Writing about the history of these border areas and their relationships with the "images of science" was, ultimately, the main goal of Rossi's epistemologically informed history.

4.5 Gargani's Knowledge without Foundations

Gargani moved in a completely different direction and, by virtue of his cultural background and intellectual stance, held a place in his own right within the field of Italian historical epistemology. As a student of Francesco Barone at the Scuola Normale in Pisa, Gargani had established himself since his youth as one of the most brilliant Italian scholars of Ludwig Wittgenstein.[35] As in the case of Geymonat and Preti, the tools developed by neo-positivist philosophy were mobilized in an

[34] In the essays contained in *I ragni e le formiche*, Rossi engaged in a dialogue with various important authors of both historical epistemology and its Anglo-American variant, related to the integration of History and Philosophy of Science (HPS). He expressed appreciation for some of their theoretical-historiographical proposals (e.g., Fleck and Enriques), whereas for others, he did not spare his criticism (e.g., Foucault).

[35] His numerous publications on Wittgenstein include *Linguaggio ed esperienza in Ludwig Wittgenstein* (1966); *Introduzione a Wittgenstein* (1973); *Wittgenstein tra Austria e Inghilterra* (1979); *Wittgenstein. Dalla verità al senso della verità* (2003); and *Wittgenstein. Musica, parola, gesto* (2008).

anti-idealistic key in the cultural context of Pisa as well (Gargani and Iofrida 2002, 31). However, these influences were also combined with those coming out of Marxist debates at Pisa thanks to the influence of Nicola Badaloni, who provided Gargani with the opportunity to engage with the historical-critical method. This line of research led to *Hobbes e la scienza* in 1971, in which Gargani directly engaged with the history of science and the history of ideas (Gargani and Iofrida 2002, 33). Here, Gargani offered a reconstruction of Hobbes's attempts to found a natural science on the basis of his ideal of logical-mathematical science and on his relations to modern theories of knowledge in the context of contemporary philosophical-scientific problems (of Wittgensteinian orientation).

Starting from his profound knowledge of contemporary science, already in the 1970s, Gargani had started to conceive of epistemology "not only as an *a posteriori* reflection on science" but as a type of research "immanent to the very structure of the theory"—that is, aimed at bringing out "the epistemology implicit" in the scientific activity itself (Gargani and Iofrida 2002, 34; Alunni 2011, 161). This assumption was one of the fundamental points of his most significant contribution to historical epistemology, *Il sapere senza fondamenti. La condotta intellettuale come strutturazione dell'esperienza comune* (1974).

As the following section explores in more detail, the 1970s were a time of high agitation and powerful intellectual clashes between different epistemological-political orientations in Italy. Even though he was independent from the parties directly involved in the dispute, Gargani participated indirectly in the debates of the "Italian Science Wars" with *Il sapere senza fondamenti*, proposing a modernization of the epistemological positions that were emerging in Italian Marxism in light of the so-called linguistic turn (Gargani 1975, 26). Gargani's uniqueness lay in his ability to use elements of the tradition of analytic philosophy as theoretical devices that were useful for historicizing epistemology.

One of the key issues for Gargani—both in this text and in subsequent works such as *Lo stupore e il caso* (1985)—was the critique of the correspondence theory of truth. As a counter to this theory, Gargani proposed an epistemological approach with a constructivist orientation based on Wittgenstein's theory of language games. With this approach, he aimed to show how knowledge is not based on the elaboration of a correspondence between words and things but rather on intersubjective linguistic practices (Gargani 1975, 48–50; Paoletti 1994). In Gargani's model, knowledge of objects is determined by the performative acts of language through which the knowing subject identifies them and constitutes them as such.

This assumption posed a series of consequences for a general theory of knowledge. For Gargani, the methods of knowledge construction could no longer be represented through the metaphor of a progressively accumulated "heritage" but rather as "a collection of tools, conceptual habits, behavioral models, operational conduct, decision-making techniques arising from forms of life, as an extension of them" (Gargani 1975, 21). In this context, Wittgenstein's concept of form of life played a strategic role because it highlighted how much scientific knowledge needs to be explained by referring to the "concrete" "organizational" and "social" practices that shaped the places in which this knowledge was produced throughout its history (Iofrida 2009, 501).

Scientific theories function as "grammatical models of objectivity" capable of constructing their objects according to "rules" and "procedures" that are immanent to the same linguistic practice that established them, therefore creating a process of the disciplinarization of the "strategy of scientific reasoning" (Gargani 1975, 33). Similarly to what Kuhn had argued, Gargani maintained that a look at the history of science clearly showed how scientific theories "manifested a mobility of categorical apparatuses, habits, decision-making and legitimation techniques, and models of objectivity" that were anything but attributable to an invariant and definitive paradigm. Indeed, the emergence of each new theory corresponded to "a change of categorical outlook" (Gargani 1975, 33). When one scientific theory supplants another, it is not just a process of adopting a model that has greater correspondence to a class of phenomena than the previous theory's explanation. Rather, there is a radical change in the grammar regulating the definition of the categorical apparatuses, of the decision-making and validation strategies of scientific propositions, which, in turn, incorporates the experimental devices (Gargani 1975, 72–74).

Among Gargani's objectives was the opening of epistemology to a sphere of investigation addressing the historical-social aspects that affect the creation of the various forms of knowledge. He remained dissatisfied with both the models that "limited themselves to outlining the historical conditions" of scientific research "in a broad sense" and those using an "economistic approach, poured into the sociology of knowledge." Instead, his objective was to investigate what remained as an epistemological blind spot—that is, the link between ordinary models of practical conduct and the processes through which "categorical structures, grammatical devices, inferential procedures, and legitimation paradigms of scientific theories" were formed (Gargani 1975, 45–46). In this sense, Gargani underlined how the "modalities of experience and organization of forms of human life constitute, so to speak, the interiority of cognitive modules" (Gargani 1975, 21).

Gargani's epistemology also took on evident political overtones. If cognitive activity was the result of linguistic practices linked to certain forms of life, the classic "object-designation" paradigm can represent only an "epistemological fetish," in the Marxian sense, the result of a series of "primitive and superstitious" "epistemological rituals"[36] (Gargani 1975, 51–70). In this passage, Gargani's antifoundationalist strategy clearly emerged, striving to form an epistemological model that did not seek its foundation in either the objective referent of reality or metaphysical hypotheses.

"Objects," "things," and "facts" are not the guarantors of our knowledge, but on the contrary, the process of their formation and crystallization needs to be explained. They are not sources in which the foundation of knowledge can be found but rather

[36] Gargani described and deconstructed a long series of epistemological rituals (e.g., the object model, the theoretical strategy from above, the logic of the double, and anticipated representation, among others) that, despite their differences, were all in solidarity with the philosophical search for a foundation (Vagelli 2015).

are the result of a form of the "fetishization"—which therefore implies alienation—of the description of the concrete cognitive structures through which we operate in the world. For Gargani, objects, things, and facts as such are constituted by the ways a grammar is used within a form of life. Therefore, "their status must assume a quality of revocability and fluidity that is relevant to a use of language that is unconstrained by arbitrary rigidities exercised by inflexible entities placed, so to speak, outside the medium of linguistic articulation" (Gargani 1975, 62–63). The correlative thesis is that if "the evocation of a class of objects functions as a technique for controlling assertions" (Gargani 1975, 62), then the epistemologies based on such objects are capable of functioning as "disciplinary models or structures" for knowledge and as "techniques for directing the processes of intellectual life within assigned and defined paths" (Gargani 1975, 60–61).

The process through which the linguistic activity constituting sense is hidden in the crystallized forms of "objects, things, and substances"—which occurs through "decisions, behaviors, and customs of an entire human form of life"—has a specific social function (Gargani 1975, 67). For Gargani, from a political point of view, this meant that the role of epistemologies based on rituals of the fetishization of the object-model was to regulate the "behavioral models of intellectual life and practical conduct" through "structures of order, disciplinary rules in which the objects, the cultural terms of reference, were distributed" (Gargani 1975, 60–61).

To illustrate exactly what he meant, Gargani engaged in a series of analogies with Marxist conceptual schemes. The process of epistemological fetishization is equivalent to the dynamic that constitutes money as an abstraction (Gargani 1975, 55), or to the ideological configuration that the capitalist system puts in place to describe the relationship between work and goods so that it shows itself as an "ineluctable" and "natural necessity" (Gargani 1975, 69). The processes through which raw materials are transformed into goods by work covers, in Gargani's analogy, the same function as the linguistic medium. The goods produced are assumed to have a value in themselves and, therefore, a natural property—as happens in the object-referent model—thus obscuring their constitutive conditions within the social division of labor and the organization of systems of production (Gargani 1975, 67).

Contrary to foundationalist epistemologies, Gargani proposed focusing attention on the analysis of the practical and social roots of intellectual habits (Gargani 1975, 89) through the historical analysis of the models producing cognitive forms as they emerge from forms of life. Historical analysis—which on the basis of his previous work on Hobbes was largely concentrated on sources from the modern era—was the primary way to observe what is humanly constructed in what we consider naturally irrevocable. In this sense, "Men have done many more things than have been historically recognized, and they have done them because they found themselves living in a world devoid of the ideal and practical foundations that have been bestowed by the imagination" (Gargani 1975, 115).

Gargani's constructivist epistemological model brought attention to the immanence of linguistic exchanges, observing how discursive orders—always linked to social forms—constitute the categorical structures and norms of practice. This proposal had an emancipatory potential—so much so that Gargani considered *Il sapere senza fondamenti* to be a philosophy of freedom—because it enabled the deconstruction of the apparent necessity commonly attributed to the categorical apparatuses of knowledge as they have been crystallized over time. Indeed, uncoincidentally, commonalities between Gargani and Foucault's perspectives have been emphasized a number of times (Davidson 2009; Vagelli 2015). In particular, the points of contact between these two authors can be found in the way they both conceived the relationship between the articulation of language games or discursive universes and the emergence of forms for the disciplinarization of life and structures of power.

The project of *Sapere senza fondamenti* was carried forward, albeit with a partially different tone, in the collection *La crisi della ragione* of 1983, edited by Gargani and including authors such as Carlo Ginzburg, Badaloni, Salvatore Veca, and Remo Bodei. This project was part of the Italian post-1968 debate on the crisis of classical rationality, which was also central to the 1988 collection *Weak Thought*, edited by Gianni Vattimo and Pier Aldo Rovatti, and other books of that period (Volonté 1988). *La crisi della ragione*, for its part, explored the process through which classical rationality was transformed into a new form of interaction between knowledge and human activities, adopting a pluralistic approach aimed at emphasizing the possible reactions to this impasse in specific research areas, such as history, psychoanalysis, Marxism, and so forth.

4.6 The Italian Science Wars

In the wake of the 1968 movements and the *autunno caldo* of 1969, the heightened climate of political tension was also reflected in the scientific field, as had happened in other nations with the birth of movements such as those that formed around the *Science for the People* journal in the United States, the *Radical Science Journal* in the United Kingdom, and the "critique of science" movement in France. Some of the Italian contributions arising from this intellectual context provided important stimuli to the process of historicizing epistemology.

The term *Italian Science Wars* refers to a broad series of disputes that developed in Italy beginning in the 1970s and that were characterized by heated debates relating to the political neutrality/non-neutrality of science and technology (Pancaldi 2010; Ienna 2020). This controversy—in both the academic community and the public sphere more broadly—was centered on an epistemological and political clash between the positions of Geymonat (and his Milanese school) and Paolo Rossi (and his school) against a large and varied group of scientists and militants of the radical left inspired by the social movements of 1968.[37] The debate hinged on the definition of the relationship between science and society and on the analysis of

[37] These positions were also joined by those expressed in more or less occasional contexts by a series of authoritative intellectuals, including Lucio Colletti, Lucio Lombardo Radice, Umberto

the role of ideology in science. From these foundations, different ways of investigating and conceptualizing the interconnections between epistemology and history implicitly emerged at the same time.

The most representative text of this period was certainly the collection of essays published in *The Bee and the Architect. Scientific Paradigms and Historical Materialism* (1976) by a group of Roman physicists made up of Giovanni Cicotti, Marcelo Cini, Michelangelo de Maria, and Giovanni Jona-Lasinio.[38] The main thesis of *The Bee and the Architect* was the integration of the Marxist theory of historical materialism with Kuhn's theses on the structure of scientific revolutions. The authors' objective was to free reflection on science from the epistemological influence of Engels's dialectical materialism, which, in their view, established a clear separation between nature and history, thus putting them in open conflict with Geymonat's Milanese group (Ciccotti et al. 2011/2024, 52–53). It was therefore a question of deconstructing the established position that held that the process through which humans appropriate nature—that is, the establishment of the natural sciences—is based on a progressively increasing and perfectible knowledge—and therefore on objective data collection—of a natural reality given in a completely independent manner with respect to social relations. The authors of *The Bee and the Architect* countered this position with a valorization of the methodological principles of historical materialism, emphasizing how science is a human and historical product subject to the social, political, and cultural conditioning of the time of its production, thus reaching more-radical conclusions than those of dialectical materialism.

The most significant contribution of this volume was its proposal of the inseparability of three levels of analysis: a historical analysis of the social and economic roots of science; a sociological analysis of the social functions covered by science in advanced capitalist societies; and, finally, an epistemological analysis of the impact of sociopolitical determinants on the activity of theoretical production in the sciences (Ciccotti et al. 2011/2024, 100–101). On the basis of the assumption of the impossibility of separating material aspects from spiritual ones, the authors believed that only historical materialism could account for the intrinsic correlations among these three factors.

Contrary to the epistemology of dialectical materialism, for this model, nature

> indicates everything which exists, not only what pre-exists, i.e., the material on which we work—as is typical of any materialism that ever existed in history—but also the person who makes transformations possible, the law that allows it, and the product of these transformations. […] nature is inseparably data [*dato*] and deed [*fatto*, i.e., fact]. (Ciccotti et al. 2011/2024, 102)

From this perspective, building a methodology on the basis of the clear separation of the analysis of human relations with nature and the analysis of social relations would be a mistake (Ciccotti et al. 2011/2024, 151). Instead, historical materialism

---

Cerroni, Giovanni Berlinguer, Valentino Gerretana, and many others. See Ciccotti et al. (2011/2024, 91–94), de Ceglia (2012), Cassata (2016), and Baracca and Del Santo (2017).

[38] For more information, see Gagliasso et al. (2015).

must be applied in an impartial manner to both the natural sphere and the historical and social sphere. In addition, to be consistent, methodological asymmetry must be avoided by considering the relations between the natural sphere and the historical-social sphere as one-way (which would lead to a mechanistic approach).

Indeed, the authors maintained that just as accounting for the historical evolution of the organization of science without resorting to factors that are external to scientific knowledge is impossible, referring to social explanations for the processes through which scientific contents are validated is also necessary. This means that the study of the adequacy of a scientific theory to its empirical context cannot help but take into consideration the ideological elements that condition it (Ciccotti et al. 2011/2024, 131).

The thesis of the non-neutrality of science derived from this; in strictly epistemological terms, it can be summarized by the expression that "there is no form of knowledge in which we can separate judgments of fact from judgments of value" (Ciccotti et al. 2011/2024, 120); that is, a clear dividing line between science and ideology cannot be identified. Therefore, the relationship that has traditionally been conceived between practices, generally understood as "a mere passive reflection of a given object," and theory needs to be redefined—which would instead represent "an active manifestation of a subjective thought"—understanding them, instead, as a dialectical unit (Ciccotti et al. 2011/2024, 153).

Precisely on this point, one of the most original elements of *The Bee and the Architect* lies. Compared to other ways of applying the analytical tools of Marxism in sociohistorical research, in this text, to use a term dear to the social studies of science, there is an opening of the "black box" of scientific knowledge. The idea that a close correlation exists between the emergence of the capitalist system, the rise of industrialization, and the birth of modern science was certainly not a new thesis in the context of historical materialism. What *The Bee and the Architect* added in terms of both epistemological and historiographical reflection was the idea that a "coherence between theory and practice of a certain society" exists, which indicates a "relative autonomy of theories in comparison with facts" (Ciccotti et al. 2011/2024, 135). In short, not only is the social organization of science conditioned by the socioeconomic structure, but so is its theoretical content.

Applied science, and its correlative technological innovation, objectively operate as productive forces—that is, they play a key role as structural determinants (Ciccotti et al. 2011/2024, 170–174). However, this is not sufficient to describe the social function of science in advanced capitalist societies. In the phase of capitalism in which "information" becomes a commodity on a large scale, the role that pure science plays as a specific cultural form—that is, as a superstructural ideological force—is left totally unexplored (Ciccotti et al. 2011/2024, 182). In this sense, it is a question of showing how the representations of the relations between humans and nature affect the ways individuals represent social relations (Ciccotti et al. 2011/2024, 188).

For the authors of *The Bee and the Architect*, the ideology of the existence of a pure neutral science is, indeed, something completely organic to the capitalist production system. From a historical-materialistic perspective, the contents of science, as generated within the capitalist mode of production, are always the reflection of the system's relations of production. In addition, the consolidation of an image of scientific results as completely neutral, based on an objective description of the relations between humans and nature and based on a function that is entirely an end in itself and not socially determined, is functional to capitalist ideology.

This kind of description cooperates in justifying an undue extension of this neutrality to the "scales of values, patterns of behavior, forms of organization, social goals" (Ciccotti et al. 2011/2024, 188) typical of the capitalist production system, which, for this reason, makes objectivity the absolute value, to the exclusion of other possible alternatives. This mechanism implements a process of technocratization—the selection of expert skills to which decision-making power is delegated—intrinsic to the capitalist system for which the scientific organization of work and social life becomes the principle for reproducing capitalist society.

Angelo Baracca and Arcangelo Rossi had also moved along a line similar to that of *The Bee and the Architect* group, publishing another significant text of this period, *Marxismo e scienze naturali*, in 1976. Contrary to orthodox interpretations, for Baracca and Rossi, dialectical materialism as formulated by Engels and Lenin could not be absolutized, but rather, it had a contextual and historically determined value with respect to the scientific debates of the era in which it had been formulated (a position, therefore, softer than the clear opposition in *The Bee and the Architect*). For this reason, the positions of scientific Marxism built on this form of absolutization were completely inadequate for understanding the rapid transformations of science (Baracca and Rossi 1976, 7–22). *Marxismo e scienze naturali* moved in the direction of historical materialism, showing its productive application as a methodology for studying the history of science. In particular, in the wake of della Volpe's influence, Baracca and Rossi suggested adopting the concept of "historically determined abstraction," which Marx had formulated as an analytical category for political economy, in the scientific context (Baracca and Rossi 1976, 93–114).

The analysis of the relationship between science and ideology had also been the subject of a specific investigation by a group composed of Elisabetta Donini, A. Rossi, and Tito Tonietti that was active for a number of years at the University of Lecce. Their collaboration gave rise to some important initiatives, such as the organization of a large conference that brought together various approaches to the issue of the non-neutrality of science in physics and mathematics (Donini et al. 1977). Donini and Tonietti prolonged their theoretical-political reflection on non-neutrality (Donini and Tonietti 1977) without discounting historical studies. In particular, in the following years, Tonietti dedicated himself to the history of mathematics, physics, and catastrophe theories (Tonietti 1983). For her part, Donini continued her research on the history of the foundations of quantum mechanics (Donini 1982) and, at the same time, developed epistemological reflections from a feminist point of view (Donini 1986), also promoting the Italian reception of authors such as Evelyn Fox Keller and Carolyn Merchant.

Baracca, together with various coauthors, also published several studies on the history of physics (and science in general), with a markedly Marxist orientation:

Natura e Storia, scienza e sviluppo del capitalismo nell'Ottocento with Roberto Livi (Baracca and Livi 1976), *Materia ed Energia* with A. Rossi (Baracca and Rossi 1978), and *Scienza e industria 1848–1915* with Stefano Ruffo and Arturo Russo (Baracca et al. 1979).

The emergence of the theme of the non-neutrality of science certainly did not go unnoticed by traditional Marxists or professional philosophers and historians of science. An intense and heated public debate ensued, which often took place in newspapers, weeklies, and popular scientific magazines, thus going beyond the canons of academic debate in the strict sense (and thus making it a question of cultural politics) (Cassata 2016, 58–62).

Immediately after the publication of *The Bee and the Architect* and *Marxismo e scienze naturali*, tones became more inflamed. The main opposition to the critique of science movements in Italy came from three key figures of the Italian academy of that period: Geymonat (and his Milanese group), Paolo Rossi, and Colletti. Notwithstanding some criticisms, Colletti's writings formed an important theoretical element in the formulation of the theses on the non-neutrality of science (Ciccotti et al. 2011/2024, 91–94; Alunni 2020). For Colletti, however, the ideas advanced by these militant scientists were completely unacceptable. Supporting the non-neutrality of science was, from his perspective, both erroneous from a theoretical point of view and useless from a political point of view. In 1976, he wrote a scathing review of *The Bee and the Architect* in the weekly *L'Espresso*, titled "La dea sragione," in which he dismissed the book's theses with disconcerting simplicity: "bodies fall in the same way under the action of gravity, both in socialist and in capitalist countries" (Ciccotti et al. 2011/2024, 367). *L'Espresso* offered the authors of *The Bee and the Architect* the opportunity to reply, and they responded with a text titled "La fisica, caro Colletti, è quella cosa…," which gave rise to further public controversy. Gerratana echoed Colletti in the article "Ma Galileo di che partito è?"—as did Paolo Rossi in "Il dialogo dei pessimi Sistemi." At the same time, the debate had also spread to the cultural pages of *La Repubblica*, where articles by Giulio Alfredo Maccacaro, Lombardo Radice, and Giovanni Battista Zorzoli appeared to support the usefulness of taking on the issue of non-neutrality. Berlinguer, instead, denounced the unilateral nature of the positions taken by both *The Bee and the Architect* group and Colletti (Cassata 2016, 62).

Even Paolo Rossi—despite having been an initial source of inspiration for some of these physicists who were interested in the history of science—attacked the radical scientists by labeling them as "Sunday epistemologists and amateur historians," claiming that their scientific training hindered, in his view, the creation of professional historical-philosophical criticism (Paolo Rossi 1982). Although he had been a student of Garin, Rossi made no secret of his skepticism of Marxist thought: "Those in Italy who refer directly to Marxism in their research have generally followed two paths [Geymonat or post-1968 culture]. Both lead very far from serious and detailed studies of the science–society relationship" (Paolo Rossi 1982, 146). As noted previously, Rossi had instead undertaken an intellectual project based on the history of ideas. In addition to directing the *Storia della scienza* book series with Feltrinelli, Rossi was also close to *Il Mulino* (a publisher from the democratic political area that tended to reconcile the Catholic wing and the center-left), collaborating for some years with the founding of the journal *Intersezioni* and the proposal and editing of several works on the history of scientific thought. Among these, Rossi edited and introduced the translation of Ludwik Fleck's *Genesis and Development of a Scientific Fact* for Il Mulino as a potential antidote to the antiscientist tendencies that the Italian reception of Kuhn was generating in far-left circles (Govoni 2015, 361).[39] In *Fra arcadia e apocalisse* (1977), Paolo Rossi labeled all criticisms of science as part of an irrationalist drift (Paolo Rossi 1977, 269–315). This position became a fixed point in his thought, even in later years with his tendency to consider critiques of science on the same level as "ramblings about a new proletarian science" and other pseudo- or antiscientific phenomena (Paolo Rossi 2002, 2004, 21–23).

For his part, Geymonat also responded to the attacks coming from supporters of the non-neutrality of science in various articles and speeches, advocating, once again, the need to renew a rationalist program and adopt dialectical materialism as an epistemological perspective (Geymonat 1978; see also Caldirola et al. 1975). Bellone, from Geymonat's Milanese group, wrote in *Il Manifesto* that "even if 200,000 metalworkers gather in Piazza San Babila [Milan], the value of gravitational acceleration does not change" (Baracca and Del Santo 2017).

As we have seen, unlike the Anglophone Science Wars, the specificity of the Italian political climate resulted in a complete inversion of the typical division of social roles. Those who criticized the idea of the neutrality of science were professional scientists, whereas those who defended the objectivity and nonpolitical nature of science were, rather, humanists. The theme of the debates in the Italian Science Wars was no longer whether epistemology and history should be integrated—that the scientific enterprise was historically determined was a concept shared by all—but rather how to outline the methodology of historical epistemology itself and which political orientation it should have.

For Geymonat and those aligned with his school, the activity of historicizing science showed a positive route of progressive rectifications that left no doubt as to the intrinsic objectivity of scientific activity. Correlatively, this position viewed objectivity as a guarantee of the development of nonideological scientific practices (against the distortions of the Lysenko affair, or the Soviet rejection of quantum mechanics and the theory of relativity) and, at the same time, the rational foundation of political practices. In this context, the history of mathematics and the history of physics were privileged sites of investigation through which to reveal a pathway progressively nearing the truth.

Paolo Rossi was convinced that an integrated historical and epistemological method should focus on showing the cultural interrelations between the elaboration of scientific theories and the sphere of intellectual production in the broad sense. The objective of his historiographical model was to reconstruct the culturally shared

[39] A slightly reworked version of his introduction to Fleck's volume was also published as a chapter in *I ragni e le formiche* (Paolo Rossi 1986, 59–94).

images of science while devaluing Marxist interpretations that emphasized the social and economic roots of intellectual practices.

What emerges from the historicization of the epistemology proposed by the radical scientists is the intricate network of relationships between the capitalist production system and the organization of research activity—which clearly orients its aims—as well as the production of scientific theories. The privileged objects of investigation in this context were the relationships between science, the economy, and society and between the activity of theoretical production and ideology. However, for these authors, supporting the idea of the impossibility of separating facts and values did not mean supporting relativistic positions but rather aiming for a political emancipation of research activity through a rigorous analysis of the relationship between science and ideology.

References

Agazzi, Evandro. 1986. *La filosofia della scienza in Italia nel '900*. Milan: Franco Angeli.

Agazzi, Evandro, Fabio Minazzi, and Ludovico Geymonat. 1989. *Filosofia, scienza e verità*. Milan: Rusconi.

Alcaro, Marco. 1977. *Dellavolpismo e la nuova sinistra*. Bari: Dedalo.

Alunni, Charles. 2011. Aldo Giorgio Gargani (1933-2009) In Memoriam. *Revue de Synthèse* 132 (2): 153–163. https://doi.org/10.1007/s11873-011-0154-3.

———. 2018. *Spectres de Bachelard. Gaston Bachelard et l'écoles surrationaliste*. Paris: Hermann.

———. 2020. Alfred Sohn-Rethel en Italie—question de contexte: L'operaismo ou de l'araignée au tisserand. *Recherches germaniques* 15 (July): 11–22. https://doi.org/10.4000/rg.2943.

Alunni, Charles, and Yves André. 2015. *Federigo Enriques o le armonie nascoste della cultura europea. Tra scienza e filosofia*. Pisa: Edizioni della Normale.

Alunni, Charles, and Éric Brian. 2002. La mémoire des gestes de science et ses enjeux. *Actes de la recherche en sciences sociales* 141-142 (1): 127–134.

Bachelard, Gaston. 1934. *Le Nouvel esprit scientifique*. Paris: Alcan. *The new scientific Spirit*, trans. Arthur Goldhammer. Boston: Beacon Press, 1984.

Badaloni, Nicola. 1962. *Marxismo come storicismo*. Milan: Feltrinelli.

Badino, Massimiliano, and Pietro Daniel Omodeo. 2020. *Cultural hegemony in a scientific world: Gramscian concepts for the history of science*. Leiden: Brill.

Baracca, Angelo, and Flavio Del Santo. 2017. La giovane generazione dei fisici e il rinnovamento delle scienze in Italia negli anni Settanta. *Altronovecento* 34: http://www.fondazionemicheletti.it/altronovecento/articolo.aspx?id_articolo=34&tipo_articolo=d_saggi&id=357#sdfootnote19anc.

Baracca, Angelo, and Roberto Livi. 1976. *Natura e Storia. Fisica e sviluppo del capitalismo nell'Ottocento*. Messina: D'Anna.

Baracca, Angelo, and Arcangelo Rossi. 1976. *Marxismo e scienze naturali. Per una storia integrale delle scienze*. Bari: De Donato.

———. 1978. *Materia e Energia*. Milan: Feltrinelli.

Baracca, Angelo, Stefano Ruffo, and Arturo Russo. 1979. *Scienza e industria, 1848–1915: gli sviluppi scientifici connessi alla seconda rivoluzione industriale*. Bari: Laterza.

Barotta, Pierluigi, and Eleonora Montuschi, eds. 2019. *Filosofia della scienza in Italia*. Rome: Armando.

Barsanti, Giulio. 2007. Storia della scienza, delle scienze, del pensiero scientifico: le scelte di Giulio Preti. In *Scienze e storia nell'Italia del Novecento*, ed. Claudio Pogliano, 129–150. Pisa: Pisa University Press.

Bellone, Enrico, Ludovico Geymonat, Giulio Giorello, and Silvano Tagliagambe. 1974. *L'attualità del materialismo dialettico*. Rome: Editori Riuniti.

Boschetti, Anna. 2024. *Benedetto Croce. Dominio simbolico e storia intellettuale*. Macerata: Quodlibet.

Caldirola, Pietro, Ludovico Geymonat, and Giuseppe Montalenti. 1975. Impegno razionalista. *Scientia* 110 (1): 5–8.

Cases, Cesare. 1958. *Marxismo e neopositivismo*. Turin: Einaudi.

Cassata, Francesco. 2016. La Scomparsa di Majorana e il dibattito sulle "due culture" in Italia. *TODOMODO Rivista internazionale di studi sciasciani. A Journal of Sciascia Studies* 6:51–66.

Castellana, Mario. 1988. Materiali per una storia dell'epistemologia italiana. In *Aspetti e problemi della filosofia italiana contemporanea*, ed. Antonio Quarta, 151–179. Lecce: Il protagora.

———. 2004. *Razionalismi senza dogmi*. Soveria Mannelli: Rubettino.

———. 2005. Federigo Enriques, Gaston Bachelard et Ferdinand Gonseth: Esquisse d'une tradition épistémologique. *Revue de Synthèse* 126 (2): 303–316. https://doi.org/10.1007/BF02965675.

———. 2016. Dal metodo storico-critico all'epistemologia genetica: Federigo Enriques e Jean Piaget. In *L'epistemologia genetica, by Jean Piaget*, ed. Mauro Ceruti, Luisa Damiano, and Mario Castellana, 173–201. Rome: Studium.

———. 2019. *Federigo Enriques e la Nuova Epistemologia*. Lecce: Pensa MultiMedia.

———. 2020. *Il contributo di Maximilien Winter alla critique des sciences. In Il metodo storico-critico per una nuova filosofia delle matematiche, 9–85*. Milan: Meltemi.

———. 2021. In *Il surrazionalismo di Gaston Bachelard con due saggi inediti*, ed. Pietro Console. Lecce: Milella.

Castelli Gattinara, Enrico. 2017. Epistemologia dell'errore in Federigo Enriques e Gaston Bachelard. *Matematica, Cultura e Società. Rivista dell'Unione Matematica Italiana* 2 (2): 195–208.

Cavazzini, Andrea. 2011. Épistémologie Historique Italienne. *Revue de Synthèse* 132 (2): 165–168. https://doi.org/10.1007/s11873-011-0147-2.

Ciccotti, Giovanni, Marcello Cini, Michelangelo de Maria, and Giovanni Jona-Lasinio. 2011. *L'Ape e l'Architetto. Paradigmi scientifici e materialismo storico*. Milan: FrancoAngeli. *The bee and the architect: Scientific paradigms and historical materialism*, trans. Giuliana Giobbi, eds. Gerardo Ienna and Pietro Daniel Omodeo. Venice: Verum Factum, 2024.

Ciliberto, Michele. 1981. Scienza, filosofia e politica: Federigo Enriques e il neoidealismo italiano. *Studi Storici* 22 (4): 861–886.

———. 2011. *Eugenio Garin. Un intellettuale nel Novecento*. Rome/Bari: Laterza.

Cimino, Guido, and Gabriella Sava. 2007. Le istituzioni italiane per la storia della scienza. In *Scienze e storia nell'Italia del Novecento*, ed. Claudio Pogliano, 237–275. Pisa: Pisa University Press.

Coniglione, Francesco. 2011. Giulio Preti, Ludovico Geymonat e la filosofia scientifica. In *Giulio Preti: intellettuale critico e filosofo attuale*, ed. Franco Cambi and Giovanni Mari, 51–83. Florence: Firenze University Press.

Croce, Benedetto. 1908. Il risveglio filosofico e la cultura italiana. *La critica. Rivista di Letteratura, Storia e Filosofia* 6:161–178.

———. 2007. *Filosofia come scienza dello spirito. Teoria e storia della storiografia*, eds. Edoardo Massimilia, Teodoro Tagliaferri, and Fulvio Tessitore. Naples: Bibliopolis. *Theory and History of Historiography*, trans. Douglas Ainslie. London: George G. Harrap, 1921.

Dalla Chiara, Maria Luisa. 1981. *Italian studies in the philosophy of science*. Dordrecht: Springer. https://doi.org/10.1007/978-94-009-8937-5.

Davidson, Arnold I. 2009. Dai giochi linguistici all'epistemologia politica. In *Il sapere senza fondamenti, by Aldo Giorgio Gargani*, 2nd ed., 7–17. Milan: Mimesis.

de Ceglia, Francesco Paolo. 2012. Il fattaccio dell'Accademia. MATEpristem. 2012. https://matematica.unibocconi.eu/articoli/il-fattaccio-dell%E2%80%99accademia.

della Volpe, Galvano. 1969. *Logica come scienza storica*, ed. Ignazio Ambrogio. Rome: Editori Riuniti. *Logic as a Positive Science*, trans. Jon Rothschild. London: New Left Books, 1980.

Dobb, Maurice. 1969. *Introduzione. In Per la critica dell'economia politica, by Karl Marx, vii-xix*. Rome: Editori Riuniti.

Donini, Elisabetta. 1982. *Il caso dei quanti: dibattito in fisica e ambiente storico, 1900–1927*. Milan: CLUP.

———. 1986. *Scienza, tecnologia e identita di genere: introduzione all'edizione italiana*. Turin: Rosenberg & Sellier.

Donini, Elisabetta, Arcangelo Rossi, and Tito Tonietti. 1977. *Matematica e fisica: struttura e ideologia*. Bari: De Donato.

Donini, Elisabetta, and Tito Tonietti. 1977. Lo scientifico è politico (gli scienziati nella crisi). *Quaderni Piacentini* 16 (62–63): 99–133.

Enriques, Federigo. 1906. *Problemi della scienza*. Bologna: Zanichelli. *Problems of science*, trans. Katharine Royce. Chicago: Open Court, 1914.

———. 1912. *Scienza e razionalismo*. Bologna: Zanichelli.

———. 1934. *Signification de l'histoire de la pensée scientifique*. Paris: Hermann.

———. 1975. Il Razionalismo sperimentale. In *Antologia, ed.* Mario Castellana. Lecce: Milella.

———. 1992. Il problema della ragione. In *Alle origini della nuova epistemologia. Il Congrès Descartes del 1937*, ed. Mario Castellana, 103–106. Lecce: Il protagora.

———. 2001. *Federigo Enriques. Matematiche e filosofia. Lettere inedite. Bibliografia degli scritti*, eds. Ornella Pompeo Faracovi and Luca M. Scarantino. Livorno: Belforte editore libraio.

Enriques, Federigo, and Giorgio De Santillana. 1932. *Storia del pensiero scientifico*. Bologna: Zanichelli.

Gagliasso, Elena, Mattia Della Rocca, and Rosanna Memoli. 2015. *Per una scienza critica: Marcello Cini e il presente: filosofia, storia e politiche della ricerca*. Pisa: Edizioni ETS.

Gambetti, Francesca. 2014. Enriques e la società filosofica italiana: Scienza, filosofia e riforma dell'università. *Rivista di storia della filosofia* 2:237–250.

Gargani, Aldo Giorgio. 1975. *Il sapere senza fondamenti*. 2nd ed. Milan: Mimesis.

Gargani, Aldo Giorgio, and Manlio Iofrida. 2002. "I vari volti della verità." Intervista biografico-teorica a cura di Manlio Iofrida. Iride 1: 21–74. doi:https://doi.org/10.1414/7701.

Garin, Eugenio. 1997. *Con Gramsci*. Rome: Editori Riuniti.

Gentile, Giovanni. 1908. Federigo Enriques—I problemi della scienza—Bologna, Zanichelli, 1906. *La critica. Rivista di Letteratura, Storia e Filosofia* 6:430–446.

Gerratana, Valentino. 1997. *Gramsci. Problemi di metodo*. Rome: Editori Riuniti.

Geymonat, Ludovico. 1977. *Scienza e realismo*. Milano: Feltrinelli.

Geymonat, Ludovico. 1945. *Studi per un nuovo razionalismo*. Turin: Chiantore.

———. 1953. *Saggi di filosofia neorazionalistica*. Turin: Einaudi.

———. 1972. Metodologia neopositivistica e materialismo dialettico. *Critica Marxista Quaderni* 6:24–40.

———. 1974. Primi lineamenti di una teoria della conoscenza materialistico dialettica. In *L'attualità del materialismo dialettico*, ed. Enrico Bellone, Ludovico Geymonat, Giulio Giorello, and Silvano Tagliagambe, 95–124. Rome: Editori Riuniti.

———. 1978. In *Contro il moderatismo. Interventi dal '45 al '78*, ed. Mario Quaranta. Milan: Feltrinelli.

———. 1983. *Riflessioni critiche su Kuhn e Popper*. Bari: Dedalo.

———. 1991. Sulle lezioni di filosofia della scienza di Giulio Preti. *Belfagor* 46 (3): 245–259.

Geymonat, Ludovico, and Silvano Tagliagambe. 1972. *Sul marxismo e le scienze*. Rome: Critica Marxista Quaderni.

Govoni, Paola. 2004. *Che cos'è la storia della scienza?* Rome: Carocci.

———. 2015. Il Mulino, la storia della scienza e la "Cultural Cold War." In Una scienza bolognese? figure e percorsi nella storiografia della scienza, eds. Annarita Angelini, Marco Beretta, and Giuseppe Olmi, 347–364. Bologna: Bononia University Press.

Gramsci, Antonio. 1977a. *Quaderni del carcere*, ed. Felice Platone. Turin: Einaudi. *Selections from the prison notebooks*, trans. And eds. Quentin Hoare and Geoffrey Nowell smith. New York: International Publishers, 1971.

———. 1977b. *Quaderni del carcere*, ed. Valentino Gerratana. Turin: Einaudi. *Further selections from the prison notebooks*, trans. And ed. Derek Boothman. Minneapolis: University of Minnesota Press, 1995.

———. 1977c. *Quaderni del carcere*, ed. Valentino Gerratana. Turin: Einaudi. *Prison notebooks*, vol. 3, trans. And ed. Joseph a. Buttigieg. New York: Columbia University Press, 2007.

———. 1977d. *Quaderni del carcere*, ed. Valentino Gerratana. Turin: Einaudi. *From Kant to Croce: Modern philosophy in Italy 1800–1950*, trans. And eds. Brian Copenhaver and Rebecca Copenhaver. Toronto: University of Toronto Press, 2011.

Hahn, Hans, Otto Neurath, and Rudolf Carnap. 1929. *Wissenschaftliche Weltauffassung. Der wiener Kreis, Veröffentlichnungen des Vereines Ernst Mach*. Vienna: Arthur wolf Verlag. *The scientific conception of the world: The Vienna circle*, in *empiricism and sociology*, trans. Paul Foulkes and Marie Neurath, eds. Marie Neurath and Robert S. Cohen, 299–318. Dordrecht: D. Reidel, 1973.

Ienna, Gerardo. 2020. Fisici italiani negli anni '70. *Fra scienza e ideologia. Physis* 55 (1–2): 415–442.

———. 2021. Métaphysique des mathématiques. Bachelard et le spinozisme mathématisant, trans. Charles Alunni. In *Métaphysique des mathématiques, by Gaston Bachelard*, ed. Charles Alunni and Gerardo Ienna, 23–48. Paris: Hermann.

Iofrida, Manlio. 2009. Ricordo di Aldo Giorgio Gargani. *Iride* 3:497–510. https://doi.org/10.1414/31432.

Lecourt, Dominique. 1976. *Lyssenko. Histoire réelle d'une science prolétarienne*. Paris: Maspero.

Maggi, Michele. 2009. Il Gramsci di Eugenio Garin. In *Nuova Antologia*, 255–268. Milan: Mondadori Education. https://doi.org/10.1400/165622.

Marx, Karl. 1932. *Ökonomisch-philosophische Manuskripte aus dem Jahre 1844*. Berlin: Marx-Engels-Verlag. *Economic and Philosophic Manuscripts of* 1944, trans. Rodney Livingstone and Gregory Benton. Moscow: Progress Publishers, 1977.

Minazzi, Fabio. 1984. Appunti per una storia del neopositivismo in Italia. Ludovico Geymonat e Giulio Preti. *Bollettino del Centro Internazionale di Storia dello Spazio e del Tempo* 3-4:119–127.

———. 2004. *Il cacodemone neoilluminista: l'inquietudine pascaliana di Giulio Preti*. Milan: Franco Angeli.

———. 2010. *Ludovico Geymonat epistemologo. Con documenti inediti e rari*. Milan: Mimesis.

———. 2011a. *Suppositio pro significato non ultimato. Giulio Preti neorealista logico studiato nei suoi scritti inediti*. Milan: Mimesis.

———. 2011b. Ludovico Geymonat le Problème de L'épistémologie Historique. *Revue de Synthèse* 132 (2): 213–232. https://doi.org/10.1007/s11873-011-0148-1.

———. 2021. *Epistemologia storico-evolutiva e neorealismo logico*. Florence: Olschki. *Historical epistemology and European philosophy of science: Rethinking critical rationalism and transcendentalism*, trans. Richard Sadleir. Cham: Springer, 2022.

Neurath, Otto. 1937. Die neue Enzyklopaedie des wissenschaftlichen Empirismus. *Scientia* 62: 309–320.

Nieto-Galan, Agustí. 2011. Antonio Gramsci revisited: Historians of science, intellectuals, and the struggle for hegemony. *History of Science* 49 (4): 453–478. https://doi.org/10.1177/007327531104900404.

Omodeo, Pietro Daniel. 2016. Egemonia e scienza. Temi gramsciani in epistemologia e storia della scienza. *Gramsciana* 2016 (2): 59–86.

———. 2018. Soggettività, strutture, egemonie. Questioni politico-culturali in epistemologia storica. *Studi culturali* 15 (2): 211–234.

———. 2020. The struggle for objectivity: Gramsci's historical-political vistas on science against the background of Lenin's epistemology. *HoST - Journal of History of Science and Technology* 14 (2): 13–49.

Pancaldi, Giuliano. 1980. The history and social studies of science in Italy. *Social Studies of Science* 10 (3): 351–374. https://doi.org/10.1177/030631278001000305.

———. 2010. Purification rituals: Reflections on the history of science in Italy. In *Impure cultures. Interfacing science, technology and humanities*, ed. Massimo Mazzotti and Giuliano Pancaldi, 233–248. Bologna: CIS.

Paoletti, Catherine. 1994. Aldo G. Gargani ou du (dé)constructivisme en philosophie. *Archives de Philosophie* 57 (1): 73–81.

Parolini, Maria Teresa. 1978. La dialettica scientifica di Galvano della Volpe. *Rivista di Filosofia Neo-Scolastica* 70 (3): 450–488.

Parrini, Paolo. 2004. *Filosofia e scienza nell'Italia del Novecento. Figure, correnti, battaglie.* Milan: Guerini e associati.

Pastore, Annibale. 1906. *Del nuovo spirito della scienza e della filosofia.* Turin: Bocca.

Pogliano, Claudio. 2007. *Scienze e storia nell'Italia del Novecento.* Pisa: Pisa University Press.

Polizzi, Gaspare. 1982. *La filosofia scientifica a congresso: Parigi 1935.* Livorno: Belforte Editore Libraio.

Pompeo Faracovi, Ornella. 2007. In *Appunti su Enriques, Giovanni Gentile jr. e la storia delle scienze*, ed. Claudio Pogliano. Pisa: Pisa University Press.

———. 2015. Federigo Enriques e il IV Congresso internazionale di filosofia. In *Federigo Enriques o le armonie nascoste della cultura europea. Tra scienza e filosofia*, ed. Charles Alunni and Yves André, 35–52. Pisa: Edizioni della Normale.

Prestipino, Giuseppe. 1973. Il marxismo nella "storia" di Geymonat. *Critica Marxista* 11 (5): 181–204.

Preti, Giulio. 1946. I limiti del neopositivismo. *Studi filosofici* 7:87–96.

———. 1950. Due orientamenti nell'epistemologia. *Rivista critica di storia della filosofia* 5 (3): 200–217.

———. 1957a. *Praxis ed empirismo.* Turin: Einaudi.

———. 1957b. *Storia del pensiero scientifico.* Milan: Mondadori.

———. 1974. *Retorica e logica. Le due culture.* Turin: Einaudi.

———. 1976a. *Saggi filosofici. Empirismo logico, epistemologia e logica.* Vol. 1. Florence: La nuova italia.

———. 1976b. *Saggi filosofici. Storia della logica e storiografia filosofica.* Vol. 2. Florence: La nuova italia.

Rivista di scienza. 1907. Presentazione. *Rivista di scienza.* 1 (1): 1–3.

Romeo, Carmelo. 2002. Galvano della Volpe e la tradizione del marxismo italiano. In *Le avanguardie della filosofia italiana nel XX secolo*, ed. Piero Di Giovanni, 250–266. Milan: FrancoAngeli.

Rossi, Paolo. 1957. *Francesco Bacone. Dalla magia alla scienza.* Bologna: Il Mulino. *Francis Bacon: From Magic to Science*, trans. Sacha Rabinovitch. London: Routledge and Kegan Paul, 2009.

———. 1960. *Clavis universalis. Arti della memoria e logica combinatoria da Lullo a Leibniz.* Milan/Naples: Riccardo Ricciardi. *Logic and the art of memory: The quest for a universal language*, trans. Stephen Clucas. London: Continuum, 2006.

———. 1962. *I filosofi e le macchine. 1400–1700.* Milan: Feltrinelli. *Philosophy, technology and the arts in the early modern era*, trans. Salvator Attanasio. New York: Harper and Row, 1970.

———. 1969. *Storia e filosofia. Saggi sulla storiografia filosofica.* Turin: Einaudi.

———. 1977. *Immagini della scienza.* Rome: Editori Riuniti.

———. 1982. Filosofia di fronte alle scienze: alcune discussioni sui rapporti scienza-società. In *La cultura filosofica italiana. Dal 1945 al 1980*, ed. Giuseppe Cantillo and Eugenio Mazzarella, 143–158. Naples: Guida Editori.

———. 1986. *I ragni e le formiche. Un'apologia della storia della scienza*. Bologna: Il Mulino.

———. 1997. *La nascita della scienza moderna in Europa*. Rome/Bari: Laterza. *The birth of modern science*, trans. Cynthia de Nardi Ipsen. Oxford: Blackwells, 2001.

———. 2002. Antiscienza: processi di globalizzazione. *Rivista di filosofia* 2:309–332. https://doi.org/10.1413/7594.

———. 2004. Prefazione alla terza edizione. In *Dalla magia alla scienza*, ed. Francesco Bacone, 9–50. Bologna: Il Mulino.

Rossi, Pietro. 2009. La stagione delle ideologie. *Rivista di filosofia* 1:117–156. https://doi.org/10.1413/28948.

Santucci, Antonio A. 1996. La "filologia vivente": Eugenio Garin e il metodo di Gramsci. *The Italianist* 16 (1): 364–375. https://doi.org/10.1179/ita.1996.16.1.364.

Snow, Charles Percy. 1959. *The two cultures*. Cambridge, UK: Cambridge University Press.

Stöltzner, Michael. 1998. Federigo Enriques e l'enciclopedia neurathiana. *Rivista di Storia della Filosofia* 53 (3): 463–494.

Tonietti, Tito. 1983. *Catastrofi: una controversia scientifica*. Bari: Dedalo.

Vacca, Giuseppe. 1970. *Scienza, stato e critica di classe. G. della Volpe e il marxismo*. Bari: De Donato.

Vagelli, Matteo. 2015. Le problème du fondement et l'épistémologie historique italienne. *Archives de Philosophie* 78 (3): 509–515.

Volonté, Paolo. 1988. La "crisi della ragione" come figura dello scietticismo nella filosofia italiana contemporanea. *Rivista di Filosofia Neo-Scolastica* 80 (3): 440–465.

The manufacturer's authorised representative in the EU is Springer Nature Customer Service Centre GmbH, Europaplatz 3, 69115 Heidelberg, Germany. If you have any concerns regarding our products, please contact ProductSafety@springernature.com

Printed and bound by CPI Group (UK) Ltd, Croydon, CR0 4YY

30/09/2026

02225589-0007